AF595032

Meat Products: From Animal (Farm) to Meal (Fork)

Meat Products: From Animal (Farm) to Meal (Fork)

Editors

Benjamin W.B. Holman

Eric Nanthan Ponnampalam

MDPI • Basel • Beijing • Wuhan • Barcelona • Belgrade • Manchester • Tokyo • Cluj • Tianjin

Editors

Benjamin W.B. Holman
Centre for Red Meat and
Sheep Development
Australia

Eric Nanthan Ponnampalam
Animal Production Sciences,
Agriculture Victoria Research
Australia

Editorial Office
MDPI
St. Alban-Anlage 66
4052 Basel, Switzerland

This is a reprint of articles from the Special Issue published online in the open access journal *Sustainability* (ISSN 2071-1050) (available at: https://www.mdpi.com/journal/foods/special_issues/Meat_Products_From_Animal_to_Meal).

For citation purposes, cite each article independently as indicated on the article page online and as indicated below:

LastName, A.A.; LastName, B.B.; LastName, C.C. Article Title. *Journal Name* **Year**, *Volume Number*, Page Range.

ISBN 978-3-0365-3879-2 (Hbk)
ISBN 978-3-0365-3880-8 (PDF)

Contents

About the Editors

Benjamin W.B. Holman is a Senior Research Scientist (Meat Science) at the NSW Department of Primary Industries. His study of meat production, fabrication, and preservation has contributed tangible benefits for stakeholders throughout the meat supply chain. Benjamin's scientific leadership has resulted in his receipt of several prestigious national awards, editorial board appointments, and an honorary position at Charles Sturt University (AUS). In all engagements, Benjamin strives to share his enthusiasm for meat quality, its nutritional enhancement, and sustainable production.

Eric Nanthan Ponnampalam is a Senior Research Scientist at Agriculture Victoria Research, Australia. His current research involves sustainable animal production and the utilization of forages and supplements to produce premium carcasses with improved nutritional and retail value. Eric is an internationally recognized expert, specializing in balancing ruminant diets with special forage feeds and secondary metabolites using abundantly available, low-nutritive materials to improve livestock productivity, animal health and wellness, and product quality. He is a member of the Editorial Boards of Meat Science and Animals.

MDPI

Editorial

Meat Products: From Animal (Farm) to Meal (Fork)

Benjamin W. B. Holman [1,*] and Eric N. Ponnampalam [2,*]

1 Centre for Red Meat and Sheep Development, NSW Department of Primary Industries, Cowra, NSW 2794, Australia
2 Animal Production Sciences, Agriculture Victoria Research, Department of Jobs, Precincts and Regions, Bundoora, VIC 3083, Australia
* Correspondence: benjamin.holman@dpi.nsw.gov.au (B.W.B.H.); eric.ponnampalam@agriculture.vic.gov.au (E.N.P.)

Citation: Holman, B.W.B.; Ponnampalam, E.N. Meat Products: From Animal (Farm) to Meal (Fork). *Foods* **2022**, *11*, 933. https://doi.org/10.3390/foods11070933

Received: 17 March 2022
Accepted: 22 March 2022
Published: 24 March 2022

Publisher's Note: MDPI stays neutral with regard to jurisdictional claims in published maps and institutional affiliations.

Meat composition and quality are not independent of the effects of animal production systems. The interactions of an animal's diet (type, form, energy content, nutritional value, etc.), an animal's physiological status or age, and an animal's environment will impact the yield, composition, quality, and sensorial appeal of its meat products. If we consider the adoption of forages, novel feeds, industry by-products and more sustainable production systems for holistic benefit, we must also consider their implications for the value of the meat product delivered for human consumption. Consequently, it is important to understand farm (animal/farm) factors that contribute to the quality, expense, and nutritional value of meat. This imperative is a response to the consumers of meat being increasingly aware of the cost, convenience, production system ethics or efficiencies, environmental impacts, food safety and the health or wellness benefits derived from a meat product.

The articles included in this Special Issue have been curated to present findings from different production systems and animal husbandry strategies, in terms of their capacity to deliver meat to consumers that matches their demands.

We begin with an overview of omega-3 and omega-6 fatty acids in red meat, with special reference made to their biochemistry, enhancement by agricultural practice, preservation until the point of consumption, and their recommended dosages within a human diet [1]. The fatty acid and nutraceutical properties of the lipids in the meat of fallow deer, produced under organic and conventional farming systems, were compared [2]. The concentrations of odd- and branched-chain fatty acids in the meat of lamb was compared between different feeding regimes (starch- versus fibre-rich), in an effort to authenticate the production system *post hoc* [3]. Different aroma compounds were identified in the roasted cuts of pre- and post-rigor mutton to understand the preference of some consumers for sheep meat, sold immediately, without chilled storage [4]. The effect of scopoletin supplementation on the quality and antioxidant status of meat from broilers, when held under stocking density stress, was investigated [5]. Breed, gender, season, and geographic region were explored as factors influencing the carcass weight and animal age at slaughter for Greek cattle [6]. The meat quality of Holstein (dairy) cattle was quantified to meet the rising demand for beef in Israel, sustainable production systems, and help address product biases [7]. The maternal prepartum dietary carbohydrate source, during mid- and late-gestation, was explored in terms of its effect on the growth, carcass, and meat quality of the offspring [8]. The supplementation of cattle with distiller's grains was explored as a means to balance the oxidative stability of beef when displayed under commercial conditions [9]. Finally, the effect of packaging type on the shelf-life, quality and oxidative status of the meat from yearling sheep and young lambs was compared against the effect of the finishing diets (standard diet; standard diet supplemented with camelina forage; or standard diet supplemented with camelina meal) [10].

From these articles, we gain an increased understanding of whole-system approaches and efficacies when managing the transition between animal and meal (i.e., from farm to fork). We are reminded that meat is the result of animal production systems and, more broadly, the agricultural sector. Furthermore, we are focused towards a future where meat production, quality and value are considered within the context of sustainable feed and supplementation selection, the management of animal genetic and husbandry practices, delivery upon market demands and trends, as well as the sustainable processing and packaging of meat products.

Author Contributions: Conceptualization, B.W.B.H. and E.N.P.; Writing—original and draft preparation, B.W.B.H. and E.N.P.; writing—review and editing, B.W.B.H. and E.N.P. All authors have read and agreed to the published version of the manuscript.

Funding: This research received no external funding.

Acknowledgments: The authors acknowledge the support from their respective organisations.

Conflicts of Interest: The authors declare no conflict of interest.

References

1. Ponnapalam, E.N.; Sinclair, A.J.; Holman, B.W.B. The sources, synthesis and biological actions of omega-3 and omega-6 fatty acids in red meat: An overview. *Foods* **2021**, *10*, 1358.
2. Kilar, J.; Kaspryzk, A. Fatty acids and nutraceutical properties of lipids in fallow deer (*Dama dama*) meat produced in organic and conventional farming systems. *Foods* **2021**, *10*, 2290. [CrossRef] [PubMed]
3. Gómez-Cortés, P.; Domenech, F.R.; Rueda, M.C.; de la Fuente, M.Á.; Schiavone, A.; Marin, A.L.M. Odd- and branched-chain fatty acids in lamb meat as potential indicators of fattening diet characteristics. *Foods* **2021**, *10*, 77. [CrossRef] [PubMed]
4. Liu, H.; Hui, T.; Fang, F.; Ma, Q.; Li, S.; Zhang, D.; Wang, Z. Characterization and discrimination of key aroma compounds in pre- and postrigor roasted mutton by GC-O-MS, GC E-Nose and aroma recombination experiments. *Foods* **2021**, *10*, 2387. [PubMed]
5. Ha, S.H.; Kang, H.K.; Hosseindoust, A.; Mun, J.Y.; Moturi, J.; Tajudeen, H.; Lee, H.; Cheong, E.J.; Kim, J.S. Effects of scopoletin supplementation and stocking density on growth performance, antioxidant activity, and meat quality of Korean native broiler chickens. *Foods* **2021**, *10*, 1505. [PubMed]
6. Nikolaou, K.; Koutsouli, P.; Bizelis, I. Evaluation of Greek cattle carcass characteristics (carcass weight and age of slaughter) based on SEUROP classification system. *Foods* **2021**, *9*, 1764.
7. Shabtay, A.; Shor-Shimoni, E.; Orlov, A.; Agmon, R.; Trofimyuk, O.; Tal, O.; Cohen-Zinder, M. The meat quality characteristics of Holstein calves: The story of Israeli 'dairy beef'. *Foods* **2021**, *10*, 2308. [PubMed]
8. Gubbels, E.R.; Block, J.J.; Salverson, R.R.; Harty, A.A.; Rusche, W.C.; Wright, C.L.; Cammack, K.M.; Smith, Z.K.; Grubbs, J.K.; Underwood, K.R.; et al. Influence of maternal carbohydrate source (concentrate-based vs. forage-based) on growth performance, carcass characteristics, and meat quality of progeny. *Foods* **2021**, *10*, 2056. [PubMed]
9. Merayo, M.; Rizzo, S.; Rossetti, L.; Pighin, D.; Grigioni, G. Effect of aging and retail display conditions on the color and oxidant/antioxidant status of beef from steers finished with DG-supplemented diets. *Foods* **2022**, *11*, 884.
10. Ha, M.; Warner, R.D.; King, C.; Wu, S.; Ponnampalam, E.N. Retail packaging affects colour, water holding capacity, texture and oxidation of sheep meat more than breed and finishing feed. *Foods* **2021**, *11*, 144.

 *foods*

MDPI

Review

The Sources, Synthesis and Biological Actions of Omega-3 and Omega-6 Fatty Acids in Red Meat: An Overview

Eric N. Ponnampalam [1,*], Andrew J. Sinclair [2] and Benjamin W. B. Holman [3]

1 Animal Production Sciences, Agriculture Victoria Research, Department of Jobs, Precincts and Regions, Bundoora, VIC 3083, Australia
2 Department of Nutrition, Dietetics and Food, Monash University, Clayton, VIC 3800, Australia; andrew.sinclair@monash.edu
3 Centre for Red Meat and Sheep Development, NSW Department of Primary Industries, Cowra, NSW 2794, Australia; benjamin.holman@dpi.nsw.gov.au
* Correspondence: eric.ponnampalam@agriculture.vic.gov.au

Abstract: The maximisation of available resources for animal production, food security and maintenance of human–animal wellbeing is important for an economically viable, resilient and sustainable future. Pasture and forage diets are common sources of short chain omega-3 (n-3) polyunsaturated fatty acids (PUFA), while grain-based and feedlot diets are common sources of short chain omega-6 (n-6) PUFA. Animals deposit n-3 and n-6 PUFA as a result of their direct consumption, as feeds or by synthesis of longer chain PUFA from short chain FA precursors in the body via desaturation and elongation processes. Research conducted over the last three decades has determined that the consumption of n-3 PUFA can improve the health and wellbeing of humans through its biological, biochemical, pathological and pharmacological effects. n-6 PUFA also play an important role in human health, but when consumed at high levels, are potentially harmful. Research shows that current consumption of n-6 PUFA by the human population is high due to their meal choices and the supplied food types. If consumption of n-3 PUFA from land- and marine-based foods improves human health, it is likely that these same food types can improve the health and wellbeing of livestock (farm animals) by likewise enhancing the levels of the n-3 PUFA in their circulatory and tissue systems. Modern agricultural systems and advanced technologies have fostered large scale animal and crop production systems. These allow for the utilisation of plant concentrate-based diets to increase the rate of animal growth, often based on economics, and these diets are believed to contribute to unfavourable FA intakes. Knowledge of the risks associated with consuming foods that have greater concentration of n-6 PUFA may lead to health-conscious consumers avoiding or minimising their intake of animal- and plant-based foods. For this reason, there is scope to produce food from plant and animal origins that contain lesser amounts of n-6 PUFA and greater amounts of n-3 PUFA, the outcome of which could improve both animal and human health, wellbeing and resilience to disease.

Keywords: agricultural practices; animal production; nutrition; human health; fatty acid profile; fat digestion and absorption; consumer guidelines; preservation

Citation: Ponnampalam, E.N.; Sinclair, A.J.; Holman, B.W.B. The Sources, Synthesis and Biological Actions of Omega-3 and Omega-6 Fatty Acids in Red Meat: An Overview. *Foods* **2021**, *10*, 1358. https://doi.org/10.3390/foods10061358

Academic Editor: Angel Cobos

Received: 19 May 2021
Accepted: 9 June 2021
Published: 11 June 2021

Publisher's Note: MDPI stays neutral with regard to jurisdictional claims in published maps and institutional affiliations.

1. Introduction

The utilisation of available resources for resilient animal production systems, food security and maintenance of the health and wellbeing of human–animal population are important for the sustainability of future agriculture and food production. The use of available feed resources from cultivated pasture and natural range lands in animal production systems may be beneficial for the wellbeing of livestock. The same is true for those humans who consume milk, meat and offal from those animals, especially when compared to products from livestock fed commercially formulated feeds or low-quality forages that are deficient in nutrients. In many livestock production systems, pasture and forage diets are common sources of omega-3 (n-3) polyunsaturated fatty acids (PUFA). Grain-based and feedlot diets are, instead, common sources of omega-6 (n-6) PUFA.

It is well known that lipids (fats from animals and oils from plants) provide energy, nutrient mediation, signal transduction, disease prevention, insulation, cell membrane structure, and organ protection upon consumption. Lipids include triglyceride, phospholipid, cholesterol, cholesterol ester, free fatty acids (FFA), sphingomyelin subgroups and glycolipids. Not many animal or plant scientists understand the complexity of the lipid fraction and the differences between animal and plant tissues. For example, fat deposits in animal tissue are mainly triacylglycerols (TAG); muscle lipids contain TAG, cholesterol and phospholipids; plant leaf tissue lipids are mainly polar lipids (glycerophospholipids); oil seed lipids are mainly TAG. The different lipid fractions in plant and animal tissues can be separated and observed by using a thin layer chromatography (TLC) technology, an example is shown in Figure 1.

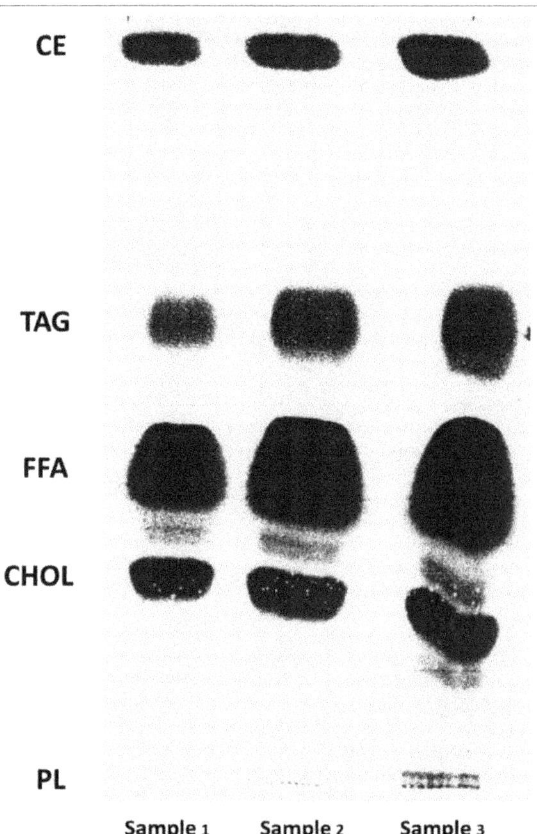

Figure 1. Lipid fractions of lyprinol (green-lipped mussel) separated by using a thin layer chromatography (TLC) technology. Lipid fractions were separated by thin layer chromatography (TLC) on silica gel plates (Silica gel 60H, Merck, Darmstadt, Germany). The solvent system for all TLC was petroleum spirit/diethyl ether/glacial acetic acid (85:15:2 by volume). Lyprinol (50 g) was made up in 1 mL of chloroform, and from this stock, 10 μL (sample 1), 20 μL (sample 2) and 30 μL (sample 3) were spotted as shown above. Lipid classes were visualised with fluorescein 5-isothiocyanate against TLC standard 18-5 (Nuchek Prep Inc, Elysian, MN). Lipid fractions identified from left top to bottom are cholesterol esters (CE), triacylglycerols (TAG), free fatty acids (FFA), cholesterols (CHOL) and phospholipids (PL).

Previous research of both humans and animals, including livestock and companion animals, demonstrate that dietary background plays a major role on lipid metabolism, fatty acid (FA) synthesis and fat accretion in the body—more so than genetic or gender associated factors alone. The effects of dietary fat are contributed by both their energy concentration and the types of lipids present. Genetic effects in FA synthesis and accretion in farm animals has been shown to be associated with desaturase and elongation activity [1]. Genetics are also known to influence desaturase and elongase activity in humans [2]. There are many studies which refer to the influence of gender on PUFA synthesis in human (higher activity in females than males).

Fatty acids are classified as saturated or unsaturated. Unsaturated FA can be further delineated as monounsaturated (MUFA) and PUFA. Common dietary sources of PUFA include leafy vegetables, oilseed, nuts, meat, eggs and seafood. Characteristics of PUFA are their low melting point and liquid state when held at room temperature. Hence, these are often referred to as oils. It is of interest, therefore, that fat melting point has been applied to estimate the unsaturated FA content of non-liquefied fat deposits from, for example, beef and sheep meat [3,4]. There is substantive evidence to support the regular consumption of n-3 PUFA, as these are beneficial for growth, development, health and the welfare of humans and animals [5–9]. The n-3 PUFA alpha-linolenic acid (ALA, C18:3n-3) and the long chain derivatives eicosapentaenoic acid (EPA, C20:5n-3), docosapentaenoic acid (DPA, C22:5n-3) and docosahexaenoic acid (DHA, C22:6n-3) have each been reported to play a role in the prevention of cardiovascular disease, diabetes, hypertension, inflammation, allergies, cancer, renal disorders, neural function and improve immune response [5,10,11].

All fish are rich in long chain n-3 PUFA, especially EPA and DHA, but this is especially true for oily fish such as salmon and mackerel. The levels of these same PUFA are comparatively moderate in red meat sourced from pasture grazed ruminants, these having levels similar to many white fish which are low in fat such as snapper, leatherjacket, flounder [12,13]. The application of grains or some feedlot rations within livestock industries to hasten animal growth rates can diminishes the level of n-3 PUFA in red meat. In addition, recent climate variation has led to prolonged drought in some parts of the world, which diminishes the availability of n-3 PUFA rich feed sources to livestock and increases reliance on concentrate and commercial feeds that are rich in n-6 PUFA.

Collective research indicates that the evolutionary aspects of modern farming (agribusiness), selection of specialised pastures for high yield, commercially oriented crop and animal production systems, and food processing have contributed to alterations in the concentrations of n-3 and n-6 PUFA in pasture and field crops [9]. This is believed to be impacting the health and wellness of animals and humans. Indeed, the ratio of n-6 and n-3 PUFA (n-6/n-3 ratio) in human and animal diets is proposed to have been nearly 1:1 during evolutionary time, but direct interventions and climate variation has led to a shift towards a ratio closer to 20:1. This is of concern because present recommendations advise that animal and human diets should have a n-6/n-3 ratio of 1–4:1 to help maintain a balanced and healthy life [14]. The nutritionally important n-3 PUFA found in meat and other products, such as milk and non-lean edible parts of a carcass, are summarised in Table 1.

Table 1. Common name, abbreviation and scientific name (IUPAC, International Union of Pure and Applied Chemistry) of omega-3 (n-3) and omega-6 (n-6) fatty acids found in dietary sources.

Common Name	Abbreviations	Systematic Name
Linoleic acid	C18:2n-6 (LA)	cis-9, cis-12-octadecatrienoic acid
Alpha-linolenic acid	C18:3n-3 (ALA)	cis-9, cis-12, cis-15-octadecatrienoic acid
Stearidonic acid	C18:4n-3 (SDA)	cis-6, -9, -12, -15-octadecatetraenoic acid
Arachidonic acid	C20:4n-6 (AA)	cis-5, -8, -11, -14-eicosatetraenoic acid
Eicosapentaenoic acid	C20:5n-3 (EPA)	cis-5, -8, -11, -14, -17-eicosapentaenoic acid
Adrenic acid	C22:4n-6 (AdA)	cis-7, -10, -13, -16-docosatetraenoic acid
Docosapentaenoic acid	C22:5n-3 (DPA)	cis-7, -10, -13, -16, -19-docosapentaenoic acid
Docosahexaenoic acid	C22:6n-3 (DHA)	Cis-4, -7, -10, -13, -16, -19-docosahexaenoic acid

This overview aims to describe the biochemical basis of n-3 and n-6 PUFA and agricultural practices unique to the modern era that are applied for their enhancement in red meat. Special reference is made to their preservation, biological actions and recommended dosages within a human diet.

2. Synthesis

2.1. Molecular Structure of Omega-3 and Omega-6 Fatty Acids

PUFA can be classified by carbon chain length, where 20–24 carbon atoms are long chain and 26 or more carbon atoms are very long chain PUFA (FAO/WHO, 2008). Researchers (i.e., nutritionists, dietitians and biochemists) often use the 'n minus' term of notation to name the naturally occurring *cis* unsaturated FA, where the 'n minus' indicates the position of first double bond of the FA closest to the methyl end of the molecule. For example, ALA is designated as C18:3n-3 since the first double bond is present 3 carbon atoms from the methyl end, but this nomenclature does not specify the position and confirmation of remaining double bonds in the molecular structure [15]. In this system, the *cis* unsaturated FA are classified as n-3 (omega-3), n-6 (omega-6) and n-9 (omega-9). Chemical structure of n-3 and n-6 PUFA naturally available in meat and other dietary sources are shown in Table 2.

Table 2. Names and chemical structures of commonly available omega-3 and omega-6 fatty acids.

Common Name	Chemical Structure
Linoleic acid	 $^{18}CH_3$ … $^{1}COOH$
Alpha-linolenic acid	 $^{18}CH_3$ … $^{1}COOH$
Stearidonic acid	 $^{18}CH_3$ … $^{1}COOH$
Arachidonic acid	 $^{20}CH_3$ … $^{1}COOH$
Eicosapentaenoic acid	 $^{20}CH_3$ … $^{1}COOH$
Adrenic acid	 $^{22}CH_3$ … $^{1}COOH$
Docosapentaenoic acid	 $^{22}CH_3$ … $^{1}COOH$
Docosahexaenoic acid	 $^{22}CH_3$ … $^{1}COOH$

2.2. Biosynthesis of Omega-3 and Omega-6 Fatty Acids

ALA is the precursor (parent) FA of the n-3 family, whereas linoleic acid (LA) is the precursor FA of n-6 family. Among the four n-3 PUFA most commonly found in animal tissues (i.e., ALA, EPA, DPA and DHA), ALA cannot be synthesised by humans and animals and is therefore referred to as an essential FA [16]. Only plants can produce essential FA, and animals and humans must obtain these FA, through dietary means, for use in the synthesis of their longer chain n-3 and n-6 PUFA derivatives *viz.* EPA or DHA, or arachidonic acid (AA) [17]. In the body, the synthesis or conversion of ALA to its longer chain derivatives is controlled by many biological factors that, according to both animal and human studies, are slow and inefficient [18,19]. The conversion efficiency is not dependent on the metabolic demand of the body but is mainly determined by the amount of ALA and (interestingly) LA, present in the diet. This is because of their competitive nature whereby the same enzymes mediate ALA and LA desaturation and elongation processes.

Due to the low conversion efficiency, it is necessary to provide substantial amounts of dietary ALA to promote higher levels of required EPA and DHA in the circulatory and tissue systems. Therefore, it is suggested that animals and humans should be fed with edible wild plant leaves or vegetable oils rich in ALA or, alternatively, with marine-based diets which are rich sources of EPA and DHA and avoid the desaturation and elongation processes required by ALA [20,21]. There are studies that have also shown significant increases in EPA, DPA and DHA concentrations in the blood of humans or muscle tissues of ruminants when terrestrial-based diets rich in ALA, such as flax(seed) or canola, are fed for long durations and/or at high doses [3,5,6,13,18–22]. Some studies indicated that there are three desaturase enzymes involved in the formation of 22 carbon long chain n-3 (DHA) and n-6 (DPAn-6) PUFA from ALA and LA in human and animal tissues. The dietary sources and biosynthetic pathways of n-3 and n-6 PUFA in mammals involving delta-6, delta-5 and delta-4 desaturase enzymes are shown below in Figure 2. It should be noted that some research has found there to be no involvement of delta-4 desaturase enzyme activity in the conversion of ALA to DHA and LA to DPAn-6, respectively. Rather, there will be a further elongation through the second use of delta-6 desaturase enzymes and then beta-oxidation processes take place in the circulatory and peripheral tissue systems for the synthesis of DHA and DPAn-6 PUFA [22]. This is illustrated in Figure 3.

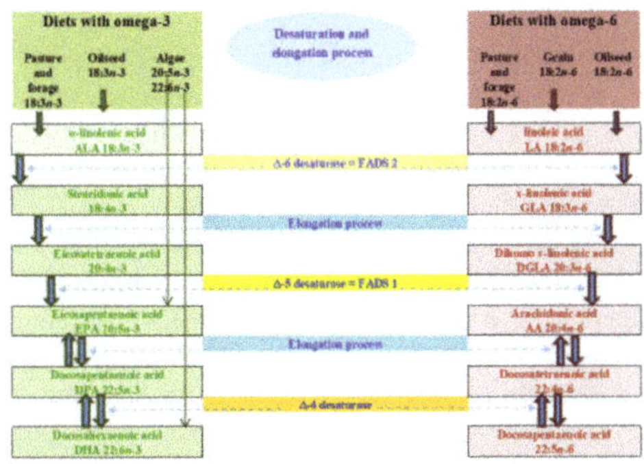

Figure 2. Dietary sources and biosynthesis of omega-3 and omega-6 fatty acids through enzymatic desaturation and elongation processes adapted from Ponnampalam et al. [23].

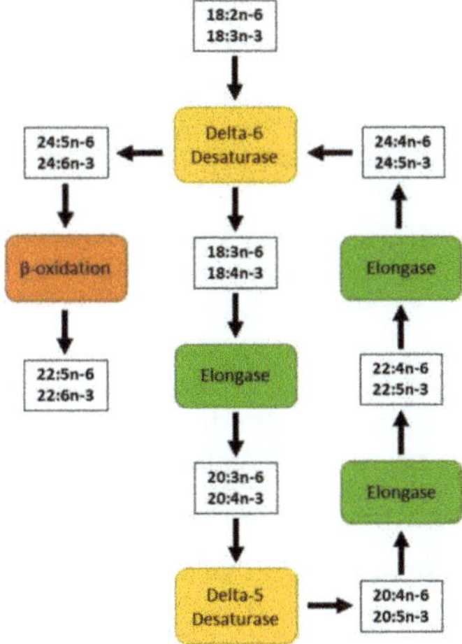

Figure 3. A diagram of omega-3 and omega-6 fatty acid elongation and desaturation to highlight the second use of delta-6 desaturase adapted from Gibson, Neumann, Lien, Boyd and Tu [22].

Long chain PUFA (LCPUFA) in animal and human muscle tissues are mainly found in phospholipids, where they play a major role in the metabolic, functional and physiological status of the body, organelles and tissues. In vivo studies conducted in animals have indicated that the relative levels of n-3 and n-6 LCPUFA in animal tissues can be regulated by altering the balance of ALA and LA in the diet. Gibson, Neumann, Lien, Boyd and Tu [22] used rats as a model species to show that feeding ALA at 1–3% and LA at 1–2% of dietary energy, while maintaining the intake of total PUFA less than 3% of dietary energy, DHA in plasma phospholipid can be positively and linearly increased. Mammals can convert ALA into LCPUFA such as EPA, DPA and DHA via a series of desaturase and elongase catalysed reactions [24]. Both the FA desaturase 1 (FADS1) and FA desaturase 2 (FADS2) prioritise ALA compared with LA. High LA intake, such as characterised by grain finishing or feedlot feeding of animals, can interfere with the desaturation of ALA and also of 24:5n-3, which is a precursor of 24:6n-3, the final precursor of DHA (Figure 3). The concentration of ALA present in the phospholipids of plasma and tissues is usually less than 0.5%. It is not known whether this level is sufficient for FADS2 to compete with LA, which is comparatively more abundant in animal tissues [25]. Past research has indicated that the conversion of ALA to DHA is not immediate, nor as effective as direct consumption of fish or a fish oil supplement [26–28].

Human studies conducted using isotope-labelled ALA have shown that males, when compared to females, are less efficient at synthesising EPA and DHA from ALA. The estimated net conversion rates of ALA to EPA is 21% for females and 8% for males, and of ALA to DHA is 9% for females and 0% for males. Sex differences in EPA and DPA content have been observed, with females having higher erythrocyte phospholipid EPA, lower adipose tissue EPA and lower plasma DPA content than males. There was a significant difference between sexes in terms of human response to increased dietary ALA, with females having a significantly greater increase in the EPA content of plasma phospholipids after six months of an ALA-rich diet compared to males [29]. A detailed study of genetically divergent sheep, raised in several disparate production regions, showed there to be a small gender effect on health claimable fatty acid content EPA and DHA such that females had higher levels than males. As female lambs approach their reproductive stage, it is possible that they synthesise more n-3 PUFA in the body for the production of series-3 eicosanoids, which is associated with the ovulation process, conception and pregnancy. Lambs from Merino dams had about 2 mg/100 g higher levels of EPA + DHA than lambs from crossbred dams when the sire breed was Poll Dorset. This is similar to Ponnampalam et al. [30], who found that the ratio of PUFA to saturated FA (SFA) in meat increased from second cross Poll Dorset to first cross Poll Dorset and from first cross Poll Dorset to purebred Merino. This same study also found this to be due to an increase in PUFA, and not due to a decrease in SFA [30].

Metabolism studies using stable isotope labelling, candidate gene single nucleotide polymorphisms (SNP), genome-wide association studies (GWAS) and metabolomics show interindividual variation in the conversion of LCPUFA precursors to LCPUFA products depends on genetic factors [31]. The FA desaturase genes (FADS1 and FADS2) code for enzymes that catalyse the introduction of double bonds at specific positions in a FA chain. FADS1 (D5-desaturase) and FADS2 (D6/D8/D4-desaturase) have specificity for several FA substrates [32]. Minor allele homozygotes (D/D) had significantly lower expression of FADS1 than the I/I major allele homozygotes. ARA is the immediate product of FADS1, leading directly to the hypothesis that individuals carrying D/D genotype have lower metabolic capacity to produce LCPUFA from precursors than I/I individuals. It was reported that individuals with I/I genotype having higher metabolic capacity to convert precursors to longer chain PUFA may be at increased risk for proinflammatory disease states as they efficiently convert LA to ARA [2] as FADS SNP was found to influence synthesis of ARA and synthesis of pro-inflammatory lipoxygenase products.

3. Sources

Twenty and 22 carbon LCPUFA, especially ARA, EPA and DHA, are ubiquitous in mammalian tissue, are bioactive components of membrane phospholipids and serve as precursors to cell signalling eicosanoids and docosanoids that are major drug targets (e.g., COX-1, COX-2 inhibitors, leukotriene receptor antagonists). LCPUFA can be obtained directly from animal foods or endogenously synthesised from 18 carbon essential FA precursors LA and ALA and their metabolites by an alternating series of desaturation and elongation reactions [32]. Vegans rely on this biochemical pathway to generate all LCPUFA from precursors. Classic carnivores (e.g., cats and most marine fish) have lost the metabolic ability to make LCPUFA and rely on consumption of animal tissue or fish to supply all their LCPUFA requirements.

3.1. Feeding Type and Digestive System of Ruminants

Cattle, sheep, goats, buffalo, yak, alpacas and deer are categorised as ruminants and are unable to digest plant material directly because they lack the enzymes needed to break down cell walls (cellulose and hemicellulose). Ruminants have a complex four-chambered stomach, comprising rumen, reticulum, omasum and abomasum, due to the nature of the high roughage feedstuffs they consume. Ruminant animals support a large population of bacteria, protozoans and fungi in their four-chambered stomach because they consume a large proportion (80–85%) of highly fibrous plant materials (roughage diets). Ruminant microorganisms play a major role in the degradation of undigestible fibrous materials, thereby making use of the dietary energy and nutrients by themselves as well as providing a medium for digestion and absorption in the small (duodenum, jejunum and ileum) and large (cecum, colon and rectum) intestines of the host animals. The important function of the salivary gland is adding saliva to the feeds to form bolus and to buffer pH levels in the rumen and reticulum so that the microbial activity and degradation process is optimised. The rumen and reticulum are home for the population of microorganisms that ferment and break down plant materials and produce volatile organic compounds and release other nutrients—both microbes and host animals use these volatile organic compounds for energy. The anatomical and functional attributes of small intestine of ruminants is similar to non-ruminants and ranges in length between approximately 12–30 times the body length of the animal [33].

3.2. Digestion, Absorption and Deposition of Dietary Lipids in Tissue of Livestock

Lipids are either consumed or synthesised de novo to contribute structure, integrity, recognition systems and energy to cells of most tissues. Not many researchers realise that the digestion and absorption of lipids (or fats) in ruminant and monogastric animals are different. This is due to their feeding nature and structure of digestive systems. In general, diets consumed by ruminants consist of 80–85% roughage and 15–20% concentrate while the diets consumed by monogastric animals are the opposite. More details on digestion, absorption and metabolism of dietary lipids can be found elsewhere [33,34]. Ruminant diets generally consist of 1–4% fat, and lipid supplements fed to ruminants above 5–6% on a dry matter basis have negative effects on rumen microbial activity, mainly on carbohydrate (fibre as cellulose and hemicellulose) degradation, particularly when PUFA are included in the diet. Supplementation of lipids in ruminant diets have some benefits to livestock industry in the following aspects: (1) it helps reducing the methane emission to environment from degradation of high fibrous diets; (2) it helps bypassing the dietary lipids (PUFA) from rumen to small intestine for absorption avoiding biohydrogenation; and (3) saving the dietary energy captured from methane emission for assimilation of tissue growth. With monogastric animals having a stomach as one organ for temporary storage of diet (fats) in the absence of rumen microbial activity, they can handle greater amounts of lipids in their diet for digestion and absorption process.

In any species, acetate (mainly cattle and sheep as ruminants) or glucose (mainly swine and poultry as monogastric animals) is absorbed in the intestine to enter FA biosynthesis via malonyl-CoA production through the acetyl-CoA carboxylase reaction and then palmitate production through FA synthase. Once palmitate is synthesised, other medium to long chain SFA and MUFA are generated by desaturation and elongation process. Since animals cannot synthesise essential PUFA (ALA and LA), these lipids have to come from consumed feeds. In monogastric animals, dietary fats are unchanged by digestion in the intestine so that tissue FA more directly reflect their present in the diet. Several steps are involved in resynthesis and transport of lipids in ruminants from the enterocyte where FA are absorbed, until they reach the peripheral tissues such as adipose and muscle tissues. The FA, monoglycerides and diglycerides reaching the jejunum from micelles are absorbed into the epithelial cells of small intestine. These FA are esterified, and triglycerides and phospholipids are assembled into lipoprotein particles (chylomicrons, very low-density lipoproteins, etc.) in the enterocyte, which are then secreted into lymph vessels and enter the bloodstream.

In monogastric animals, the liver plays a major role in FA synthesis. In ruminant animals, the contribution of liver is minimal and, instead, FA synthesis is very extensive in adipose tissue. Upon entry to the blood, chylomicrons and very low-density lipoproteins acquire apoproteins apo-C and apo-E provided by high-density lipoprotein. Apo-C inhibits liver removal of chylomicrons and very low-density lipoproteins and this enhances the extent of diversion of these entities to other tissues. One of the apo-C components activates the lipoprotein lipase enzyme, which is situated primarily on the surface of the endothelium of skeletal muscle, adipose and mammary tissue sites. FAs and partial glycerides are apportioned to triglycerides, phospholipids and other lipids in the organs or oxidation for energy according to the metabolic demands of the body either in skeletal muscles, adipose and/or mammary tissues. The state of dietary lipids rich in n-3 PUFA from digestion in the intestine to deposition in the peripheral tissues through the circulatory systems is shown in Figure 4.

In the circulatory or tissue systems, diacylglycerol (DAG) is produced from phosphatidylcholine or from other phospholipids. Phospholipase C (PLC) cleaves the membrane phospholipid phosphatidylinositol-4,5-bisphosphate (PIP_2) to generate inositol-1,4,5-trisphosphate (IP_3) and DAG. Another type of phospholipase, phospholipase D (PLD), is activated by various stimuli in the cell. PLD hydrolyses phosphatidylcholine, which is abundant in the cell plasma membrane, producing phosphatidic acid and choline. Phosphatidic acid is hydrolysed by phosphatidic acid hydrolase to release DAG and phosphate. This is a second pathway that generates DAG. While this intermediate is the product of the action of both PLC and PLD, cellular responses in both cases are usually not identical due to differences in the cellular localisation of enzymes or the fatty acid composition of the DAG produced. The stimulation of specific cell-surface receptors activates phospholipase A2, leading to the release of arachidonic acid from the cell membrane.

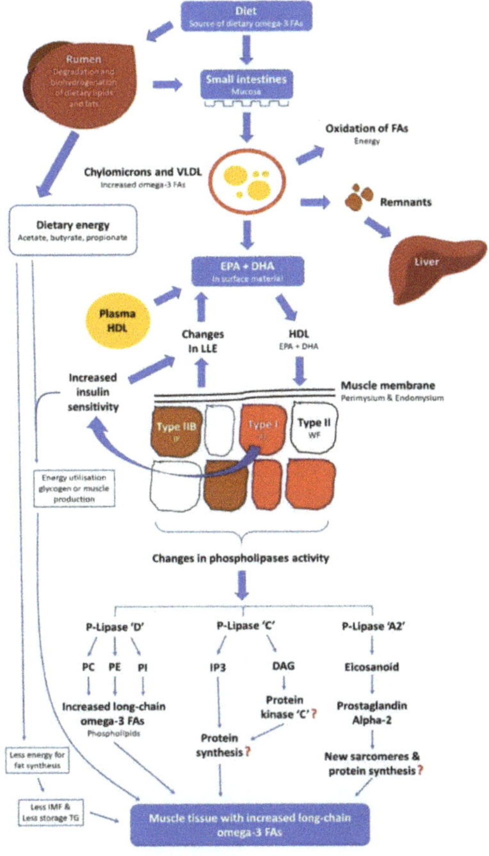

Figure 4. The mechanisms of dietary omega-3 fatty acid digestion and metabolism in ruminants, adapted from Ponnampalam [34]. Abbreviations include fatty acid (FA); very low-density lipoproteins (VLDL); eicosapentaenoic acid (EPA); docosahexaenoic acid (DHA); high-density lipoproteins (HDL); lipoprotein lipase enzyme (LLE); intermediate fibres (IF); red fibres (RF); white fibres (WF); phosphatidylcholine (PC); phosphatidylethanolamine (PE); phosphatidylinositol (PI); Inositol triphosphate (IP3); diacylglyceride (DAG); and triglyceride (TG or triacylglycerol (TAG)).

3.3. Factors Affecting Omega-3 and Omega-6 Fatty Acid Deposition in Muscle Tissues (Meat) of Ruminants

The modern era has brought advancement in agricultural practices and food processing. However, large scale commercially oriented crop and animal production has, in general, decreased n-3 PUFA concentrations, increased n-6 PUFA concentrations and increased n-6/n-3 ratios in meat, eggs and milk when sourced from intensively farmed animals compared to animals living in range lands. Within this context, there are three things to recognise. The amount of a specific n-3 PUFA deposited in the cells and tissues (1) are not directly related to the amounts of n-3 PUFA present in the animal diets; (2) are related to the amount of n-6 PUFA available in the feeds for consumption and amount already deposited in the peripheral tissues; and (3) are dependent on the interference of desaturation and elongation enzymes reactions of both FA families that generate their LCPUFA in tissues. Many different factors influence the n-3 and n-6 PUFA concentrations of meat in ruminants, which will be briefly discussed below.

A. Type of based diet: It has been well established that pasture feeding systems or fodder feeds containing silage produce meat with more n-3 PUFA than the grain feeding or lot feeding system. Green leafy materials contain more ALA in the chloroplasts, whereas the grain-based or cereal-based diets often contain more LA and MUFA. For example, a high concentration of LA and AA is found in meat from ruminants consuming grain-based feeds in contrast to ruminants consuming brassica-based feeds or lucerne (alfalfa) pasture-based feeds which have more ALA, EPA and DHA in their meat [35].
B. Type of supplement (ingredient) used in the diet: Supplementation of oilseeds and meals from sunflower, safflower, corn and cotton to ruminant diets could elevate n-6 PUFA whilst flax, canola, chia and camelina can increase the concentration of n-3 PUFA within meat [36]. LA is found in most plant and cereal seeds and is abundantly available naturally for ruminant consumption. Therefore, it is expected that the concentration of n-6 PUFA would be greater than n-3 PUFA and lead to a higher n-6/n-3 ratio in the meat from these animals.
C. Form of lipid present in the diet: It has been reported that the efficacy of direct absorption of product FAs (AA, EPA and DHA) were greater than parent FA (LA and ALA). For example, when EPA and DHA were directly consumed by ruminants as algae or fish oil supplementation, the deposition of EPA and DHA were greater than when feeding diet containing ALA (flaxseed) to increase the concentration of EPA and DHA within cells and tissues [6,20,21], and the same outcomes can be applied to other n-6 PUFA. Moreover, the ratio of EPA, DHA and ALA to the total PUFA in the diet can determine the amounts of EPA, DHA or ALA absorbed at the enterocyte and further deposition at the peripheral tissue sites [22].
D. Competition between desaturation and elongation enzymes: The affinity of FADS2 for ALA is greater than LA. Nevertheless, high intake of LA or high concentration of dietary LA can adversely affect the conversion of ALA within tissues. The nature of the current animal production systems prefers the application of concentrate feeding for fast growth and quick turn over to market, offering large quantity of dietary LA available for animal consumption and this results in greater deposition of LA and AA than their counterparts of ALA and its derivatives EPA, DHA or DPA [37]. A recent study indicated that the application of forages, such as lucerne hay around 50%, in the mixed ration, compared with a ration containing 50% of grain made of barley and oats, significantly altered the concentration of ALA, LA, and EPA of muscle fat in two sheep types of diverse genetics namely pure Merino and Crossbred sheep. This same study showed a diet by breed interaction that was proposed to be the result of different concentration of ALA and LA being deposited in muscle fat of those two genetics, allowing for domination of one type of FA on another when deposited at or above certain concentrations in the muscle fat [38]. When the animals were fed lucerne hay diet, the crossbred lambs produced higher ALA in the tissues and had 20 mg higher ALA/100 g tissue than the Merino lambs (ALA concentrations for Crossbred and Merino animals were 50 vs. 31 mg/100 g muscle). This was not observed in animals fed the grain-based diet, and the ALA concentrations for crossbred and Merino animals were 30 vs. 15 mg/100 g muscle. The greater ALA concentration in muscle tissues from crossbred lambs fed the lucerne hay diet might have suppressed the LA deposition whilst enhancing the elongation process of ALA to its longer chain EPA. The LA concentrations for crossbred and Merino animals fed lucerne hay diet were 103 vs. 95 mg/100 g muscle while the values for grain fed sheep were 168 vs. 138 mg/100 g muscle, respectively.
E. Availability of secondary metabolites: Animal feeds contain a vast range of secondary metabolites called phytonutrients. Pastures, fodder crops and higher plant species produce (secrete) these phytonutrients in their body for their protection, survival and establishment against disease, pest and harsh climate under various conditions. These phytonutrients have health enhancing compounds and animals may selectively

consume these pastures, fodders and other by-products of field crops to support their good health and wellbeing. Phytonutrients are classified as alkaloids, polyphenols, organosulfur compounds and so on. It is likely that animal feeds containing some types of polyphenols, such as tannins, phenolic acids and flavonoids, can protect dietary PUFA from the hydrolysis and biohydrogenation in the rumen resulting in beneficial effects. Hence, increased n-3 LCPUFA would be available for absorption across enterocytes and, therefore, have increased deposition within tissue and meat. This is possibly due to these phytonutrients having low bioavailability and long retention times within the rumen, causing a slow degradation of fibrous diets by the microflora and allowing the PUFA and other nutrients present in the diet to bypass the rumen and be available for intestinal absorption by host animals.

F. Level of antioxidants and carotenoids in muscle tissues: Ruminants are specialised to consume 80–85% of diet as forage (fibrous materials) such as green pastures, fodders, silage and other forage materials. From these diets, they ingest adequate amounts of antioxidants, such as vitamins, minerals and carotenoids. Monogastric animals grown under intensive systems consume 80–85% concentrated diets and they receive carotenoids and antioxidants from the ingredients of cereal grains, protein meal and oilseeds. Carotenoids, such as carotenes and xanthophylls, are pigments present in leaves, seeds, fruits and animal products of blood, meat and milk. Carotenoids have the ability to act as antioxidants as they are quenchers of singlet oxygen ($^1O_2$) and other reactive oxygen species (ROS) or substances that causes oxidative damage in the body from cell to tissue level. The biological roles of carotenoids and polyphenols in the ruminant digestive system and their metabolism are not yet fully understood. It is speculated that increased level of antioxidant potential in the circulatory and tissue systems can protect the oxidation of n-3 LCPUFA from the tissues. This improves the health and wellbeing of individuals. For example, several studies [7,21,35–37] described the relationship among antioxidants, n-3 PUFA and lipid oxidation in muscle tissues in sheep and goats.

4. Biological Actions

In the context of human health and wellness, ALA, EPA, DPA, DHA and their secondary metabolites have been the focus of attention throughout the previous 50 years. It seems reasonable to suggest that the mode of actions and effects would be similar between farm and companion animals. The dramatic advancement in the analytical technologies of n-3 PUFA, identification of their intermediate metabolites and understanding of their important role in human growth, development and disease prevention has facilitated the introduction of new dietary regulations and recommendations for foods high in PUFA and *trans* FAs, particularly n-3 PUFA and vaccenic acid. In this context, long chain PUFA are considered essential and/or health enhancing nutrients that impact on growth and development in early life as well as metabolic disorders and chronic diseases in later life.

DHA is a major constituent of cardiomyocytes, sperm, grey matter of the brain and the retina. Several studies have indicated DHA is necessary for central nervous system functionality as well as the visual activity of infants. The 20 and 22 carbon chain-length PUFA (i.e., EPA, DHA, and AA) can be converted to a series of hormone-like substances called eicosanoids and docosanoids, respectively, including prostaglandins (PGs), thromboxanes (TXs), prostacyclin (PGI2), leukotrienes (LTs), resolvins (RVD) and other lipid mediators (Figure 5). These eicosanoids and docosanoids contain many intermediary metabolites and isoforms. These agents play major roles in the regulation of diverse pathophysiological functions, including blood pressure, platelet aggregation, blood clotting, blood lipid profiles, immune response, the inflammation response to injury and infections and the resolution of inflammation [15,39]. A large proportion of research conducted in laboratory animals and humans has been devoted to the pathophysiological functions and properties of EPA, DPA and DHA and the roles of the derived lipid mediators.

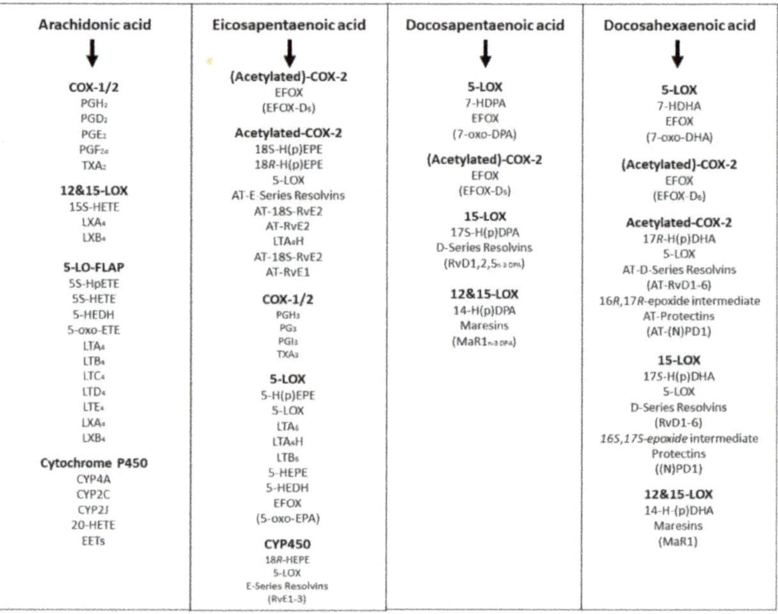

Figure 5. The formation of lipid mediators (intermediary metabolites and isoforms) from eicosanoids and docosanoids derived from long chain omega-3 (EPA, DPA, DHA) and omega-6 (AA) fatty acids in animals and human tissues or body.

The ARA is then rapidly converted into two major classes of enzymes, called cyclooxygenases (COX) and lipoxygenases (LOX). COX enhance the production of prostaglandins, prostacyclin and thromboxanes, while lipoxygenase enhance the production of leukotrienes. The physiological actions of these metabolites are widespread and diverse. Briefly, prostaglandins and prostacyclin are potent vasodilators whilst thromboxanes are potent vasoconstrictors, whereas leukotrienes produce bronchoconstriction. Lipoxygenases in plants and animals are heme-containing dioxygenases that oxidise PUFA at specific carbon sites to give enantiomers of hydroperoxide derivatives with conjugated double bonds. The number in specific enzyme names such as 5-LOX, 12-LOX, or 15-LOX refers to the ARA site that is predominantly oxidised. Of these, 5-LOX is best known for its role in the biosynthesis of leukotrienes A4, B4, C4, D4 and E4. The oxidised metabolites generated by 5-LOX were found to alter the intracellular redox balance and to induce signal transduction pathways and gene expression. The enzyme 5-LOX has been identified as an inducible source of ROS production in lymphocytes [40]. Cyclooxygenase-1 has been implicated in ROS production through formation of endoperoxides, which are susceptible to scavenging by some antioxidants in cells stimulated with TNF-α, interleukin-1, bacterial lipopolysaccharide, or the tumour promoter 4-otetradecanoylphorbol-13-acetate [41].

DPA is another n-3 LCPUFA which has potential in maintaining health and wellness of animals and humans. Its applicability and efficacy in terms of metabolic activity and disease prevention have not been fully investigated. However, it deserves attention for various reasons, including that it is the intermediate substrate of EPA conversion to DHA within the cell or tissue systems of animals and humans, and its tissue concentration is dependent on the balance between EPA and DHA. There is emerging evidence that DPA levels are positively correlated with the expression of certain enzymes involved in inflammatory processes of the cardiovascular system [42]. Research indicates that DHA and its metabolites are used for tissue-based metabolic activities such as insulin-stimulated energy disposal, phospholipid-induced signal transduction towards gene expression, active autoimmune systems towards cell defence, etc. Within these contexts, the overall n-3 PUFA metabolic process must be efficient in converting EPA to DHA via DPA and the reverse reaction of DPA to EPA; and the availability of DPA in the tissue system is important for transitional processes to maintain both EPA and DHA levels. It is noteworthy that the concentration of DPA in red meat is equal or greater than that of EPA and DHA. Therefore, the contribution of DPA from red meat should not be neglected or ignored in terms of its role in the maintenance of health and wellness of people who regularly consume more red meat than fish or vegetables.

5. Dietary Recommendation

Past research has determined adults (aged 18 years and older) to have no upper intake limit for n-3 PUFA to ensure their safety, *viz.* ALA, EPA, DHA and DPA [43,44]. Instead, it is apparent that diets which fail to provide the minimum requirements of these FA are a greater health concern. In response, many authorities and organisations from around the world have proposed guidelines that define the daily recommendable intakes for n-3 PUFA, n-6 PUFA and total LCPUFA (Table 3). From these, we can observe that male and female adults have different requirements and, furthermore, the requirement for females will depend on their physiological status (e.g., every day, during pregnancy, during lactation). This complements previous knowledge that age will impact on dietary requirements for these PUFA, with children aged less than 18 years proposed to require a diet that contains more n-3 PUFA and n-6 PUFA than necessary for an adult [45].

The examples included in Table 3 also show that dietary FA guidelines differ between organisations and, sometimes, these differences are substantial. A possible basis for this disparity could be the basal diets typical to the populations represented by these organisations. For example, MHLW [49] identify the diet of the Japanese population to be comparatively lower in n-6 PUFA and, therefore, a lower requirement for n-3 PUFA is necessary to achieve an acceptable n-6/n-3 ratio. This is reliant on the n-6/n-3 ratio's importance to human health, an observation previously made and resulting in ratio recommendations that range from 5:1–10:1 for adults [46]. That said, the FAO [54] report that if n-3 PUFA and n-6 PUFA intakes adhere to their individual guidelines then there is no rationale to support a recommendation for n-6/n-3 ratio intake. This premise does depend on their being no biochemical competition or inhibition between the functionalities of n-6 PUFA and n-3 PUFA that affects their bioavailability.

Table 3. Examples of omega-3 (n-3), omega-6 (n-6), and total long chain polyunsaturated fatty acid (LCPUFA) dietary recommendations for a healthy adult (18 years and older). Abbreviations include male (M); female (F); alpha-linolenic acid (ALA); linoleic acid (LA); percentage total energy (%E); eicosapentaenoic acid (EPA); docosahexaenoic acid (DHA); and acceptable macronutrient distribution range (AMDR). Please note that for female adults, recommendations are categorised using physiological status (as everyday, during pregnancy and during lactation).

Organisation	n-3	n-6	LCPUFA
NHMRC [46] Australia and New Zealand	M: 1.3 g/day ALA F: 0.8 ǀ 1.0 ǀ 1.2 g/day ALA	M: 13 g/day LA F: 8 ǀ 10 ǀ 12 g/day LA	M: 160 mg/day LCPUFA F: 90 ǀ 110–115 ǀ 140–145 mg/day LC-PUFA
Health Canada [47] Canada	M: 1.6 g/day ALA F: 1.1 ǀ 1.4 ǀ 1.3 g/day ALA	M: 14–17 g/day LA F: 11–12 ǀ 13 ǀ 13 g/day LA	-
EFSA [44] Europe Union	4.0%E	0.5%E	250 mg/day EPA + DHA +100–200 mg/day DHA (during pregnancy or lactation)
[48] France	0.8%E ALA	2.0%E LA	Min. physiological requirement: DHA 0.1 E%
MHLW [49] Japan	M: 2.0–2.4 g/day (total) F: 1.1 ǀ 1.4 ǀ 1.3 g/day (total)	M: 8–11 g/day (total) F: 7–8 ǀ 9 ǀ 9 g/day (total)b	
MVO [50] The Netherlands	1%E ALA	2%E LA	450 mg/day EPA + DHA
Bartrina and Majem [51] Spain	1–2%E ALA	5%E LA	500–1000 mg/day EPA + DHA
Otten [52] United States of America	5–10%E	20–35%E	10%E EPA + DHA
Food and Nutrition Board [53] United States of America	M: 1.2–1.6 g/day ALA F: 1.1–1.2 ǀ 1.3 ǀ 1.3 g/day ALA	M: 14–17 g/day (total) F: 21–26 ǀ 28 ǀ 29 g/day (total)b	2 g/day U-AMDR
FAO [54] United Nations	0.5%E ALA	2.5%E LA	-

6. Concentration Range in Red Meats

Approximately 80% of Australians are not meeting the recommended n-3 LCPUFA intake for optimum health [55]. The same is likely true for majority of the population in other Western countries such as United States of America and the United Kingdom. The general advice from dieticians and health professionals is to consume 2–3 fish meals weekly to elevate the n-3 LCPUFA levels in the body. Fish and other seafood are rich sources of n-3 LCPUFA. Nevertheless, the consumption of LCPUFA (i.e., EPA and DHA) from marine-based foods is low to very low in many Western countries, and alternative food sources for these FA may therefore be advantageous in these populations.

Unlike human and other monogastric animals, ruminants are mainly dependent on the microbial population in the rumen for the digestion and absorption of dietary lipids. This microbial activity is responsible for the hydrolysis of dietary lipids and further isomerisation and conversion of unsaturated FAs into MUFA and SFA intermediates. The latter process leads to an increase in stearic acid (C18:0) concentration for small intestine absorption. However, several studies have reported that PUFA content in red meat from ruminants can be significantly modified by feeding systems. Pasture- and silage-fed animal deliver meat with higher PUFA, particularly in terms of n-3 PUFA content, when compared with their grain-fed and concentrate feedlot fed counterparts [56–58]. Particular secondary metabolites found in pasture and forage diets may exert a greater protection against microbial biohydrogenation of PUFA in the rumen and, therefore, facilitate increased absorption and deposition of PUFA in ruminant tissues [59]. Previous research has also shown considerable differences on animal growth performances, carcass characteristics and meat quality attributes in sheep and cattle fed forage-based diet versus concentrate diet and these areas have been discussed in detail by others [60].

Western populations consume more meat and processed meat products than marine-based foods due to preference, availability and affordability. As a consequence, red meat can contribute up to 20% of their n-3 LCPUFA requirements [61]. An earlier study in humans consuming lean red meat showed that 2 weeks consumption of 500 g lean meat per day was sufficient to raise plasma DPAn-3 levels [62]. This may therefore be considered an alternative or complementary source for those with poor fish consumption. The enrichment of n-3 PUFA levels in meat through dietary management has been a focus in the animal production systems for the past 20 years. This aims to improve n-3 PUFA consumption for those who consume lower amounts of meat [20,34,63,64]. For this reason, many studies have investigated feeding lipid sources such as marine-based oils, and grains and oilseeds in ruminant and monogastric animals. To our knowledge, the studies conducted with algae, fish oil, flaxseed and canola seed supplementation and, likewise, using specialised forage or grazing options have shown prominent outcomes in increasing the n-3 LCPUFA levels in ruminants [65–69]. This information provides insight into management practices that can optimise the nutritional value of the meat products. That said, ALA is the primary FA source from plant-based diets. A good understanding of the efficacy of the elongation process of ALA to EPA, DPA and DHA in ruminants is long overdue [17].

7. Preservation until the Point of Consumption

Management systems have been adopted that enrich animal tissues so that they become a source of n-3 PUFA and n-6 PUFA. These efforts are often implemented without first considering the interim between processing and consumption. This is important as longer carbon chain-length FA with double C-H bonds (e.g., EPA or DHA or AA) are more susceptible to oxidation than shorter or more hydrogenated FA, such as MUFA [70,71]. Manifestations of this effect are observed in Adeyemi et al. [72], with results showing that n-3 PUFA and n-6 PUFA concentrations in goat meat declined across 12 days of a chilled storage period at 4 °C; in Muino et al. [73], with findings that lamb PUFA decreased in a linear trend with increased chilled storage period when held in oxygen-rich modified atmospheric packaging; and in Diaz et al. [74], with conclusions that 6 days of chilled storage at 2 °C was sufficient to degrade the PUFA content of lamb meat. It is interesting, therefore, that Holman et al. [75] observed no change in beef PUFA composition across a 12 week chilled storage period—although the authors suggest this was an outcome of anaerobic storage, low initial levels of PUFA and a relatively high concentration of vitamin E within the samples.

Vitamin E (α-tocopherol) has been widely acknowledged as practical and intrinsic means to preserve the FA profile of meat. Indeed, for lamb meat, Ponnampalam et al. [76] proposed a tissue concentration of greater than 3.45 mg/kg vitamin E as sufficient to inhibit excessive peroxidation. A similar recommendation was made for beef, with Arnold et al. [77] concluding that concentrations of 3.3 mg/kg vitamin E were appropriate. The concentrations of tissue vitamin E have been reportedly improved with animal supplementation [78,79] and functions as a result of its inhibition of the production of reactive oxygen species and propagation of free radical reactions [80]. Alternatively, meat may be stored within anaerobic packaging conditions or with embedded antioxidants to inhibit peroxidation. Examples have been described in Holman et al. [81] with review of different patents for smart packaging devices and antioxidant coatings that can scavenge specific gases, including oxygen, from an in-pack atmosphere to preserve against oxidation, and can be implemented within packaging systems to assure anaerobic conditions. That said, temperature controls (cold-chain, frozen storage) and vacuum packaging alone may be enough to prevent excessive peroxidation if their consistency and efficacy can be confirmed across the interim. From these findings, it is recommended that the preservation of FA composition beyond its immediate enhancement should be considered when seeking to enrich the composition of different food types.

8. Conclusions

We report here that animal and human foods in the modern era are composed of higher n-6 PUFA levels and n-6/n-3 ratio compared to foods consumed by humans and animals during early evolutionary periods. These variations are primarily the consequence of changes to agricultural practice, animal production systems and food processing during the last 100–200 years. Changes in ecosystems and climate variability also contribute to these variations. Taking Australian production systems as an example, it is perceived that the application of low-nutritive or low-quality roughage diets (haylage), crop residues, senesced hay materials in the ruminant production systems is vital for sustainable and resilient future animal industries, but this will further reduce the n-3 PUFA and vitamin consumption and as a consequence in red meat. Commercially based animal industries using proportionately high concentrate diets in their animal feeds may also be attributable to increased consumption of n-6 PUFA and reduced n-3 PUFA by livestock and, thus, elevated n-6/n-3 ratio in red meat. It is likely that animals grazing single stand pasture (monoculture) receive lower amounts of essential FA, vitamins and minerals than those grazing mixed pastures. This is due to limited selection of herbage materials that are rich in nutrient values. It is known that ruminants consuming feeds rich in lipids mainly n-3 PUFA (oils and fats—e.g., diets containing brassica family members such as canola, camelina, or flax) also emit lower amounts of methane to the ecosystem than those consuming diets with highly fibrous structural carbohydrates, such as diets high in cellulose. Forages high in secondary metabolites such as polyphenols (tannins, flavonoids and phenolic acids), alkaloids and carotenoids may protect n-3 PUFA against microbial fermentation and biohydrogenation in the rumen due to their low bioavailability, allowing PUFA to reach the intestine for the absorption by host animals. This observation notwithstanding, additional research is necessary in this area to better understand the biological pathways and mechanisms of actions.

A sustainable animal and plant production system is essential for economic viability and the health and welfare of animals and humans, reinforcing the consideration of n-3 PUFA and n-6 PUFA in animal feeding systems equivalent to range feeding. The literature clearly indicates that animal grazing diets high in essential FA and vitamins have better metabolic conditions and oxidative status than those consuming diets of low nutritive value, contributing to improved wellness and lower veterinary care. It is likely that ruminant animal feeding systems will, in the future, utilise more concentrate-based specialised diets, which consist of less n-3 PUFA, to tackle the extended dry seasons and shortage in green pasture with climate variation. This scenario requires that the producers and researchers identify forage diets and supplements high in n-3 PUFA, vitamins and phytonutrients whilst low in n-6 PUFA and structural carbohydrates (cellulose, lignin) so that the health and wellbeing of animals and humans can be advantaged. The capacity to maintain the essential PUFA, vitamins and trace elements in meat from farm to fork and throughout processing and preservation must also be considered. Taken together, we state that offering n-3 PUFA rich diets to animals has many advantages economically, environmentally and socially, not only for animals but also for those humans who consume red meat in moderate to high quantities.

Author Contributions: Conceptualisation, E.N.P.; writing—original draft preparation, E.N.P., A.J.S. and B.W.B.H. All authors have read and agreed to the published version of the manuscript.

Funding: This research received no external funding.

Acknowledgments: The authors are grateful to the support of their corresponding organisations, specifically Agriculture Victoria and NSW Department of Primary Industries. The first author (Eric N Ponnampalam) wishes to acknowledge Andrew J Sinclair's contribution as an adviser and mentor in his early scientific research career development (year 1994–2000) in the area covering the nutritional management of omega-3 fatty acids in ruminants for better production aspects and animal health and wellbeing. Sinclair is internationally recognised and published his research in *Nature* four decades ago (Rivers, Sinclair and Crawford [16]) reporting on the importance of omega-3 fatty acids for a healthy life.

Conflicts of Interest: The authors declare no conflict of interest.

References

1. Pewan, S.B.; Otto, J.R.; Huerlimann, R.; Budd, A.M.; Mwangi, F.W.; Edmunds, R.C.; Holman, B.W.B.; Henry, M.L.E.; Kinobe, R.T.; Adegboye, O.A.; et al. Genetics of omega-3 long-chain polyunsaturated fatty acid metabolism and meat eating quality in Tattykeel Australian White lambs. *Genes* **2020**, *11*, 587. [CrossRef] [PubMed]
2. Kothapalli, K.S.D.; Ye, K.; Gadgil, M.S.; Carlson, S.E.; O'Brien, K.O.; Zhang, J.Y.; Park, H.G.; Ojukwu, K.; Zou, J.; Hyon, S.S. Positive selection on a regulatory insertion–deletion polymorphism in FADS2 influences apparent endogenous synthesis of arachidonic acid. *Mol. Biol. Evol.* **2016**, *33*, 1726–1739. [CrossRef] [PubMed]
3. Malau-Aduli, A.E.O.; Edriss, M.A.; Siebert, B.D.; Bottema, C.D.K.; Pitchford, W.S. Breed differences and genetic parameters for melting point, marbling score and fatty acid composition of lot-fed cattle. *J. Anim. Physiol. Anim. Nutr.* **2000**, *83*, 95–105. [CrossRef]
4. Holman, B.W.B.; Flakemore, A.R.; Kashani, A.; Malau-Aduli, A.E.O. Spirulina supplementation, sire breed, sex and basal diet effects on lamb intramuscular fat percentage and fat melting points. *Int. J. Vet. Med.* **2014**, *2014*, 263951. [CrossRef]
5. Benatti, P.; Peluso, G.; Nicolai, R.; Calvani, M. Polyunsaturated fatty acids: Biochemical, nutritional and epigenetic properties. *J. Am. Coll. Nutr.* **2004**, *23*, 281–302. [CrossRef]
6. Ponnampalam, E.N.; Sinclair, A.J.; Egan, A.R.; Blakeley, S.J.; Li, D.; Leury, B.J. Effect of dietary modification of muscle long-chain n-3 fatty acid on plasma insulin and lipid metabolites, carcass traits, and fat deposition in lambs. *J. Anim. Sci.* **2001**, *79*, 895–903. [CrossRef]
7. Ponnampalam, E.N.; Vahedi, V.; Giri, K.; Lewandowski, P.; Jacobs, J.L.; Dunshea, F.R. Muscle antioxidant enzymes activity and gene expression are altered by diet-induced increase in muscle essential fatty acid (α-linolenic acid) concentration in sheep used as a model. *Nutrients* **2019**, *11*, 723. [CrossRef]
8. Simopoulos, A.P. Omega-3 fatty acids in health and disease and in growth and development. *Am. J. Clin. Nutr.* **1991**, *54*, 438–463. [CrossRef]
9. Simopoulos, A.P. Importance of the omega-6/omega-3 balance in health and disease: Evolutionary aspects of diet. In *Healthy Agriculture, Healthy Nutrition, Healthy People: World Review of Nutrition and Dietetics*; Simopoulos, A.P., Ed.; Karger Publishers: Basel, Switzerland, 2011; Volume 102, pp. 10–21.
10. Aranceta, J.; Pérez-Rodrigo, C. Recommended dietary reference intakes, nutritional goals and dietary guidelines for fat and fatty acids: A systematic review. *Br. J. Nutr.* **2012**, *107*, 8–22. [CrossRef]
11. Sinclair, A.J. Docosahexaenoic acid and the brain—What is its role? *Asia Pac. J. Clin. Nutr.* **2019**, *28*, 675–688.
12. Dunstan, G.A.; Sinclair, A.J.; O'Dea, K.; Naughton, J.M. The lipid content and fatty acid composition of various marine species from southern Australian coastal waters. *Comp. Biochem. Physiol. Part B Comp. Biochem.* **1988**, *91*, 165–169. [CrossRef]
13. Sinclair, A.J.; O'Dea, K. The lipid levels and fatty acid compositions of the lean portions of Australian beef and lamb. *Food Technol. Aust.* **1987**, *39*, 228–231.
14. Simopoulos, A.P. An increase in the omega-6/omega-3 fatty acid ratio increases the risk for obesity. *Nutrients* **2016**, *8*, 128. [CrossRef]
15. Ratnayake, W.M.N.; Galli, C. Fat and fatty acid terminology, methods of analysis and fat digestion and metabolism: A background review paper. *Ann. Nutr. Metab.* **2009**, *55*, 8–43. [CrossRef]
16. Rivers, J.P.W.; Sinclair, A.J.; Crawford, M.A. Inability of the cat to desaturate essential fatty acids. *Nature* **1975**, *258*, 171. [CrossRef] [PubMed]
17. Sinclair, A.J.; Attar-bashi, N.M.; Li, D. What is the role of alpha-linolenic acid for mammals? *Lipids* **2002**, *37*, 1113–1123. [CrossRef] [PubMed]
18. De Gomez Dumm, I.N.T.; Brenner, R.R. Oxidative desaturation of α-linolenic, linoleic, and stearic acids by human liver microsomes. *Lipids* **1975**, *10*, 315–317. [CrossRef]
19. Emken, E.A.; Adlof, R.O.; Rakoff, H.; Rohwedder, W.K.; Gulley, R.M. Metabolism in vivo of deuterium-labelled linolenic and linoleic acids in humans. *Biochem. Soc. Trans.* **1990**, *18*, 766–769. [CrossRef] [PubMed]
20. Simopoulos, A.P. New products from the agri-food industry: The return of n-3 fatty acids into the food supply. *Lipids* **1999**, *34*, 297–301. [CrossRef]
21. Simopoulos, A.P. Omega-3 fatty acids and antioxidants in edible wild plants. *Biol. Res.* **2004**, *37*, 263–277. [CrossRef]

22. Gibson, R.A.; Neumann, M.A.; Lien, E.L.; Boyd, K.A.; Tu, W.C. Docosahexaenoic acid synthesis from alpha-linolenic acid is inhibited by diets high in polyunsaturated fatty acids. *Prostaglandins Leukot. Essent. Fat. Acids* **2013**, *88*, 139–146. [CrossRef] [PubMed]
23. Ponnampalam, E.N.; Hopkins, D.L.; Jacobs, J.L. Increasing omega-3 levels in meat from ruminants under pasture-based systems. *Rev. Sci. Tech.* **2018**, *37*, 57–70. [CrossRef] [PubMed]
24. Sprecher, H. Metabolism of highly unsaturated n-3 and n-6 fatty acids. *Biochem. Biophys. Acta* **2000**, *1486*, 219–231. [CrossRef]
25. Burdge, G.C.; Calder, P.C. Conversion of alpha-linolenic acid to longer-chain polyunsaturated fatty acids in human adults. *Reprod. Nutr. Dev.* **2005**, *45*, 581–597. [CrossRef]
26. Arterburn, L.M.; Hall, E.B.; Oken, H. Distribution, interconversion, and dose response of n-3 fatty acids in humans. *Am. J. Clin. Nutr.* **2006**, *83*, 1467–1476. [CrossRef]
27. Innis, S.M.; Hansen, J.W. Plasma fatty acid responses, metabolic effects, and safety of microalgal and fungal oils rich in arachidonic and docosahexaenoic acids in healthy adults. *Am. J. Clin. Nutr.* **1996**, *64*, 159–167. [CrossRef]
28. Park, Y.; Harris, W. EPA, but not DHA, decreases mean platelet volume in normal subjects. *Lipids* **2002**, *37*, 941–946. [CrossRef]
29. Childs, C.E.; Kew, S.; Finnegan, Y.E.; Minihane, A.M.; Leigh-Firbank, E.C.; Williams, C.M.; Calder, P.C. Increased dietary α-linolenic acid has sex-specific effects upon eicosapentaenoic acid status in humans: Re-examination of data from a randomised, placebo-controlled, parallel study. *Nutr. J.* **2014**, *13*, 1–5. [CrossRef]
30. Ponnampalam, E.N.; Hopkins, D.L.; Butler, K.L.; Dunshea, F.R.; Sinclair, A.J.; Warner, R.D. Polyunsaturated fats in meat from Merino, first- and second-cross sheep slaughtered as yearlings. *Meat Sci.* **2009**, *83*, 314–319. [CrossRef]
31. Illig, T.; Gieger, C.; Zhai, G.; Römisch-Margl, W.; Wang-Sattler, R.; Prehn, C.; Altmaier, E.; Kastenmüller, G.; Kato, B.S.; Mewes, H.-W. A genome-wide perspective of genetic variation in human metabolism. *Nat. Genet.* **2010**, *42*, 137–141. [CrossRef] [PubMed]
32. Park, W.J.; Kothapalli, K.S.D.; Lawrence, P.; Tyburczy, C.; Brenna, J.T. An alternate pathway to long-chain polyunsaturates: The FADS2 gene product Δ8-desaturates 20: 2n-6 and 20: 3n-3. *J. Lipid Res.* **2009**, *50*, 1195–1202. [CrossRef] [PubMed]
33. Swanson, K.C. Small intestinal anatomy, physiology, and digestion in ruminants. In *Reference Module in Food Science;* Elsevier: Amsterdam, The Netherlands, 2019.
34. Ponnampalam, E.N. Nutritional Modification of Muscle Long Chain Omega-3 Fatty Acids in Lamb: Effects on Growth and Composition and Quality of Meat. Ph.D. Thesis, The University of Melbourne, Melbourne, Australia, 1999.
35. Ponnampalam, E.N.; Plozza, T.; Kerr, M.G.; Linden, N.; Mitchell, M.; Bekhit, A.E.D.; Jacobs, J.L.; Hopkins, D.L. Interaction of diet and long ageing period on lipid oxidation and colour stability of lamb meat. *Meat Sci.* **2017**, *129*, 43–49. [CrossRef] [PubMed]
36. Ponnampalam, E.N.; Lewandowski, P.A.; Fahri, F.T.; Burnett, V.F.; Dunshea, F.R.; Plozza, T.; Jacobs, J.L. Forms of n-3 (ALA, C18:3n-3 or DHA, C22:6n-3) fatty acids affect carcass yield, blood lipids, muscle n-3 fatty acids and liver gene expression in lambs. *Lipids* **2015**, *50*, 1133–1143. [CrossRef]
37. Ponnampalam, E.N.; Norng, S.; Burnett, V.F.; Dunshea, F.R.; Jacobs, J.L.; Hopkins, D.L. The synergism of biochemical components controlling lipid oxidation in lamb muscle. *Lipids* **2014**, *49*, 757–766. [CrossRef]
38. Ponnampalam, E.N.; Dunshea, F.R.; Warner, R.D. Use of lucerne hay in ruminant feeds to improve animal productivity, meat nutritional value and meat preservation under a more variable climate. *Meat Sci.* **2020**, *170*, 108235. [CrossRef]
39. Fard, S.G.; Cameron-Smith, D.; Sinclair, A.J. n–3 Docosapentaenoic acid: The iceberg n–3 fatty acid. *Curr. Opin. Clin. Nutr. Metab. Care* **2021**, *24*, 134–138. [CrossRef]
40. Bonizzi, G.; Piette, J.; Merville, M.-P.; Bours, V. Cell type-specific role for reactive oxygen species in nuclear factor-kappaB activation by interleukin-1. *Biochem. Pharmacol.* **2000**, *59*, 7–11. [CrossRef]
41. Dröge, W. Free radicals in the physiological control of cell function. *Physiol. Rev.* **2002**, *82*, 47–95. [CrossRef]
42. Solakivi, T.; Jaakkola, O.; Kalela, A.; Pispa, M.; Salomäki, A.; Lehtimäki, T.; Höyhtyä, M.; Jokela, H.; Tnikkari, S. Lipoprotein docosapentaenoic acid is associated with serum matrix metalloproteinase-9 concentration. *Lipids Health Dis.* **2005**, *4*, 1–8.
43. EFSA. Scientific opinion on the tolerable upper intake level of eicosapentaenoic acid (EPA), docosahexanenoic acid (DHA) and docosapentaenoic acid (DPA). *Eur. Food Saf. Auth. EFSA J.* **2012**, *10*, 2815.
44. EFSA. *Dietary Reference Values for Nutrients: Summary Report;* European Food Safety Authority (EFSA): Parma, Italy, 2017; p. 92. Available online: www.efsa.europa.eu/sites/default/files/2017_09_DRVs_summary_report.pdf (accessed on 24 September 2019).
45. Elmadfa, I.; Kornsteiner, M. Fats and fatty acid requirements for adults. *Ann. Nutr. Metab.* **2009**, *55*, 56–75. [CrossRef] [PubMed]
46. NHMRC. *Nutrient Reference Values for Australia and New Zealand Including Recommended Dietary Intakes;* Australian Government Department of Health and Ageing, New Zealand Ministry of Health: Canberra, Australia, 2019.
47. Health Canada. Dietary Reference Intake Tables; Government of Canada, Health Canada. 2010. Available online: www.canada.ca/content/dam/hc-sc/migration/hc-sc/fn-an/alt_formats/hpfb-dgpsa/pdf/nutrition/dri_tables-eng.pdf (accessed on 24 September 2019).
48. FAFEOHS. Actualisation des Apports Nutritionnels Conseillés pour les Acides Gras. *Rapport D'expertise Collective 2011; French Agency for Food, Environmental and Occupational Health & Safety (FAFEOHS).* 2011. Available online: www.anses.fr/fr/system/files/NUT2006sa0359Ra.pdf (accessed on 24 September 2019).
49. MHLW. *Dietary Reference Intakes for Japanese;* Health Service Bureau, Ministry of Health, Labour and Welfare (MHLW): Chiyoda-ku, Japan, 2015; Available online: www.mhlw.go.jp/file/06-Seisakujouhou-10900000-Kenkoukyoku/Full_DRIs2015.pdf (accessed on 24 September 2019).

50. MVO. Dutch Recommendations for Adults and Children as from 1 Year of Age; The Netherlands Oils and Fats Industry (MVO): 2011. Available online: www.mvo.nl/en-vetzuursamenstelling (accessed on 24 September 2019).
51. Bartrina, J.A.; Majem, L.S. Objetivos nutricionales para la población española: Consenso de la Sociedad Española de Nutrición Comunitaria 2011. *Rev. Española Nutr. Comunitaria Span. J. Community Nutr.* **2011**, *17*, 178–199.
52. Otten, J.J. *Dietary Reference Intakes: The Essential Guide to Nutrient Requirements*; National Academy of Sciences: Washington, DC, USA, 2006; p. 1323. Available online: www.nal.usda.gov/sites/default/files/fnic_uploads/DRIEssentialGuideNutReq.pdf (accessed on 24 September 2019).
53. Food and Nutrition Board. *Dietary Reference Intakes for Energy, Carbohydrate, Fiber, Fat, Fatty Acids, Cholesterol, Protein, and Amino Acids*; National Academy of Sciences: Washington, DC, USA, 2005.
54. FAO. *Fats and Fatty Acids in Human Nutrition. Report of an Expert Consultation*; Food and Agriculture Organisation of the United Nations (FAO): Rome, Italy, 2010; Available online: www.fao.org/3/a-i1953e.pdf (accessed on 24 September 2019).
55. Meyer, B.J. Australians are not meeting the recommended intakes for omega-3 long chain polyunsaturated fatty acids: Results of an analysis from the 2011–2012 national nutrition and physical activity survey. *Nutrients* **2016**, *24*, 111. [CrossRef] [PubMed]
56. Dewhurst, R.J.; Shingfield, K.J.; Lee, M.R.F.; Scollan, N.D. Increasing the concentrations of beneficial polyunsaturated fatty acids in milk produced by dairy cows in high-forage systems. *Anim. Feed Sci. Technol.* **2006**, *131*, 168–206. [CrossRef]
57. Nuernberg, N.; Dannerberger, D.; Nuernberg, G.; Ender, K.; Voigt, J.; Scollan, N.D.; Wood, J.D.; Nute, G.R.; Richardson, R.I. Effect of a grass-based and a concentrate feeding system on meat quality characteristics and fatty acid composition of longissimus muscle in different cattle breeds. *Livest. Prod. Sci.* **2005**, *94*, 137–147. [CrossRef]
58. Realini, C.E.; Duckett, S.K.; Brito, G.W.; Dalla Rizza, M.; De Mattos, D. Effect of pasture vs. concentrate feeding with or without antioxidants on carcass characteristics, fatty acid composition, and quality of Uruguayan beef. *Meat Sci.* **2004**, *66*, 567–577. [CrossRef]
59. Lourenço, M.; Van Ranst, G.; Vlaeminck, B.; De Smet, S.; Fievez, V. Influence of different dietary forages on the fatty acid composition of rumen digesta as well as ruminant meat and milk. *Anim. Feed Sci. Technol.* **2008**, *145*, 418–437. [CrossRef]
60. De Brito, G.F.; Ponnampalam, E.N.; Hopkins, D.L. The effect of extensive feeding systems on growth rate, carcass traits, and meat quality of finishing lambs. *Compr. Rev. Food Sci. Food Saf.* **2017**, *16*, 23–38. [CrossRef]
61. Russo, G.L. Dietary n−6 and n−3 polyunsaturated fatty acids: From biochemistry to clinical implications in cardiovascular prevention. *Biochem. Pharmacol.* **2009**, *77*, 937–946. [CrossRef]
62. Sinclair, A.; Johnson, L.; O'Dea, K.; Holman, R. Diets rich in lean beef increase the eicosatrienoic, arachidonic, eicosapentaenoic and docosapentaenoic acid content of plasma phospholipids. *Lipids* **1994**, *29*, 337–343. [CrossRef]
63. Scollan, N.; Hocquette, J.F.; Nuernberg, K.; Dannenberger, D.; Richardson, I.; Moloney, A. Innovations in beef production systems that enhance the nutritional and health value of beef lipids and their relationship with meat quality. *Meat Sci.* **2006**, *74*, 17–33. [CrossRef]
64. Simopoulos, A.P. Overview of evolutionary aspects of ω3 fatty acids in the diet. In *The Return of ω3 Fatty Acids into the Food Supply*; Simopoulos, A.P., Ed.; Karger Publishers: Basel, Switzerland, 1998; Volume 83, pp. 1–11.
65. Kashani, A.; Holman, B.W.B.; Nichols, P.D.; Malau-Aduli, A.E.O. Effect of level of spirulina supplementation on the fatty acid composition of adipose, muscle, heart, kidney and liver tissues in Australian dual-purpose lambs. *Ann. Anim. Sci.* **2015**, *15*, 945–960. [CrossRef]
66. De Brito, G.F.; Holman, B.W.B.; McGrath, S.R.; Friend, M.A.; van de Ven, R.J.; Hopkins, D.L. The effect of forage-types on the fatty acid profile, lipid and protein oxidation, and retail colour stability of muscles from White Dorper lambs. *Meat Sci.* **2017**, *130*, 81–90. [CrossRef]
67. Cooper, S.L.; Sinclair, L.A.; Wilkinson, R.G.; Hallett, K.G.; Enser, M.; Wood, J.D. Manipulation of the n-3 polyunsaturated fatty acid content of muscle and adipose tissue in lambs. *J. Anim. Sci.* **2004**, *82*, 1461–1470. [CrossRef]
68. Kronberg, S.L.; Barcelo-Coblijn, G.; Shin, J.; Murphy, E.J. Bovine muscle n-3 fatty acid content is increased with flaxseed feeding. *Lipids* **2006**, *41*, 1059–1068. [CrossRef]
69. Ponnampalam, E.N.; Burnett, V.F.; Norng, S.; Hopkins, D.L.; Plozza, T.; Jacobs, J.L. Muscle antioxidant (vitamin E) and major fatty acid groups, lipid oxidation and retail colour of meat from lambs fed a roughage based diet with flaxseed or algae. *Meat Sci.* **2016**, *111*, 154–160. [CrossRef] [PubMed]
70. Porter, N.A.; Caldwell, S.E.; Mills, K.A. Mechanisms of free radical oxidation of unsaturated lipids. *Lipids* **1995**, *30*, 277–290. [CrossRef]
71. Cosgrove, J.P.; Church, D.F.; Pryor, W.A. The kinetics of the autoxidation of polyunsaturated fatty-acids. *Lipids* **1987**, *22*, 299–304. [CrossRef]
72. Adeyemi, K.D.; Sabow, A.B.; Shittu, R.M.; Karim, R.; Karsani, S.A.; Sazili, A.Q. Impact of chill storage on antioxidant status, lipid and protein oxidation, color, drip loss and fatty acids of semimbranosus muscle in goats. *J. Food* **2016**, *14*, 405–414.
73. Muino, I.; Apeleo, E.; Fuente, J.D.L.; Perez-Santaescolastica, C.; Rivas-Canedo, A.; Perez, C.; Diaz, M.T.; Caneque, V.; Lauzurica, S. Effect of dietary supplementation with red wine extract or vitamin E, in combination with linseed and fish oil, on lamb meat quality. *Meat Sci.* **2014**, *98*, 116–123. [CrossRef]
74. Diaz, M.T.; Caneque, V.; Sanchez, C.I.; Lauzurica, S.; Perez, C.; Fernandez, C.; Alvarez, I.; Fuente, J.D.L. Nutritional and sensory aspects of light lamb meat enriched in n-3 fatty acids during refrigerated storage. *Food Chem.* **2011**, *124*, 147–155. [CrossRef]

75. Holman, B.W.B.; Bailes, K.L.; Kerr, M.J.; Hopkins, D.L. Point of purchase fatty acid profile, oxidative status and quality of vacuum-packaged grass fed Australian beef held chilled for up to 12 weeks. *Meat Sci.* **2019**, *158*, 107878. [CrossRef] [PubMed]
76. Ponnampalam, E.N.; Butler, K.L.; Pearce, K.M.; Mortimer, S.I.; Pethick, D.W.; Ball, A.J.; Hopkins, D.L. Sources of variation of health claimable long chain omega-3 fatty acids in meat from Australian lamb slaughtered at similar weights. *Meat Sci.* **2014**, *96*, 1095–1103. [CrossRef] [PubMed]
77. Arnold, R.N.; Arp, S.C.; Scheller, K.K.; Williams, S.N.; Schaefer, D.M. Tissue equilibration and subcellular distribution of vitamin E relative to myoglobin and lipid oxidation in displayed beef. *J. Anim. Sci.* **1993**, *71*, 105–118. [CrossRef]
78. Holman, B.W.B.; Baldi, G.; Chauhan, S.S.; Hopkins, D.L.; Seymour, G.R.; Dunshea, F.R.; Collins, D.; Ponnampalam, E.N. Comparison of grain-based diet supplemented with synthetic vitamin E and lucerne hay-based diet on blood oxidative stress biomarkers and lamb meat quality. *Small Rumin. Res.* **2019**, *177*, 146–152. [CrossRef]
79. Alvarez, I.; Fuente, J.D.L.; Caneque, V.; Lauzurica, S.; Perez, C.; Diaz, M.T. Changes in the fatty acid composition of M. longissimus dorsi of lamb during storage in a high-oxygen modified atmosphere at different levels of dietary vitamin E supplementation. *J. Agric. Food Chem.* **2009**, *57*, 140–146. [CrossRef]
80. McDowell, L.R.; Williams, S.N.; Hidiroglou, N.; Njeru, C.A.; Hill, G.M.; Ochoa, L.; Wilkinson, N.S. Vitamin E supplementation for the ruminant. *Anim. Feed Sci. Technol.* **1996**, *60*, 273–296. [CrossRef]
81. Holman, B.W.B.; Kerry, J.P.; Hopkins, D.L. A review of patents for the smart packaging of meat and muscle-based food products. *Recent Pat. Food Nutr. Agric.* **2018**, *9*, 3–13. [CrossRef]

Article

Fatty Acids and Nutraceutical Properties of Lipids in Fallow Deer (*Dama dama*) Meat Produced in Organic and Conventional Farming Systems

Janusz Kilar [1] and Anna Kasprzyk [2,*]

[1] Institute of Agricultural and Forest Economy, Jan Grodek State University in Sanok, 21 Mickiewicza, 38-500 Sanok, Poland; janusz.kilar@wp.pl
[2] Institute of Animal Breeding and Biodiversity Conservation, University of Life Sciences in Lublin, 13 Akademicka, 20-950 Lublin, Poland
* Correspondence: anna.kasprzyk@up.lublin.pl

Abstract: The aim of the study was to assess the fatty acid profile and nutraceutical properties of lipids contained in fallow deer (*Dama dama*) meat produced in organic and conventional farming systems. *Longissimus lumborum* (LL) and *semimembranosus* (SM) muscles from 24 fallow deer carcasses were selected for the study. The fallow deer meat from the organic farming system was characterized by significantly lower intramuscular fat content. The fatty acid profile in the organic meat was characterized by a particularly high proportion ($p < 0.0001$) of conjugated linoleic acid—CLA (LL—2.29%, SM—2.14%), alpha-linolenic acid—ALA (LL—4.32%, SM—3.87%), and docosahexaenoic acid—DHA (LL—2.83%, SM—2.60%). The organic system had a beneficial effect ($p < 0.0001$) on the amount of polyunsaturated fatty acids (PUFAs), including n-3 PUFAs, which resulted in a more favorable n-6 PUFA (polyunsaturated fatty acid)/n-3 PUFA ratio. The significantly higher nutritional quality of organic meat lipids was confirmed by such nutraceutical indicators as the thrombogenic index (TI), Δ9-desaturase C16, elongase, and docosahexaenoic acid+eicosapentaenoic acid (DHA+EPA) in the LL and SM and cholesterol index (CI), and the cholesterol-saturated fat index (CSI) indices in the SM. LL was characterized by higher overall quality.

Keywords: venison; feeding system; muscles; intramuscular fat; cholesterol; fatty acids

Citation: Kilar, J.; Kasprzyk, A. Fatty Acids and Nutraceutical Properties of Lipids in Fallow Deer (*Dama dama*) Meat Produced in Organic and Conventional Farming Systems. *Foods* **2021**, *10*, 2290. https://doi.org/10.3390/foods10102290

Academic Editors: Benjamin W. B. Holman and Eric Nanthan Ponnampalam

Received: 6 August 2021
Accepted: 23 September 2021
Published: 27 September 2021

1. Introduction

The dynamic changes in nutrition in terms of food quality and health safety requirements prompt the consumer to look for food produced in strictly defined systems, e.g., in the organic farming system [1,2]. This is confirmed by the continuous increase in the global organic food market [3–5] and the frequency of purchase of such products [6]. There is a common belief that organic food has higher nutritional values, is healthier and safer [7–10], and reduces the risk of overweight and obesity [11]. Recent literature reviews and meta-analyses have revealed significant differences in the nutritional composition of organic and conventional food [5,8]. Organic meat has better nutritional properties than industrially produced meat, which is reflected in a considerably improved composition of animal fats (higher concentrations of n-3 fatty acids), higher content of biologically active compounds, and lower cholesterol levels [1,12]. A higher frequency of consumption of organic meat in the diet can significantly reduce the prevalence of various lifestyle diseases [13].

Meat from game animals is a highly valued culinary and processing raw material due to its flavor, excellent nutritional properties, and high pH stability in the maturation process [14–16]. It can be classified as very lean meat [16–18]. Over the last few years, the interest in this meat has been increasing as it is regarded as "natural and sustainable" [19]. Wild-living animals experience a high standard of welfare and eat mainly natural food. Their meat is free of antibiotics and hormones [20]. Meat of various cervid species is also produced on farms. In Poland, the red deer (*Cervus elaphus*), the sika deer (*Cervus nippon*), and especially the fallow deer (*Dama dama*) are reared quite commonly [21]. Compared with other cervid species, fallow deer meat is regarded as a healthier product due to the higher content of n-3 polyunsaturated fatty acids (PUFAs) [17]. It is high in protein, heme iron, copper, zinc, and potassium and low in saturated fatty acids [17,18,22–25]. The meat of cervids is also a good source of nutritionally important conjugated linoleic acid (CLA) [25]. Due to the essential role of fat in the human diet, the quantity and quality of lipids in meat from various animal species are the focus of considerable research interest [7,26–31].

The aim of the study was to evaluate the fatty acid profile and nutraceutical properties of lipids in fallow deer (*Dama dama*) meat produced in organic and conventional farming systems. Additionally, an attempt was made to estimate the nutritional contribution of this type of meat in the diet of adults in terms of the levels of total fat, fatty acids, and cholesterol.

2. Materials and Methods

2.1. Meat Samples, Animals, and Treatments

Given the special culinary preferences of Polish consumers for two parts of cervid meat, e.g., the loin and leg [32,33], the research material consisted of longissimus lumborum (LL) and semimembranosus (SM) muscles dissected from 12 carcasses of fallow deer reared on an organic farm and 12 carcasses of fallow deer kept in a conventional farming system. In both groups of animals, there were six does and six bucks with an equal age ratio of ca. 18 and 30 months. The farms were located in the area of the Beskid Niski Mts., Podkarpackie Province. The organic farm followed the requirements of Regulation (EU) 2018/848 of the European Parliament and Council of 30 May 2018 [34] and the Act on organic farming (Journal of Laws 2009, No. 116, item 975) [35]. The organic farming certificate covered the grazing area, animals, fodder, and all other rearing procedures. The staple food for the animals was provided by a natural grazing ground with a density of 0.42 LU (large livestock units)/ha. The 191 plant species identified in the pasture [36] represented the following floristic groups: grasses—10.99%, legumes—22.51%, sedges—2.09%, rushes—2.62%, dicotyledonous herbs—44.50%, deciduous trees—6.28%, coniferous trees—1.05%, blackberries—1.05%, shrubs—4.71%, shrublets—2.62%, ferns—0.53%, and horsetails—1.05%. In winter, the animals received hay, straw, cereal grains, and carrots ad libitum.

The animals from the conventional farming system were reared at a density of 0.67 LU/ha following the DEFRA [37] and FEDFA [38] recommendations. The pasture comprised 72 plant species [36] from the following floristic groups: grasses—22.22%, legumes—11.11%, sedges—1.39%, herbaceous dicotyledons—43.05%, deciduous trees—5.56%, coniferous trees—2.78%, blackberries—1.39%, shrubs—6.94%, shrublets—4.17%, and horsetails—1.39%. In winter, the animals received hay, haylage, straw, and fodder beets ad libitum.

The animals were slaughtered by shooting from October to December upon the consent and supervision of veterinary services. The post-slaughter treatments, e.g., evisceration, skinning, and veterinary inspection of carcasses, were carried out after transporting the carcasses in a refrigeration truck to an authorized processing plant. During the dissection of cooled carcasses (cooling for 48 h at a temperature of 4 °C), three equal-sized samples of LL and SM were collected and an approx. 500-g pooled sample placed in polyethylene bags was transported in an isothermal container to the laboratory. The meat was stored at −200 C until laboratory analyses.

2.2. Chemical Analysis

Fat was extracted with the method proposed by Folch [39]. Approximately 5 g of meat were homogenized with 5 mL of methanol. The samples were extracted using an automated Soxhlet extractor (Soxtec Avanti, Tecator). The extracted lipids were converted into fatty acid methyl esters (FAME). Fatty acids were saponified with 0.5 N KOH in methanol at

80 °C and then esterified with boron trifluoride/methanol in accordance with the PN-ISO 1444:2000 standard [40]. Separation and quantification of the fatty acid methyl esters was carried out using a gas chromatograph (Varian 450-GC with an FID detector) equipped with a flame ionization detector and fitted with a Select™ Biodiesel for FAME capillary column (30 m × 0.32 mm internal diameter, and 0.52 µm film thickness, Shinwa Inc.) A split/splitless injection system (split ratio of 1:50) and helium as a carrier gas at a flow rate of 1.5 mL/min were used. The injection port and the detector were maintained at 250 and 270 °C, respectively. The column oven temperature was programmed at 100 °C, and finally held at 240 °C for 20 min. The identification of individual FAMEs was based on a standard mixture of 37 Component FAME Mix-CRM47885, St. Louis, MO, USA. The Galaxie ™ Chromatography Data System software was used to convert the results. All samples were analyzed in triplicate. The results were expressed as g/100 g of total identified fatty acids.

The cholesterol content in the muscles was determined with the SOP M.023a method (2011) [41]. The analysis was performed with a gas chromatograph (GC—2010 Shimadzu) equipped with an on-column capillary injector and a flame ionization detector. A capillary column (Zebron ZB-5, L = 30 mm, I.D. = 0.25 mm; df. = 0.5 µm) and a ramped oven temperature were used (increased to 150 °C from 100 °C at 30 °C/min, then increased to 360 °C at 15 °C/min). The cholesterol content was expressed as mg/100 g of fresh meat.

Using the content of individual fatty acids (FA), the following parameters were calculated: SFA—saturated fatty acids, UFA—unsaturated fatty acids, MUFA—monounsaturated fatty acids, PUFA—polyunsaturated fatty acid, PUFA n-6—polyunsaturated fatty acid n-6, PUFA n-3—polyunsaturated fatty acid n-3, OFA—hypercholesterolemic fatty acids = (C12:0 + C14:0 + C16:0), and SFA/UFA; MUFA/SFA; PUFA/SFA; PUFA n-6/PUFA n-3.

The assessment of the nutritional quality of lipids was based on calculation of the following parameters: DFA—desirable fatty acids = (MUFA + PUFA + C18:0), AI—atherogenic index = [C12:0 + 4×C14:0 + C16:0]/[MUFA + PUFA] [42], TI—thrombogenic index = [C14:0 + C16:0 + C18:0]/[0.5×MUFA + 0.5×n-6 + 3×n-3 + n-3/n-6] [42], h/H—ratio of hypo- and hypercholesterolemic fatty acids = [C18:1c9 + C18:2n-6 + C18:3n-3 + C20:3n-6 + C20:4n-6 + C20:5n-3 + C22:5n3]/[C12:0 + C14:0 + C16:0] [7], NV—nutritional value = [C12:0 + C14:0 + C16:0]/[C18:1c9 + C18:2n-6] [43], DHA + EPA—sum of docosahexaenoic acid and eicosapentaenoic acids, HPI—health-promoting index = UFA/[C12:0 +(4 × C14:0) + C16:0] [44], CI—cholesterol index = 1.01 (g of SFA 100 g^{-1} of fresh matter—0.5 × g of PUFA 100 g−1 of fresh matter) + (0.06 × mg of cholesterol 100 g^{-1} of fresh matter) [45], and CSI—cholesterol-saturated fat index = (1.01 × g of SFA 100 g^{-1} of fresh matter) + (0.05 × mg of cholesterol 100 g−1 fresh matter) [46]. The activities of Δ9-desaturase C16 = [C16:1/(16:0 + C16:1)] × 100, Δ9-desaturase C18 [C18:1n9c/(18:0 + C18:1n9c)] × 100, and elongase = [C18:0 + C18:1n9c]/[C16:0 + C16:1+ C18:0 + C18:1n9c] × 100 [29] were calculated as well.

The nutritional contribution of 100 g of fallow deer meat in the diet for adults was estimated by comparison of the total fat and fatty acid content to the recommendations of FAO (2010) [47] and EU (2011) [48]. The cholesterol level was compared with to the recommendations of WHO/FAO (2003) [49].

2.3. Statistical Analysis

The numerical data were analyzed with methods of descriptive statistics and statistical hypothesis testing. The arithmetic mean ($\overline{x}$) and standard error (SE) were calculated. The statistical hypothesis testing was preceded by the examination of the normality of selected empirical distributions. The χ^2 chi-squared test showed that the empirical distributions were consistent with the normal distribution; hence, the hypotheses were verified with the one-way analysis of variance (ANOVA) and the F-test (Fisher–Snedecor). Differences were considered significant at $p \leq 0.05$. The data were analyzed using Statistica software (v. 13.3, TIBCO Software Inc., Palo Alto, CA, USA).

3. Results

Both muscles (LL, SM) of the fallow deer from the organic farm had significantly lower content of intramuscular fat (Figure 1). It was shown that the farming system had a significant impact on the cholesterol content only in the SM muscle (Figure 2). The cholesterol content in the SM muscle of the fallow deer from the organic farm was 3.47 mg/g lower ($p \leq 0.022$) than in the muscle of the conventionally farmed fallow deer. The cholesterol content in the LL muscle was 2.01 mg/g lower in the fallow deer from the conventional farm (Figure 2).

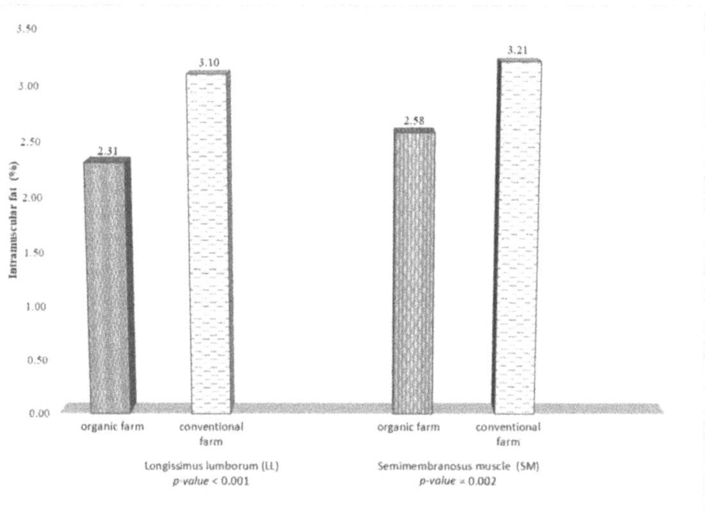

Figure 1. Intramuscular fat content (%) in the meat of fallow deer from organic and conventional farms.

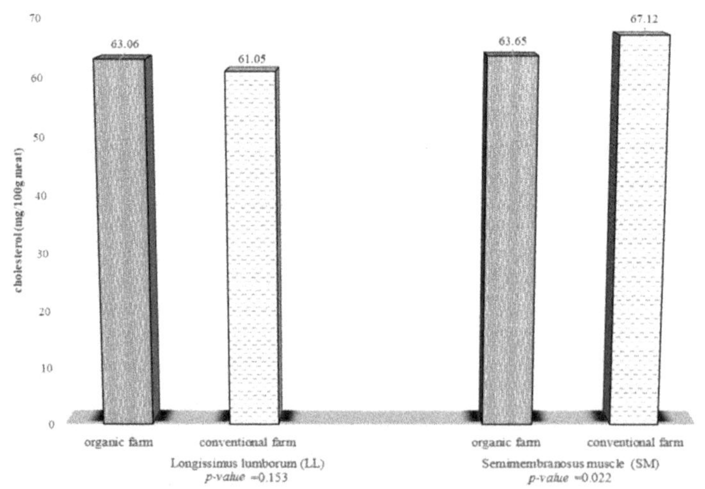

Figure 2. Cholesterol content (mg/100 g meat) in the meat of fallow deer from organic and conventional farms.

Table 1 shows the analyzed fatty acids. In comparison with the muscle samples from the conventional farm, the LL muscles of the organically reared fallow deer had a lower concentration of C12:0, C15:0, C18:0, and C21:0 and a higher level of C24:0 ($p \leq 0.0001$). In turn, significantly higher levels of C12:0, C15:0, C17:0, and C18:0 and lower contents of C21:0 and C24:0 were determined in the SM of the conventionally farmed fallow deer, compared to the SM muscles from the animals reared on the organic farm. The LL muscles of the conventionally farmed fallow deer were characterized by higher concentrations of C15:1, C20:1, and C24:1n-9 than those in the muscles of the other animals. In turn, the SM muscle of the organically farmed animals had lower ($p \leq 0.0001$) concentrations of C14:1, C15:1, and C20:1 and a higher level of C16:1, compared to the meat produced in the organic farming system. Increased levels of C18:2n-6t, C18:2c9t11, C18:3n-6, C18:3n-3, and C22:6n-3 ($p \leq 0.0001$) as well as C20:5n-3 ($p \leq 0.005$) were found in the LL muscle of the fallow deer from the organic farm, whereas the content of C20:3n-6 ($p \leq 0.0001$) was higher in the LL muscle of the conventionally farmed animals. The organic SM muscle had significantly higher levels of C18:2n-6c, C18:2n-6t, C18:2c9t11, C18:3n-3, C20:3n-6, and C22:6n-3.

In both muscles, the SFA sum was higher in the meat of the conventionally farmed fallow deer, but the differences ($p \leq 0.011$) were significant only in the SM muscle. The higher SFA sum was reflected in the higher SFA/UFA value (Table 2). The LL and SM muscles of the organically reared fallow deer had higher total PUFA content, including n-6 PUFAs and n-3 PUFAs. Additionally, the organic SM had higher ($p \leq 0.011$) UFA content than the muscle from the conventional farming system. A higher value of the PUFA/SFA ratio and a lower n-6 PUFA/n-3 PUFA ratio were recorded in the LL and SM muscles of the organically farmed fallow deer. In terms of the nutraceutical properties, the LL and SM muscles of the fallow deer from the organic farm had a higher TI value and higher EPA + DHA content. The cholesterol-saturated fat index (CSI) was significantly higher ($p \leq 0.008$) in the MS muscle from the conventional system. Significantly higher activity of Δ9-desaturase C16 and lower activity of elongase were determined in both types of muscle of the organically farmed animals (Table 2).

Table 3 shows the FAO (Food and Agriculture Organization) recommended levels of fat and FAs as an energy source in a diet for adults [47] and the relevant EU recommendations for the dietary fat, FA (Fatty acid), and cholesterol intake [48]. In a 2000-kcal daily diet, the consumption of 100 g of fresh fallow meat from the organic farm covered the largest percentage of the EPA + DHA demand (over 28% in SM and 29% in LL). The recommended daily intake of other components was realized as follows: cholesterol (over 21%), n-3 PUFAs (mean 9.15% in SM and 8.81% in LL), fat (mean 4.58% in SM and 4.11% in LL), SFAs (3.81% in SM and 3.49% in LL), and n-6 PUFAs (mean 2.85% of SM and 2.40% of LL). Consumption of the conventionally farmed fallow deer meat was found to cover a slightly higher percentage of fat demand (mean 5.70% in SM and 5.51% in LL) and SFA demand (5.14% and 5.02%).

Table 1. Fatty acid profile (g·100 g^{-1} FA; mg·100 g^{-1} fresh meat) in the meat of fallow deer from organic and conventional farms.

Fatty Acids	Muscle													
	longissimus lumborum (LL)							*semimembranosus* (SM)						
	Organic Farm			Conventional Farm				organic Farm			Conventional Farm			
	$\bar{x}$	SE	mg·100 g^{-1}	$\bar{x}$	SE	mg·100g^{-1}	*p*-Value	$\bar{x}$	SE	mg·100g^{-1}	$\bar{x}$	SE	mg·100g^{-1}	*p*-Value
C10:0	0.09	0.022	1.45	0.07	0.013	1.66	0.123	0.09	0.026	1.64	0.08	0.013	1.71	0.180
C12:0	0.16	0.020	2.65	0.23	0.082	5.10	0.006	0.16	0.033	2.94	0.23	0.083	5.23	0.019
C13:0	0.19	0.223	3.09	0.18	0.261	3.87	0.967	0.17	0.207	2.96	0.24	0.344	5.31	0.571
C14:0	3.44	1.316	58.38	3.84	1.531	83.02	0.503	3.39	1.534	63.57	3.72	1.045	87.25	0.551
C15:0	0.87	0.393	13.90	1.49	0.492	33.11	0.002	0.38	0.256	7.33	1.74	0.253	40.14	<0.0001
C16:0	23.96	2.424	398.55	22.49	2.376	500.72	0.146	21.66	2.732	399.67	20.46	3.288	467.10	0.341
C17:0	1.10	0.506	17.33	0.86	0.281	18.70	0.154	0.51	0.238	9.59	1.23	0.418	28.99	<0.0001
C18:0	15.52	4.780	254.27	19.07	3.491	472.42	0.050	17.80	2.896	328.77	20.53	3.405	477.84	0.046
C20:0	0.30	0.220	5.12	0.50	0.298	11.51	0.069	0.21	0.067	3.83	0.23	0.103	5.33	0.578
C21:0	0.23	0.119	4.02	0.75	0.110	6.44	<0.0001	0.33	0.106	5.97	0.22	0.073	4.96	0.006
C22:0	0.07	0.029	1.18	0.08	0.013	1.75	0.593	0.13	0.041	2.34	0.22	0.144	5.00	0.055
C24:0	0.18	0.033	2.91	0.09	0.023	1.90	<0.0001	0.17	0.034	3.07	0.10	0.027	2.32	<0.0001
C14:1	2.57	0.232	42.65	2.87	0.839	64.09	0.248	1.63	0.409	29.60	3.29	1.186	76.82	<0.0001
C15:1	1.23	0.386	20.93	2.50	0.816	54.28	<0.0001	1.08	0.896	19.87	3.59	1.277	84.18	<0.0001
C16:1	3.56	0.927	57.83	1.90	1.155	41.53	0.001	3.89	0.942	70.99	2.11	1.138	49.58	<0.0001
C17:1	1.31	0.360	21.19	0.67	0.336	14.61	<0.0001	1.04	0.302	18.94	0.86	0.342	20.22	0.186
C18:1n9c	19.95	4.833	336.86	22.86	3.039	513.16	0.092	23.26	3.239	432.12	22.77	3.227	522.55	0.716
C18:1n9t	0.48	0.250	8.28	0.70	0.282	15.88	0.057	0.75	0.994	13.04	0.62	0.390	14.89	0.683
C20:1	0.23	0.119	4.01	0.75	0.110	16.82	<0.0001	0.16	0.073	2.98	0.64	0.088	14.80	<0.0001
C24:1n9	0.19	0.034	3.13	0.23	0.036	5.17	0.004	0.19	0.050	3.41	0.16	0.041	3.69	0.183
C18:2n6c (LA)	7.64	0.966	128.31	6.99	1.419	156.99	0.208	7.24	1.022	133.61	5.97	1.659	142.21	0.035
C18:2n6t	0.63	0.052	10.54	0.36	0.061	7.95	<0.0001	0.60	0.036	10.99	0.33	0.068	7.88	<0.0001
C18:2c9t11 (CLA)	2.29	0.123	37.97	1.60	0.078	35.86	<0.0001	2.14	0.343	39.22	1.66	0.073	38.50	<0.0001
C20:2n6	0.16	0.031	2.63	nd	nd	nd	-	0.15	0.046	2.74	nd	nd	nd	-
C22:2n6	0.32	0.236	5.02	nd	nd	nd	-	0.16	0.065	2.89	nd	nd	nd	-
C18:3n6 (GLA)	1.31	0.157	21.58	0.36	0.196	8.21	<0.0001	1.43	0.461	25.99	1.38	0.235	32.39	0.745
C18:3n3 (ALA)	4.32	0.691	71.93	2.15	0.694	48.92	<0.0001	3.87	0.742	70.81	1.96	0.737	46.63	<0.0001
C20:3n6	0.91	0.237	15.49	1.65	0.263	39.94	<0.0001	0.85	0.170	15.64	0.45	0.221	10.43	<0.0001
C20:3n3	0.64	0.142	10.60	0.73	0.114	16.38	0.079	0.91	0.547	16.35	0.81	0.102	18.69	0.544
C20:4n6 (AA)	1.68	0.146	28.04	1.71	0.065	38.21	0.545	1.75	0.077	32.31	1.74	0.068	40.36	0.781
C20:5n3 (EPA)	1.56	0.464	26.17	1.14	0.037	25.34	0.005	1.23	0.129	22.88	1.19	0.028	25.50	0.240
C22:6n3 (DHA)	2.83	0.093	47.22	1.63	0.072	36.42	<0.0001	2.60	0.083	47.01	1.50	0.029	34.82	<0.0001

nd = not detected.

Table 2. Sums of FA groups (g·100 g^{-1} FA), fatty acid ratios, and nutraceutical indices of lipids in the meat of fallow deer from organic and conventional farms.

Specification	Muscle									
	Longissimus lumborum (LL)					*Semimembranosus* (SM)				
	Organic Farm		Conventional Farm		*p*-Value	Organic Farm		Conventional Farm		*p*-Value
	$\bar{x}$	SE	$\bar{x}$	SE		$\bar{x}$	SE	$\bar{x}$	SE	
					Sums of FA groups					
SFA	46.12	5.742	49.19	4.790	0.169	44.98	3.673	48.96	3.332	0.011
UFA	53.88	5.742	50.81	4.791	0.169	55.02	3.673	51.04	3.332	0.011
MUFA	29.62	5.096	32.49	3.390	0.119	32.10	3.610	34.04	3.294	0.183
PUFA	24.26	1.822	18.33	2.256	<0.0001	22.92	2.350	17.00	2.502	<0.0001
PUFA n-6	12.59	1.457	11.05	1.624	0.023	12.18	1264	9.88	1.991	0.003
PUFA n-3	9.34	0.565	5.65	0.744	<0.0001	8.60	1.180	5.46	0.740	<0.0001
OFA	27.56	1.631	26.56	2.988	0.318	25.21	3.090	24.40	3.421	0.548
					FA Ratios					
SFA/UFA	0.88	0.199	0.99	0.218	0.204	0.83	0.123	0.97	0.139	0.014
MUFA/SFA	0.66	0.208	0.67	0.115	0.938	0.72	0.129	0.70	0.099	0.647
PUFA/SFA	0.54	0.094	0.38	0.073	<0.0001	0.51	0.081	0.35	0.064	<0.0001
PUFA n-6/PUFA n-3	1.35	0.123	1.96	0.176	<0.0001	1.42	0.124	1.81	0.269	<0.0001
					Nutraceutical indices					
DFA	69.40	1.921	69.88	3.261	0.666	72.81	3.265	71.57	3.284	0.361
AI	0.54	0.090	0.54	0.132	0.979	0.47	0.116	0.50	0.090	0.561
TI	0.73	0.153	0.92	0.209	0.022	0.72	0.112	0.90	0.118	<0.0001
h/H	2.53	0.227	2.68	0.426	0.310	2.94	0.470	3.00	0.530	0.771
Δ9-desaturase C16	13.01	3.692	7.59	3.978	0.002	15.24	3.527	9.24	4.082	<0.0001
Δ9-desaturase C18	56.27	13.044	54.61	6.913	0.701	56.68	4.607	52.62	7.262	0.116
Elongase	56.31	3.369	63.25	3.638	<0.0001	61.52	5.237	65.88	3.909	0.030
NV	1.04	0.224	0.91	0.221	0.186	0.84	0.180	0.86	0.142	0.835
DHA + EPA	4.38	0.477	2.77	0.076	<0.0001	3.83	0.137	2.69	0.027	<0.0001
HPI	1.37	0.190	1.34	0.304	0.747	1.55	0.354	1.42	0.226	0.277
CI	4.35	0.221	4.56	0.218	0.420	4.44	0.224	4.96	0.270	<0.0001
CSI	3.92	0.194	4.15	0.221	0.067	4.02	0.207	4.49	0.276	<0.0001

Table 3. Contribution of a 100-g portion of fallow deer meat in the diet of adults in terms of total fat, fatty acids, and cholesterol.

Specification	Percent (%) of Energy Requirements Recommended by FAO [a]	g/day (in a 2000-Kcal Diet) [b]	Mean Content (g/100g of Fresh Meat) [c]				Percent (%) of Contribution to a 2000-Kcal Diet			
			longissimus lumborum (LL)		*semimembranosus* (SM)		*longissimus lumborum* (LL)		*semimembranosus* (SM)	
			organic Farm	conventional Farm	Organic Farm	Conventional Farm	Organic Farm	Conventional Farm	Organic Farm	Conventional Farm
Total fat	20.0–35.0	44.0–78.0	2.31	3.10	2.58	3.21	2.96–5.25	3.97–7.04	3.30–5.86	4.11–7.29
∑ SFA	<10.0	<22.0	0.768	1.100	0.837	1.133	≥3.49	≥5.02	≥3.81	≥5.14
∑ MUFA	15.0–20.0	33.0–44.0	0.493	0.726	0.597	0.788	1.12–1.49	1.65–2.20	1.36–1.81	1.79–2.39
∑ PUFA	6.0–11.0	13.0–24.0	0.404	0.410	0.426	0.393	1.68–3.11	1.71–3.15	1.77–3.27	1.64–3.02
∑ PUFA n-6	2.5–9.0	5.6–20.0	0.210	0.247	0.226	0.228	1.05–3.75	1.23–4.42	1.13–4.03	1.14–4.07
∑ PUFA n-3	0.5–2.0	1.1–4.4	0.155	0.126	0.161	0.126	3.52–14.09	2.86–11.45	3.66–14.64	2.86–11.54
EPA + DHA [d]	250 mg	0.250	0.073	0.062	0.071	0.062	29.23	24.81	28.41	24.81
Cholesterol [e]	<300 mg	<0.300	0.063	0.061	0.063	0.067	≥21.02	≥20.35	≥21.22	≥22.37

[a] FAO (2010); [b] European Union (EU) (2010); [c] FA content expressed in g/100 g of muscle tissue calculated from total intramuscular fat (IMF) using a conversion factor of (F_{CON}) 0.721 [50]; [d] The upper level of EPA + DHA consumption should not exceed 2 g/day; [e] WHO/FAO (2003).

4. Discussion

The samples of muscles were taken during the production process, where the animals were slaughtered to meet the economic needs of the farm owners. The fallow deer were culled in the autumn period in agreement with hunting regulations. The animals were in the age range reported in the literature [51,52]. Fallow deer are usually slaughtered between the 16th and 24th months, due to the highest body weight gains, the most effective feed conversion, low subcutaneous fat cover, and the highest meat quality. In turn, Volpelli et al. [52] highlighted the economic benefits of the extension of fallow deer breeding from 18 to 30 months due to higher dressing proportions, higher amounts of first quality cuts, and better carcass conformation.

An important indicator of the quality of meat is its fat content. Due to its various functions, fat has an impact on human health; therefore, both excessive levels of total fat in the diet and an imbalance in the fat composition are associated with various diseases. The present study showed that the farming system had an impact on the value of this parameter. Lower fat content was determined in the organic meat, as in the study on beef conducted by Ribas-Augusti et al. [1]. The differences in the content of this component in muscles of wild and farmed deer in Lithuania were reported by Razmaitė et al. [53]. In a study conducted by Daszkiewicz et al. [54], meat from farmed fallow deer had lower intramuscular fat content than meat from wild fallow deer (0.24% vs. 0.50%). Noteworthy, the fat content in the fallow deer meat from the organic and conventional farms analyzed in the present study (Figure 1) was at the optimal level (2–3%) [28]. This is extremely important given the well-documented role of intramuscular fat in the sensory properties of meat [55]. An increase in the intramuscular fat content to the optimal level improves the intensity of meat flavor, juiciness, and tenderness [54]. Joo et al. [56] show a positive correlation between the type IIB fiber and IMF content in Hanwoo steer cattle. The higher fat content determined in the MS muscle compared to the LL muscle (Table 1) may be related to the higher amount of type IIB fibers.

Although there have been no upper limits on cholesterol intake since 2015 (previously <300 mg per day), the dietary guidelines for Americans still recommend the lowest cholesterol intake possible [44]. Thus, for recommended health reasons, the LL was characterized by lower cholesterol content. SM from the conventional vs. organic rearing system had higher cholesterol content. Similarly, Ribas-Augusti et al. [1] reported higher cholesterol content in conventionally farmed beef than in organic beef. Since there are no similar data, comparison with other results of studies of these two cervid production systems is difficult. The higher muscle cholesterol content may be a result of the differences in the diet and the slightly higher total fat content in the meat of the conventionally farmed fallow deer. Chung et al. [57] reported a strong relationship between cholesterol levels and marbling scores. It was found in the present study that the muscles of the analyzed fallow deer generally had lower cholesterol content than the muscles of wild-living deer examined by Polak et al. [30] in the Republic of Slovenia. A higher level of cholesterol was determined in chicken meat (89–129 mg100 g^{-1}) [31] and *longissimus dorsi* (LD) and *semitendinosus* (ST) muscle of lambs (99.4–223.28 mg100 g^{-1} and 68.7– 166.2 mg100 g^{-1}, respectively) [58].

The biological value of fat is determined primarily by the amount and type of FA contained therein. The fatty acid composition is essential, as it may influence the development of vascular and coronary diseases in humans [59]. As shown by Simopoulos [60], SFAs have been identified as a risk factor for human health. No significant differences in the SFA content were noted in the LL samples from the fallow deer reared in the two farming systems analyzed in the present study. The higher SFA level determined in the SM muscle of the conventionally farmed fallow deer was mainly a consequence of the higher contents of C18:0, C15:0, C17:0, and C12:0. In turn, Revilla et al. [61] reported a higher concentration of SFA in the meat of cattle reared in the organic system compared with the conventional farming system. C16:0 and C18:0 acids are the dominant SFAs in red meat [23,24,52], which was demonstrated in the present study as well. No effect of stearic acid C18:0 on total cholesterol levels has been reported. This is most likely related to its desaturation to oleic acid in the liver [28,54]. However, the thrombogenic properties of C18:0 have been demonstrated [62]. Myristic (C14: 0) and lauric (C12: 0) acids, which were detected in substantially smaller amounts (Table 1), and palmitic acid (C16:0) probably exert atherogenic effects. They inhibit the expression of the LDL (low-density lipoprotein) receptor gene, thus increasing the synthesis of LDL cholesterol and the level of total cholesterol [63]. However, the potential of C14:0 to raise total serum cholesterol is fourfold or even sixfold higher than that of C16:0 [23]. A meta-analysis of results from several studies on the effect of dietary fatty acids has demonstrated that lauric acid increases the level of high-density lipoprotein (HDL) as well [63]. An overall effect of C12:0 is the reduction of the TC-to-HDL ratio, which is associated with desirable cardiovascular outcomes. However, a meta-analysis of prospective epidemiological studies [64] has provided no compelling evidence for the correlation of dietary saturated fat with an increased risk of coronary artery disease (CHD) or cardiovascular disease (CVD).

Stearic acid contained in meat plays a significant role in meat tenderness and juiciness. As reported by Wood et al. [55,65], there is a positive correlation of meat flavor with the content of saturated and monounsaturated fatty acids and a negative correlation with the level of unsaturated fatty acids. Similar SFA levels to the values reported in the present study were shown by Daszkiewicz et al. [54] in a wild population of fallow deer from northeastern Poland and by Bures et al. [66] for LL of fallow deer from the Czech Republic. A higher level was reported by Daszkiewicz et al. [54] in muscles from a farm-raised population and Ivanović et al. [24] in LT muscles of fallow deer from Serbia.

No differences in the content of monounsaturated fatty acids (MUFAs) were observed between the muscles from the conventional and organic farming systems. However, the C16:1 concentration was higher in the LL and SM muscles from the organically than conventionally farmed animals, as in the study on beef conducted by Ribas-Augusti et al. [1]. The high content of this FA may be related to the increased activity of stearoyl-CoA desaturase Δ9 (Δ9-desaturase C16) (Table 2). It is one of the most important endoplasmic reticulum (ER)-associated enzymes catalyzing the generation of monounsaturated fatty acids (MUFAs; C16:1 n-7) from saturated fatty acids (SFAs; C16: 0) synthesized *de novo* or supplied with food [67]. This is an important issue since increasing attention is being paid to the possibility of using stearoyl-CoA desaturase in the treatment of circulatory diseases and cancers [68,69].

Essential PUFAs are not synthesized in the human organism, and their deficiency in the diet causes metabolic and health disorders. PUFAs are classified into four families: n-3, n-6, n-9, and n-7. In terms of nutrition, α-linolenic acid (ALA; 18:3 n-3) and linoleic acid (LA; 18:2 n-6), which are precursors of long-chain polyunsaturated fatty acids (LCPUFA) are the most important PUFAs. The process of biochemical transformations of ALA leads to formation of EPA (20:5n-3) and DHA (22:6n-3), whereas LA is converted into arachidonic acid (ARA; 20:4n-6). The synthesis is possible due to the presence of appropriate enzymes [70]. LA and ALA compete for the same desaturase and elongase enzymes involved in the synthesis of LCPUFA [44].

The present study showed that the muscles from the organically farmed animals had a higher content of n-3 PUFAs (ALA and DHA) and CLA and C18:2n-6t compared with the muscles of fallow deer from the conventional system. The organic LL samples had a higher concentration of EPA and GLA, and the SM muscles contained higher amounts of LA. High PUFA content in meat is desirable due to their nutritional value and health-enhancing properties. A higher level of n-3 PUFAs and CLA was also noted by Revilla et al. [61] and Turner et al. [71] in organic beef as well as Kamihiro et al. [27], who observed differences only in the content of n-3 PUFAs. PUFAs act as carriers of fat-soluble vitamins (A, D, E, and K) and play a key role in the immune response in humans and animals [72]. They are involved in vital metabolic processes such as brain development [73], endocytosis and exocytosis, and cellular signal transduction [74]. ALA is involved in prevention of cardiovascular diseases, and LA has 2–3-fold higher efficiency in lowering the LDL-C level than oleic acid [75].

EPA and DHA exert a hypolipidemic effect by reducing the concentration of triglycerides (TG) in blood plasma via inhibition of their resynthesis in the intestinal wall and liver and activate anti-inflammatory, anticoagulant, and other anti-atherosclerotic mechanisms [44,70,76]. EPA and DHA play a beneficial role in many human diseases, including autoimmune diseases, diabetes, cancer, and Alzheimer's disease (AD) [76,77]. EPA exerts an effect mainly on the cardiovascular system through the synthesis of eicosanoids. At the same time, DHA is an important structural component of nervous cell membranes, especially in the brain cortex and retina [44]. Moreover, a beneficial role of DHA in counteracting depression and stress has been indicated [70,73]. It is also suggested that omega-3 acids reduce the severity of viral and bacterial inflammatory processes [70]. These data suggest that organic fallow deer meat characterized by the significantly higher level of DHA + EPA (Table 2) should be included in the human diet more frequently.

Recently, considerable attention has been paid to CLA due to its proven bioactivity in the prevention of obesity, cancer, diabetes, atherosclerosis, and osteoporosis [44,78,79]. Conjugated linoleic acid is synthesized by the bacterium *Butyrivibrio fibrisolvens* in the rumen of ruminants through incomplete hydrogenation of linoleic acid to stearic acid [80]. The present study shows that the LL and SM muscles of fallow deer from the organic farming system were characterized by significantly higher CLA levels, which may be associated with the great floristic diversity of the pasture and the consumption of a higher number of grass, legume, and herb species. Higher CLA concentrations in the meat of grazing animals than those receiving feed concentrates were observed in other studies as well [7,61]. As reported by Budimir et al. [26], consumption of grass and grazing significantly increased the concentration of CLA, n-3 PUFAs, and MUFA in lamb meat.

The PUFA/SFA, n-6 PUFA/n-3 PUFA, and h/H ratios are commonly used for the assessment of the IMF nutritional value and consumer health. All dietary PUFAs are believed to lower the levels of low-density lipoprotein cholesterol (LDL-C) and total cholesterol in serum, while all SFAs contribute to the elevation of serum cholesterol levels. Thus, higher ratios are correlated with more positive effects [44]. According to the relevant nutritional recommendations, the PUFA/SFA ratio in the human diet should be 0.4 or higher [55]. Some authors [81] suggest that this ratio should be in the range from ≥ 0.45 to 1.0. In the present study, these criteria were fulfilled only by the meat from the organically farmed fallow deer. Foods in the human diet with a PUFA/SFA ratio below 0.45 are regarded as undesirable due to their potential effect of increasing blood cholesterol levels [82]. In the present study, the PUFA/SFA ratio in the meat of the conventionally reared fallow deer was similar to the value reported by Ivanović et al. [24] for the musculus longissimus thoracis (LT) of fallow deer from Serbia. In turn, the ratio in the muscles of the organically farmed animals was higher than the value in meat from farmed and wild fallow deer (0.27) calculated by Daszkiewicz [52] and in beef and lamb muscles (0.11–0.37) [44]. As suggested by Simopoulos [60], a balanced n-6/n-3 ratio of 1–2/1 is one of the most important dietary factors in the prevention of obesity, whereas dieticians claim that the desired n-6/n-3 ratio should be 5. As reported by Harris [83], there are many indications that the ratio of dietary n-3 and n-6 acids is irrelevant, and its role in prevention of many diseases is unreliable. Based on the available literature [47,83], it can be concluded that achievement of an appropriate threshold of consumption of n-3 and n-6 fatty acids has fundamental importance. The diet of Western societies has been shown to be deficient in n-3 acids [60,73]. It is now known that n-3 fatty acids are highly important for the proper growth and development of the human organism. The present study showed a higher concentration of n-6 PUFAs and n-3 PUFAs and a lower n-6 PUFA/n-3 PUFA ratio in the organic meat; nevertheless, the concentration of these FAs was high in the meat from both production systems, which undoubtedly proves that fallow deer meat is a good source of n-6 PUFAs and n-3 PUFAs. Therefore, it can be recommended for the prevention and treatment of such diseases as hypertension, diabetes, arthritis, inflammatory diseases, coronary heart disease, and cancers [70,80]. Similarly, lower n-6/n-3 values in organic than conventional beef were reported by Revilla et al. [61] and Turner et al. [71].

Ulbricht and Southgate [42] have found that AI and TI indices are better indicators of atherogenicity and thrombogenicity than the PUFA/SFA ratio. In general, their lower value is more beneficial for health. This is associated with the fact that not all SFAs are hypercholesterolemic, and MUFAs exert a protective effect, likewise PUFAs. It is assumed that an AI value lower than 0.5 is beneficial for human health [84], although some authors recommend an atherogenicity index for animal lipids in the range from 0.5 to 1.0 [81,85]. In the present study, the values of AI of the organically and conventionally produced meat did not differ significantly. They were consistent with the dietary recommendations [60,73] and agreed with the results reported by Švrčula et al. [85]. This is extremely important from the health point of view, as consumption of products with a lower AI value may contribute to the reduction of the level of total cholesterol and LDL-C in human blood plasma [44].

The TI values calculated for the LL and SM muscles of the organically farmed fallow deer were lower and more favorable than in the conventional farming system. However, they were within the recommended range for a healthy diet (<1.0) in both systems [73]. Noteworthy, the fat in the organic meat appeared to have very low thrombogenicity and atherogenicity indices similar to those of seafood [86]. The TI values recorded in this study were lower than the indices for fallow deer meat (1.42–1.83) from the Czech Republic [85] as well as lamb and heifer meat (1.1–1.34) [44]. The TI value reflects the thrombogenic potential of FAs, indicating a tendency to form clots in blood vessels, and reveals the contribution of various FAs and the relationship between pro-thrombogenic FAs (C12:0, C14:0, and C16:0) and anti-thrombogenic FAs (MUFAs and the n-3 and n-6 families). Therefore, the consumption of meat with a lower TI value is beneficial for CVH [44]. FAs with lower AI and TI values have better nutritional quality, and consumption of such fatty acids may reduce the risk of coronary heart disease (CHD). Nevertheless, no specific AI and TI values have been recommended to date [44].

The hypercholesterolemic and atherosclerotic potential of meat is associated with the content of cholesterol and saturated fats. Hence, the CI and CSI indices depending mainly on the cholesterol content and, to a lesser extent, on the level of fat and SFA concentration, were analyzed in this study. These indices have been proposed as useful elements of evaluation and design in low-fat diets, and offer quick and easy assessments of daily cholesterol intake. Their low value reflects low contents of saturated fat and cholesterol and thus low atherogenicity [46]. The SM muscles from the organically farmed fallow deer were characterized by significantly reduced CI and CSI values, indicating that fallow deer meat has great potential to reduce hyperlipidemia. Therefore, the meat of fallow deer kept in organic systems can be regarded as a functional food due to its content of fatty acids and the reduced risk of cardiovascular and autoimmune diseases [70,77]. The EPA and DHA fatty acids are essential parameters for recognizing the organic meat analyzed in this study as a functional food [76,77]. It was estimated that a 100-g portion of fresh organic fallow deer meat ensured nearly 30% of the recommended daily intake of EPA+DHA for adults; therefore, it can be labeled as meat with high PUFA n-3 content. Venison is currently an exclusive product and, although Poland is one of the leading game producers in Europe, its average annual consumption per capita in our country ranges from 50 g/person (2013) to 138 g/person (2016) [87].

The present investigations of the quantity and quality of lipids in fallow deer meat show that the natural conditions of animal husbandry following the organic farming principles can provide food with a higher health-enhancing value.

5. Conclusions

The organic fallow deer meat was characterized by a significantly lower content of intramuscular fat. The fatty acid profile in the organic meat exhibited a particularly high proportion (p <0.0001) of CLA (LL—2.29%, SM—2.14%), ALA (LL—4.32%, SM—3.87%), and DHA (LL—2.83%, SM—2.60%). The beneficial effect ($p < 0.0001$) of the organic farming system on the amount of PUFAs, including n-3 PUFAs, was shown to result in a more favorable n-6 PUFA/n-3 PUFA ratio. The significantly higher nutritional quality of organic meat lipids was confirmed by such nutraceutical indicators as TI, Δ9-desaturase C16, elongase, and DHA + EPA for the LL and SM muscles and the CI and CSI indices for the SM muscle. A 100-g portion of fresh organic fallow deer meat was found to provide nearly 30% of the recommended daily intake of EPA + DHA acids in the adult diet. Regardless of the farming system, the longissimus lumborum muscle turned out to be more attractive to consumers in terms of its health-enhancing properties.

Author Contributions: Conceptualization, J.K.; methodology, J.K. and A.K.; formal analysis, J.K. and A.K.; investigation, J.K. and A.K.; data curation, J.K. and A.K.; writing—original draft preparation, J.K. and A.K.; writing—review and editing, J.K. and A.K.; supervision, A.K. Both authors have read and agreed to the published version of the manuscript.

Funding: This research received no external funding.

Institutional Review Board Statement: Ethical review and approval were waived for this study, as the Polish legislation and institutional requirements of the University Life Sciences of Lublin do not require it, as the assay was carried out as part of common zootechnical procedures and the animals did not suffer any intervention beyond those typical in such farming conditions. Muscle samples were collected for the analyses during the routine carcass cutting procedure at the meat processing plant.

Informed Consent Statement: Not applicable.

Data Availability Statement: The data presented in this study are available in the article.

Acknowledgments: The authors are grateful to the owners of the fallow deer farms and the processing plant for their help in the research.

Conflicts of Interest: The authors declare no conflict of interest.

References

1. Ribas-Agustí, A.; Díaz, I.; Sárraga, C.; García-Regueiro, J.A.; Castellari, M. Nutritional properties of organic and conventional beef meat at retail. *J. Sci. Food Agric.* **2019**, *99*, 4218–4225. [CrossRef]
2. Rizzo, G.; Borrello, M.; Dara Guccione, G.; Schifani, G.; Cembalo, L. Organic Food Consumption: The Relevance of the health attribute. *Sustainability* **2020**, *12*, 595. [CrossRef]
3. Sahota, A.; Willer, H.; Lernoud, J. (Eds.) *The world of Organic Agriculture. Statistics and Emerging Trends 2018*; Research Institute of Organic Agriculture (FiBL) and IFOAM–Organics International: Bonn, Germany, 2018; pp. 145–150.
4. Tandon, A.; Jabeen, F.; Talwar, S.; Sakashita, M.; Dhir, A. Facilitators and inhibitors of organic food buying behavior. *Food Qual. Prefer.* **2021**, *88*, 104077. [CrossRef]
5. Hurtado-Barroso, S.; Tresserra-Rimbau, A.; Vallverdú-Queralt, A.; Lamuela-Raventós, R.M. Organic food and the impact on human health. *Crit. Rev. Food Sci. Nutr.* **2019**, *59*, 704–714. [CrossRef]
6. Anisimova, T.; Mavondo, F.; Weiss, J. Controlled and uncontrolled communication stimuli and organic food purchases: The mediating role of perceived communication clarity, perceived health benefits, and trust. *J. Mark. Commun.* **2017**, *25*, 180–203. [CrossRef]
7. Santos-Silva, J.; Bessa, R.J.; Santos-Silva, F. Effect of genotype, feeding system and slaughter weight on the quality of light lambs. II. Fatty acid composition of meat. *Livest. Product. Sci.* **2002**, *77*, 187–194. [CrossRef]
8. Barański, M.; Rempelos, L.; Iversen, P.O.; Leifert, C. Effects of organic food consumption on human health; the jury is still out! *Food Nutr. Res.* **2017**, *61*, 1287333. [CrossRef] [PubMed]
9. Ditlevsen, K.; Sandøe, P.; Lassen, J. Healthy food is nutritious, but organic food is healthy because it is pure: The negotiation of healthy food choices by Danish consumers of organic food. *Food Qual. Prefer.* **2019**, *71*, 46–53. [CrossRef]
10. Kilar, M.; Kilar, J.; Ruda, M. Rolnictwo ekologiczne jako źródło żywności funkcjonalnej. In *Innowacyjne Rozwiązania w Technologii Żywności i Żywieniu Człowieka*; Tarko, T., Ed.; Wyd. Naukowe Polskiego Towarzystwa Technologów Żywności: Kraków, Poland, 2016; pp. 195–204. ISBN 978-83-937001-8-9.
11. Kesse-Guyot, E.; Péneau, S.; Méjean, C.; De Edelenyi, F.S.; Galan, P.; Hercberg, S.; Lairon, D. Profiles of organic food consumers in a large sample of French adults: Results from the Nutrinet-Santé cohort study. *PLoS ONE* **2013**, *8*, e76998. [CrossRef]
12. Rembiałkowska, E.; Wiśniewska, K. Jakość mięsa z produkcji ekologicznej. *Med. Wet.* **2010**, *66*, 188–191.
13. Gadomska, J.; Sadowski, S.; Buczkowska, M. Ekologiczna żywność jako czynnik sprzyjający zdrowiu. *Probl. Hig. I Epidemiol.* **2014**, *95*, 556–560.
14. Vargas-Ramella, M.; Munekata, P.E.S.; Gagaoua, M.; Franco, G.; Campagnol, P.C.B.; Pateiro, M.; Da Silva Barretto, A.C.; Domínguez, R.; Lorenzo, J. Inclusion of healthy oils for improving the nutritional characteristics of dry-fermented deer sausage. *Foods* **2020**, *9*, 1487. [CrossRef] [PubMed]

15. Żochowska-Kujawska, J. *Mięso Zwierząt Łownych Jako Potencjalne Źródło Surowca do Produkcji Surowych Wędzonek Dojrzewających*; Wyd. Zachodniopomorski Uniwersytet Technologiczny: Szczecin, Poland, 2017; pp. 1–110. ISBN 978-83-7663-233-9.
16. Kudrnáčová, E.; Bartoň, L.; Bureš, D.; Hoffman, L.C. Carcass and meat characteristics from farm-raised and wild fallow deer (*Dama dama*) and red deer (*Cervus elaphus*): A review. *Meat Sci.* **2018**, *141*, 9–27. [CrossRef] [PubMed]
17. Mexia, I.A.; Quaresma, M.A.G.; Coimbra, M.C.P.; Dos Santos, F.A.; Alves, S.P.A.; Bessab, R.J.B.; Antunes, I.C. The influence of habitat and sex on feral fallow deer meat lipid fraction. *J. Sci. Food Agric.* **2020**, *100*, 3220–3227. [CrossRef]
18. Cawthorn, D.M.; Fitzhenry, L.B.; Kotrba, R.; Bureš, D.; Hoffman, L.C. Chemical composition of wild fallow deer (*Dama dama*) meat from South Africa: A preliminary evaluation. *Foods* **2020**, *9*, 598. [CrossRef] [PubMed]
19. Kilar, J. Dziczyzna w jadłospisie człowieka–walory technologiczne, kulinarne i prozdrowotne. In *Nowoczesna Produkcja Świń i Stojące Przed nią Wyzwania*; Zachodniopomorski Uniwersytet Technologiczny: Szczecin, Poland, 2018; pp. 24–28. ISBN 978-83-7663-247-6.
20. Soriano, A.; Sánchez-García, C. Nutritional Composition of Game Meat from Wild Species Harvested in Europe. *Intech Open* **2021**. Available online: https://www.intechopen.com/online-first/nutritional-composition-of-game-meat-from-wild-species-harvested-in-europe (accessed on 17 May 2021). [CrossRef]
21. Kilar, J.; Ruda, M.; Kilar, M.; Grych, K. Dziczyzna. Szanse i bariery zwiększenia spożycia. In *Surowce Pochodzenia Zwierzęcego Jako Źródło Składników Bioaktywnych*; Słupski, J., Tarko, T., Drożdż, I., Eds.; Wyd. Oddział Małopolski Polskiego Towarzystwa Technologów Żywności: Kraków, Poland, 2018; pp. 53–63. ISBN 978-83-946796-2-0.
22. Stanisz, M.; Skorupski, M.; Ślósarz, P.; Bykowska-Maciejewska, M.; Składanowska-Baryza, J.; Stańczak, Ł.; Krokowska-Paluszak, M.; Ludwiczak, A. The seasonal variation in the quality of venison from wild fallow deer (*Dama dama*)—A pilot study. *Meat Sci.* **2019**, *150*, 56–64. [CrossRef]
23. Daszkiewicz, T.; Mesinger, D. Fatty acid profile of meat (*Longissimus lumborum*) from female roe deer (*Capreolus capreolus* L.) and red deer (*Cervus elaphus* L.). *Int. J. Food Prop.* **2018**, *21*, 2276–2282. [CrossRef]
24. Ivanović, S.; Pisinov, B.; Pavlović, M.; Pavlović, I. Quality of meat from female fallow deer (*Dama dama*) and roe deer (*Capreolus capreolus*) hunted in Serbia. *Ann. Anim. Sci.* **2020**, *20*, 245–262. [CrossRef]
25. Bykowska, M. Influence of selected factors on meat quality from farm-raised and wild fallow deer (*Dama dama*): A review. *Can. J. Anim. Sci.* **2018**, *98*, 405–415. [CrossRef]
26. Budimir, K.; Mozzon, M.; Toderi, M.; D'Ottavio, P.; Trombetta, M.F. Effect of breed on fatty acid composition of meat and subcutaneous adipose tissue of light lambs. *Animals* **2020**, *10*, 535. [CrossRef] [PubMed]
27. Kamihiro, S.; Stergiadis, S.; Leifert, C.; Eyre, M.D.; Butler, G. Meat quality and health implications of organic and conventional beef production. *Meat Sci.* **2015**, *100*, 306–318. [CrossRef] [PubMed]
28. Kasprzyk, A.; Tyra, M.; Babicz, B. Fatty acid profile of pork from a local and a commercial breed. *Arch. Anim. Breed.* **2015**, *58*, 379–385. [CrossRef]
29. Nogales, S.; Bressan, M.C.; Delgado, J.V.; Telo daGama, L.; Barba, C.; Camacho, M.E. Fatty acid profile of feral cattle meat. *Ital. J. Anim. Sci.* **2017**, *16*, 172–184. [CrossRef]
30. Polak, T.; Rajar, A.; Gaperlin, L.; Lender, B. Cholesterol concentration and fatty acid profile of red deer (*Cervus elaphus*) meat. *Meat Sci.* **2008**, *80*, 864–869. [CrossRef]
31. Rozbicka-Wieczorek, A.J.; Więsyk, E.; Brzóska, F.; Śliwiński, B.; Kowalczyk, J.; Czauderna, M. Fatty acid profile and oxidative stress of thigh muscles in chickens fed the ration enriched in lycopene, selenium compounds or fish oil. *Ann. Anim. Sci.* **2014**, *14*, 595–609. [CrossRef]
32. Kilar, J.; Ruda, M.; Kilar, M. *Dziczyzna. Co o Niej Wiedzą i Czy ją Jedzą Mieszkańcy Podkarpacia*; Wyd. PWSZ im. S. Pigonia: Krosno, Poland, 2016; pp. 1–101. ISBN 978-83-64457-21-0.
33. Kwiecińska, K.; Kosicka-Gębska, M.; Gębski, J. Ocena preferencji konsumentów związanych z wyborem dziczyzny. *Handel Wewnętrzny* **2016**, *1*, 53–64.
34. Regulation (EU) 2018/848—Rules on Organic Production and Labelling of Organic Products. EU Rules on Producing and Labelling Organic Products. Available online: https://eur-lex.europa.eu/legal-content/EN/LSU/?uri=CELEX%3A32018R0848 (accessed on 4 January 2021).
35. Dz.U. 2009 nr 116 poz. 975. In *Ustawa z Dnia 25 Czerwca 2009 r. o Rolnictwie Ekologicznym*; Organic Agriculture Law; Kancelaria Sejmu: Warsaw, Poland. Available online: https://isap.sejm.gov.pl/isap.nsf/download.xsp/WDU20091160975/U/D20090975Lj.pdf (accessed on 4 January 2021). (In Polish)
36. Braun-Blanguet, J. *Pflanzensoziologie, Grundzuge der Vegetationskundle*; 3. Aufl; Springer: Vienna, Austria, 1964.
37. DEFRA Code of Recommendations for the Welfare of Farmed Deer. 2006. Available online: http://www.defra.gov.uk/animalh/welfare/farmed/othersps/deer/pb0055/deercode.htm (accessed on 4 January 2021).
38. FEDFA Federation of European Deer Farmers Associations. Available online: https://www.fedfa.com/ (accessed on 4 January 2021).
39. Folch, J.; Lees, M.; Sloane-Stanley, G.H. A simple method for the isolation and purification of total lipids from animal tissues. *J. Biol. Chem.* **1957**, *226*, 497–509. [CrossRef]
40. PN-ISO 1444:2000. *Meat and Meat Products. Determination of Fat Content*; Polish Committee for Standardization: Warsaw, Poland, 2000. (In Polish)
41. SOP. *Cholesterol in meat by GC method*; M.023a wersja 1 z 20.10.2011; SOP: Kraków, Poland, 2011.

42. Ulbricht, T.L.V.; Southgate, D.A.T. Coronary heart disease: Seven dietary factors. *Lancet* **1991**, *338*, 985–992. [CrossRef]
43. Domaradzki, P.; Żółkiewski, P.; Litwińczuk, A.; Florek, M.; Dmoch, M. Profil i wartość odżywcza kwasów tłuszczowych w wybranych mięśniach szkieletowych buhajków rasy polskiej holsztyńsko-fryzyjskiej. *Med. Weter.* **2019**, *75*, 310–315.
44. Chen, J.; Liu, H. Nutritional indices for assessing fatty acids: A mini-review. *Int. J. Mol. Sci.* **2020**, *21*, 5695. [CrossRef] [PubMed]
45. Zilversmit, D.B. Cholesterol index of foods. *J. Am. Diet. Assoc.* **1979**, *74*, 562–565. [PubMed]
46. Connor, S.L.; Gustafson, J.R.; Artaud-Wild, S.M.; Favell, D.P.; Classick-Kohn, C.J.; Hatcher, L.F.; Connor, W.E. The cholesterol/saturated-fat index an indication of the hypercholesterolemic and atherogenic potential of food. *Lancet* **1986**, *327*, 1229–1232. [CrossRef]
47. FAO. Fats and Fatty Acids in Human Nutrition: Report of an Expert Consultation. *FAO Food Nutr. Pap.* **2010**, *91*, 1–180.
48. European Food Safety Authority. Scientific opinion on dietary reference values for fats, including saturated fatty acids, polyunsaturated fatty acids, monounsaturated fatty acids, trans fatty acids, and cholesterol. *EFSA J.* **2010**, *8*, 1461. [CrossRef]
49. WHO/FAO. Diet, nutrition and the prevention of chronic diseases. In *Report of a Joint WHO/FAO Expert Consultation*; World Health Organization Technical Report Series; 916. Report of a Joint WHO/FAO Expert Consultation; World Health Organization: Geneva, Switzerland, 2003; pp. 54–60.
50. Domaradzki, P.; Florek, M.; Skałecki, P.; Litwińczuk, A.; Kędzierska-Matysek, M.; Wolanciuk, A.; Tajchman, K. Fatty acid composition, cholesterol content and lipid oxidation indices of intramuscular fat from skeletal muscles of beaver (*Castor fiber* L.). *Meat Sci.* **2019**, *150*, 131–140. [CrossRef]
51. Janiszewski, P.; Bogdaszewska, Z.; Bogdaszewski, M.; Bogdaszewski, P.; Ciululko-Dołęga, J.; Nasiadka, P.; Steiner, Ż. *Rearing and Breeding of Cervids on Farms*; Wydawnictwo UWM: Olsztyn, Poland, 2014; p. 133.
52. Volpelli, L.A.; Valusso, R.; Piasentier, E. Carcass quality in male fallow deer (*Dama dama*): Effects of age and supplementary feeding. *Meat Sci.* **2002**, *60*, 427–432. [CrossRef]
53. Razmaitė, V.; Pileckas, V.; Šiukščius, A.; Juškiene, V. Fatty acid composition of meat and edible offal from free-living red deer (*Cervus elaphus*). *Foods* **2020**, *9*, 923. [CrossRef]
54. Daszkiewicz, T.; Hnatyk, N.; Dąbrowski, D.; Janiszewski, P.; Gugołek, A.; Kubiak, D.; Śmiecińska, K.; Winarski, R.; Koba-Kowalczyk, M. A comparison of the quality of the *Longissimus lumborum* muscle from wild and farm-raised fallow deer (*Dama dama* L.). *Small Rumin. Res.* **2015**, *129*, 77–83. [CrossRef]
55. Wood, J.D.; Enser, M.; Fisher, A.V.; Nute, G.R.; Sheard, P.R.; Richardson, R.I.; Hughes, S.I.; Whittington, F.M. Fat deposition, fatty acid composition and meat quality: A review. *Meat Sci.* **2008**, *78*, 343–358. [CrossRef]
56. Joo, S.H.; Lee, K.W.; Hwang, Y.H.; Joo, S.T. Histochemical characteristics in relation to meat quality traits of eight major muscles from Hanwoo steers. *Korean J. Food Sci. Anim. Resour.* **2017**, *37*, 716–725. [CrossRef] [PubMed]
57. Chung, K.Y.; Lunt, D.K.; Choi, C.B.; Chae, S.H.; Rhoades, R.D.; Adams, T.H. Lipid characteristics of subcutaneous adipose tissue and *M. Longissimus thoracis* of Angus and Wagyu steers fed to US and Japanese endpoints. *Meat Sci.* **2006**, *73*, 432–441. [CrossRef]
58. Aksoy, Y.; Çiçek, Ü.; Sen, U.; Sirin, E.; Ugurlu, M.; Önenç, A.; Kuran, M.; Ulutas, Z. Meat production characteristics of Turkish native breeds: II. meat quality, fatty acid, and cholesterol profile of lambs. *Arch. Anim. Breed.* **2019**, *62*, 41–48. [CrossRef] [PubMed]
59. Calder, P.C.; Deckelbaum, R.J. Fat as a physiological regulator: The news gets better. *Curr. Opin. Clin. Nutr. Metab. Care* **2003**, *6*, 127–131. [CrossRef]
60. Simopoulos, A. An increase in the omega-6/omega-3 fatty acid ratio increases the risk for obesity. *Nutrients* **2016**, *8*, 128. [CrossRef] [PubMed]
61. Revilla, I.; Plaza, J.; Palacios, C. The effect of grazing level and ageing time on the physicochemical and sensory characteristics of beef meat in organic and conventional production. *Animals* **2021**, *11*, 635. [CrossRef]
62. Attia, Y.A.; Al-Harthi, M.A.; Korish, M.A.; Shiboob, M.M. Fatty acid and cholesterol profiles, hypocholesterolemic, atherogenic, and thrombogenic incides of broiler meat in the retail market. *Lipids Health Dis.* **2017**, *16*, 1–11. [CrossRef]
63. Dayrit, F.D. The properties of lauric acid and their significance in coconut oil. Review. *J. Am. Oil Chem. Soc.* **2015**, *92*, 1–15. [CrossRef]
64. Siri-Tarino, P.W.; Sun, Q.; Hu, F.B.; Krauss, R.M. Meta-analysis of prospective cohort studies evaluating the association of saturated fat with cardiovascular disease. *Am. J. Clin. Nutr.* **2010**, *91*, 535–546. [CrossRef]
65. Wood, J.D.; Richardson, R.I.; Nute, G.R.; Fisher, A.V.; Campo, M.M.; Kasapidou, E.; Sheard, P.R.; Enser, M. Effects of fatty acids on meat quality: A review. *Meat Sci.* **2004**, *66*, 21–32. [CrossRef]
66. Bureš, D.; Bartoň, L.; Kotrba, R.; Hakl, J. Quality attributes and composition of meat from red deer (*Cervus elaphus*), fallow deer (*Dama dama*) and Aberdeen Angus and Holstein cattle (*Bos taurus*). *J. Sci. Food Agric.* **2015**, *95*, 2299–2306. [CrossRef] [PubMed]
67. Kucharski, M.; Kaczor, U. Stearoyl-CoA desaturase—The lipid metabolism regulator. *Postepy Hig. Med. Dosw.* **2014**, *68*, 334–342. [CrossRef] [PubMed]
68. Mayneris-Perxachs, J.; Guerendiain, M.; Castellote, A.I.; Estruch, R.; Covas, M.I.; Fitó, M.; Salas-Salvadó, J.; Martinez-González, M.A.; Aros, F.; Lamuela-Raventós, R.M.; et al. Plasma fatty acid composition, estimated desaturase activities, and their relation with the metabolic syndrome in a population at high risk of cardiovascular disease. *Clin. Nutr.* **2014**, *33*, 90–97. [CrossRef] [PubMed]
69. Von Roemeling, C.A.; Marlow, L.A.; Wei, J.J.; Cooper, S.J.; Caulfield, T.R.; Wu, K.; Tan, W.W.; Copland, J.A. Stearoyl-CoA desaturase 1 is a novel molecular therapeutic target for clear cell renal cell carcino ma. *Clin. Cancer Res.* **2013**, *19*, 2368–2380. [CrossRef]

70. Kolanowski, W. Long chain polyunsaturated omega-3 fatty acids and their role in reducing the risk of life-style related diseases. *Bromat. Chem. Toksykol.* **2007**, *11*, 229–237, (In Polish, In English abstract).
71. Turner, T.D.; Jensen, J.; Jessica, L.; Pilfold, J.; Prema, D.; Donkor, K.K.; Cinel, B.; Thompson, D.J.; Dugan, M.E.R.; Church, J.S. Comparison of fatty acids in beef tissues from conventional, organic and natural feeding systems in western Canada. *Can. J. Anim. Sci.* **2015**, *95*, 4958. [CrossRef]
72. Webb, E.C.; O'Neill, H.A. The animal fat paradox and meat quality. *Meat Sci.* **2008**, *80*, 28–36. [CrossRef]
73. Simopoulos, A.P. Evolutionary aspects of diet. The omega-6/omega-3 ratio and the brain. *Mol. Neurobiol.* **2011**, *44*, 203–215. [CrossRef]
74. Kim, W.; Khan, N.A.; McMurray, D.N.; Prior, I.A.; Wang, N.; Chapkin, R.S. Regulatory activity of polyunsaturated fatty acids in T-cell signaling. *Prog. Lipid Res.* **2010**, *49*, 250–261. [CrossRef]
75. Cichosz, G.; Czeczot, H. Trans fatty acids in the human diet. *Bromat. Chem. Toksykol.* **2012**, *2*, 181–190, (In Polish, In English abstract).
76. Bałasińska, B.; Jank, M.; Kulasek, G. Properties and the role of polyunsaturated fatty acids in health protection of human and animal. *Życie Wet.* **2010**, *85*, 749–756, (In Polish, In English abstract).
77. Li, X.; Bi, X.; Wang, S.; Zhang, Z.; Li, F.; Zhao, A.Z. Therapeutic potential of n-3 polyunsaturated fatty acids in human autoimmune diseases. *Front. Immunol.* **2019**, *10*, 2241. [CrossRef] [PubMed]
78. Kritchevsky, D.; Tepper, S.A.; Wright, S.; Czarnecki, S.K.; Wilson, T.A.; Nicolosi, R.J. Conjugated linoleic acid isomer effects in atherosclerosis: Growth and regression of lesions. *Lipids* **2004**, *39*, 611. [CrossRef]
79. Benjamin, S.; Prakasan, P.; Sreedharan, S.; Wright, A.D.; Spener, F. Pros and cons of CLA consumption: An insight from clinical evidences. *Nutr. Metab.* **2015**, *21*, 4. [CrossRef]
80. Pogorzelska-Nowicka, E.; Atanasov, A.G.; Horbańczuk, J.; Wierzbicka, A. Bioactive compounds in functional meat products. *Molecules* **2018**, *23*, 307. [CrossRef]
81. Tocci, R.; Sargentini, C. Meat quality of Maremmana young bulls. *Acta Sci. Anim. Sci.* **2020**, *42*, e46515. [CrossRef]
82. Hmso, U. *Nutritional Aspects of Cardiovascular Disease (Report on Health and Social Subjects No. 46)*; HMSO: London, UK, 1994.
83. Harris, W.S. The omega-6:omega-3 ratio: A critical appraisal and possible successor. *Prostaglandins Leukot. Essent. Fatty Acids* **2018**, *132*, 34–40. [CrossRef]
84. Popova, T.; Ignatova, M.; Petkov, E.; Stanišić, N. Difference in fatty acid composition and related nutritional indices of meat between two lines of slow-growing chickens slaughtered at different ages. *Arch. Anim. Breed.* **2016**, *59*, 319–327. [CrossRef]
85. Švrčula, V.; Košinová, K.; Okrouhlá, M.; Chodová, D.; Hart, V. The effect of sex on meat quality of fallow deer (*Dama dama*) from the farm located in the Middle Bohemia. *Ital. J. Anim. Sci.* **2019**, *18*, 498–504. [CrossRef]
86. Valfre, F.; Caprino, F.; Turchini, G.M. The health benefit of sea food. *Vet. Res. Commun.* **2003**, *27*, 507–512. [CrossRef] [PubMed]
87. Kilar, J. Zasoby zwierząt łownych w Polsce w aspekcie bazy surowcowej przetwórstwa mięsnego. In *Ochrona Środowiska Produkcji Rolniczej*; Uniwersytet Rzeszowski: Rzeszów, Poland, 2020; pp. 60–74. ISBN 978-83-7996-818-3.

MDPI

Article

Odd- and Branched-Chain Fatty Acids in Lamb Meat as Potential Indicators of Fattening Diet Characteristics

Pilar Gómez-Cortés [1], Francisco Requena Domenech [2], Marta Correro Rueda [3], Miguel Ángel de la Fuente [1], Achille Schiavone [4] and Andrés L. Martínez Marín [3,*]

1. Instituto de Investigación en Ciencias de la Alimentación (CSIC-UAM), Universidad Autónoma de Madrid, Nicolás Cabrera 9, 28049 Madrid, Spain; p.g.cortes@csic.es (P.G.-C.); mafl@if.csic.es (M.Á.d.l.F.)
2. Departamento de Biología Celular, Fisiología e Inmunología, Universidad de Córdoba, Ctra. Madrid-Cádiz km 396, 14071 Córdoba, Spain; v02redof@uco.es
3. Departamento de Producción Animal, Universidad de Córdoba, Ctra. Madrid-Cádiz km 396, 14071 Córdoba, Spain; martacorrero@gmail.com
4. Dipartimento di Scienze Veterinarie, Università degli studi di Torino, 10124 Torino, Italy; achille.schiavone@unito.it

* Correspondence: pa1martm@uco.es; Tel.: +34-957-218-746; Fax: +34-957-218-746

Abstract: There is a growing interest of researchers in meat authentication in terms of geographical and dietary background of animals, and several analytical methods have been proposed for the purpose of investigating this. We hypothesized that the odd- and branched-chain fatty acid (OBCFA) profile in intramuscular fat (IMF) might suffice to distinguish lamb meat entering the food chain supply on the basis of the type of diet fed to lambs during the fattening period. A total of 30 individual OBCFA profiles, quantified by gas chromatography, of IMF of Manchego lambs were used. During the fattening period (42 days), the lambs were fed three diets differing in concentrate composition: (i) Control, concentrate typical of commercial fattening rations, rich in starch and based on cereals and soybean meal; (ii) Camelina, similar to Control but replacing 50% of the soybean meal with camelina meal; and (iii) Fibrous, concentrate rich in neutral detergent fiber (NDF), based on fibrous by-products and not including cereals nor soybean meal. The OBCFA were grouped into three classes (linear odd, iso and anteiso fatty acids) and were then submitted to a linear discriminant analysis, using the feeding treatments as grouping variable and the OBCFA class contents in IMF as quantitative variables. The results suggested that a high NDF to starch ratio of the concentrate, being the lowest for Control (CON) treatment and the highest for Fibrous (FIB) treatment, would be negatively related to the odd/anteiso ratio and positively related to the iso/(anteiso+odd) FA ratio in IMF. Determination of OBCFA profile in lamb meat would be useful to monitor the feeding regime (starch- or NDF-rich) of lambs entering the food chain supply.

Keywords: fatty acids; meat; lambs; feeding; discriminant analysis

Citation: Gómez-Cortés, P.; Requena Domenech, F.; Correro Rueda, M.; de la Fuente, M.Á.; Schiavone, A.; Martínez Marín, A.L. Odd- and Branched-Chain Fatty Acids in Lamb Meat as Potential Indicators of Fattening Diet Characteristics. *Foods* **2021**, *10*, 77. https://doi.org/10.3390/foods10010077

Received: 12 December 2020
Accepted: 24 December 2020
Published: 3 January 2021

1. Introduction

Europe has very different sheep production systems in the Northern and Mediterranean areas as feeding and husbandry are adapted to local environmental conditions and agricultural practices. Moreover, there are great differences in the acceptability of lamb meat by consumers across regions due to flavor variations [1], which can be directly ascribed to the feeding background of the lambs [2]. Again, consumers are concerned about information about sheep production systems and, specifically, the type of diet the sheep are fed because they are aware of its effects on lamb meat quality [3]. In this regard, there has been a growing research interest in meat authentication in terms of the geographical and dietary background of animals, and several analytical methods have been proposed for that purpose [4]. Those methods are based on more or less complicated laboratory techniques intended to identify reliable predictors such as fatty acids (FA), volatile compounds, stable isotopes or a variety of metabolites in meat [5–9]. The FA profile of intramuscular fat (IMF) has proven to efficiently discriminate between diets fed to fattening lambs [5]. Furthermore, intramuscular FA are directly linked to organoleptic and nutritional attributes of ruminant meat [10,11].

Odd- and branched-chain fatty acids (OBCFA) are quantitatively minor FA that are almost exclusively of ruminant fat. These FA have attracted great attention not only because they may serve as biomarkers of rumen function [12,13] but also because of their putatively beneficial effects on human health [14]. OBCFA found in ruminant fat mainly come from intestinal digestion of bacteria washed out from the rumen in solid and liquid phases [15,16]. In turn, available data show that the contents of OBCFA in the lipids of rumen bacteria differ between species, i.e., cellulolytic bacteria usually contain more iso FA (although some strains are rich in anteiso FA), whereas amylolytic bacteria are mainly rich in linear odd-chain FA, with a few strains displaying high contents of anteiso FA [12,13]. The relative abundance of rumen bacterial populations heavily depends on diet composition [17,18]. Likewise, the relationship between diet characteristics, the relative abundance of rumen bacterial species and the OBCFA profile of rumen contents has been demonstrated [19,20].

Therefore, we hypothesized that OBCFA contents in IMF might suffice to distinguish lamb meat entering the food chain supply on the basis of the characteristics of the diet fed to lambs during the fattening period.

2. Materials and Methods

This work was carried out with the data from a study whose results have been published elsewhere [21,22]. Briefly, the animal protocol was fully in compliance with the European Union and Spanish regulations on animal welfare and experimentation. A total of 105 uncastrated male lambs of the Manchega breed with an initial bodyweight (BW) of 13.9 ± 1.74 kg and an age of 35 ± 7 days old were randomly allocated to 15 straw-bedded pens. The pens were randomly allocated to one of three treatments (five replicates per treatment): Control (CON), Camelina (CAM), and Fibrous (FIB). The concentrate of CON treatment was cereal-soybean meal based, which is similar to the concentrates commonly used in the feedlot of light lambs (11.4 MJ metabolizable energy/kg, 15.7% crude protein, as fed). The concentrate of CAM treatment was similar to CON but replaced 50% of soybean meal with solvent-extracted camelina meal (11.3 MJ metabolizable energy/kg, 15.6% crude protein, as fed). Finally, the concentrate of FIB treatment did not contain cereals or soybean meal, included several fibrous by-products and was very rich in neutral detergent fiber (NDF) (9.8 MJ metabolizable energy/kg, 15.8% crude protein, as fed). As a result, the NDF to starch ratio was 0.31, 0.40 and 1.54 in CON, CAM and FIB concentrates, respectively. Barley straw was offered as roughage in all experimental treatments. On day 42 of the trial, two lambs per pen (i.e., 10 lambs from each experimental treatment), with the final BW closest to the average pen BW, were tagged to track their carcasses. Then, all animals were sent to a commercial abattoir for slaughter. Samples of Longissimus thoracis muscle were obtained for intramuscular FA analysis after 6 days of aging at 4 °C. Analyses of IMF were made in duplicate. Firstly, muscle was homogenized with a mixture of chloroform and methanol and BHT added as antioxidant. Dilution with chloroform and water separated the homogenate into two layers. The chloroform layer containing all the lipids was collected, and, after removal of the solvent, the total lipids were preserved in amber vials frozen at −17 °C until derivatization. Fatty acids were derivatized to methyl esters by base-catalyzed methanolysis. The OBCFA were quantified by gas chromatography with an Agilent model 6890 N network system (Palo Alto, CA, USA) equipped with an autoinjector and a flame ionization detector (FID) and fitted with a CP-Sil 88 fused silica capillary column (100 m 0.25 mm i.d., Varian, Middelburg, The Netherlands). Injector and detector temperatures were set to 250 °C. Helium was used as the carrier gas, and samples were injected with a split ratio of 1:100. Initial oven temperature was 45 °C. After 4 min, it was increased at a rate of 13° C/min to 165 °C and held for 35 min. Then, oven temperature was increased to 215 °C at 4 °C/min and maintained for 30 min. Individual FA were identified by comparison with standards distributed by Nu-Chek (Elysian, MN, USA).

Statistical analyses were carried out with SAS University Edition 3.8 (SAS Institute, Cary, NC). OBCFA were grouped into three classes (linear odd, iso and anteiso FA) and the requirements of homoscedasticity, absence of outliers and normality were verified with Box, Grubb and Mardia tests, respectively, at $p > 0.05$; multicollinearity was also discarded (variance inflation factor lower than 1.3). Then a linear discriminant analysis (LDA) was performed with the DISCRIM procedure. LDA is a multivariate statistical technique that can be used to differentiate experimental groups and to determine the meaningful variables that contribute most to such differences [23]. In the analysis, the grouping variable were the feeding treatments, and the quantitative variables were the OBCFA class contents in IMF samples. Wilks' test, Bartlett's test, squared Mahalanobis distances and cross-validation of generated Fisher's functions were obtained by means of MANOVA, CAN, DISTANCE and CROSSVALIDATE options, respectively, included in the DISCRIM procedure. The GLM procedure, using the experimental treatments as the fixed effect and the Tukey's test for mean separation at $p < 0.05$, was also used to help in explaining the LDA results.

3. Results

Final BW (25.8 ± 1.12 kg), average daily gain (290 ± 23.8 g/day) and IMF level (1.25 ± 0.22%) did not differ between treatments ($p > 0.05$), but feed conversion ratio was 23% higher in the FIB treatment ($p < 0.05$) [21,22]. Total contents of linear odd, iso and anteiso FA in the Longissimus thoracis muscle of lambs under the experimental treatments are presented in Table 1. The quantitatively main FA within each class of OBCFA were C17:0 (~ 66% of linear odd FA), C17:0 anteiso (~ 82% of anteiso FA) and C14:0 iso (~ 42% of iso FA).

Table 1. Minimum (Min), maximum (Max) and average contents (mean ± standard deviation, g per 100 g of total fat) of several fatty acid groups in Longissimus muscle from lambs fed the experimental treatments.

Fatty Acids	Treatments [1]								
	CON			CAM			FIB		
	Mean	Min	Max	Mean	Min	Max	Mean	Min	Max
Total saturated	32.6 ± 1.76	29.2	34.8	33.3 ± 1.58	30.9	35.3	31.1 ± 2.04	27.1	33.7
Odd	2.99 ± 0.57	2.08	3.90	2.66 ± 0.39	2.25	3.25	1.68 ± 0.22	1.42	2.12
Iso	0.70 ± 0.09	0.56	0.83	0.59 ± 0.07	0.48	0.65	0.65 ± 0.06	0.54	0.75
Anteiso	0.37 ± 0.06	0.30	0.45	0.54 ± 0.07	0.42	0.66	0.55 ± 0.09	0.43	0.76
Monounsaturated	36.1 ± 2.56	31.6	41.5	40.1 ± 2.42	37.3	44.7	38.1 ± 2.59	35.6	43.5
Trans 18:1	4.47 ± 1.31	2.14	6.18	6.49 ± 0.74	5.70	8.08	8.86 ± 1.38	6.80	11,0
Polyunsaturated	17.3 ± 1.30	14.7	18.8	15.4 ± 1.81	12.1	17.9	18.3 ± 2.60	14.1	21.2
CLA	0.46 ± 0.07	0.36	0.62	0.57 ± 0.07	0.44	0.64	1.07 ± 0.24	0.75	1.51

[1] CON: typical commercial concentrate rich in starch and based on cereals and soybean meal. CAM: typical commercial concentrate which replaces 50% of soybean meal with camelina meal. FIB: concentrate rich in neutral detergent fiber based on fibrous by-products and not including cereals or soybean meal.

Results from the canonical discriminant analysis are shown in Table 2. Eigenvalues, which provide information about the relative efficacy of each discriminant function, were significant according to Bartlett's test. Canonical discriminant functions (DF) explained 90.92% and 9.08% of total variance, respectively. Canonical correlation of the quantitative variables and the grouping variable was higher in DF1. Wilks' lambda test, which checks how well each function separates cases into groups, supported the validity of the model. Linear odd and anteiso FA showed the greatest discriminating ability and correlation in DF1, while iso FA had a higher contribution to group separation in DF2 (Table 2 and Figure 1). The centroid of CON treatment showed a positive value in both functions, while the centroid of FIB treatment was negative in DF1 and positive in DF2. The centroid of CAM treatment was negative in both functions (Table 2 and Figure 1). The largest distance was between CON and FIB treatments with CAM treatment in an intermediate position.

Table 2. Canonical discriminant analysis results.

	Standardized Canonical Coefficients		Canonical Structure	
	DF1 [1]	DF2	DF1	DF2
Odd FA [2]	1.00	−0.47	0.81	−0.56
Iso FA	0.40	0.60	0.36	0.76
Anteiso FA	−0.85	−0.34	−0.76	−0.46
Eigenvalues	5.37	0.54		
Variance explained (%)	90.92	9.08		
Bartlett test	$p < 0.001$	$p < 0.01$		
Canonical correlation	0.92	0.59		
Wilk's lambda test	$p < 0.001$	$p < 0.01$		
Centroids [3]				
CON	2.88	0.37		
CAM	−0.43	−0.97		
FIB	−2.45	0.61		

[1] DF: discriminant function. [2] FA: fatty acids. [3] CON: typical commercial concentrate rich in starch and based on cereals and soybean meal. CAM: typical commercial concentrate which replaces 50% of soybean meal with camelina meal. FIB: concentrate rich in neutral detergent fiber based on fibrous by-products and not including cereals or soybean meal.

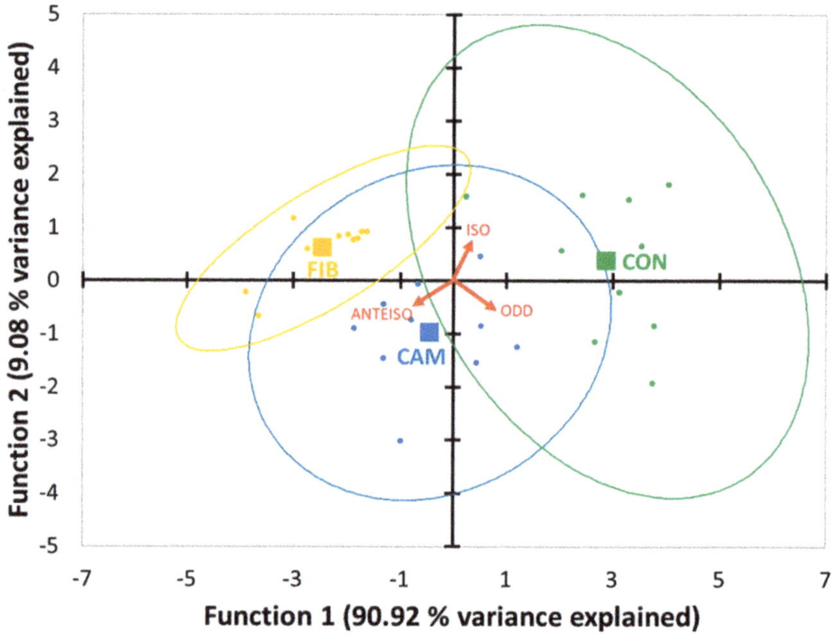

Figure 1. Canonical discriminant plot. Observations correspond to three type of diets: a typical commercial concentrate rich in starch and based on cereals and soybean meal (CON, green dots); a typical commercial concentrate which replaces 50% of soybean meal with camelina meal (CAM, blue dots); and a concentrate rich in neutral detergent fiber based on fibrous by-products and not including cereals or soybean meal (FIB, yellow dots). Centroids are indicated by squares of the corresponding color. Variable correlations with the discriminant functions are represented by red arrows.

Fisher's classification functions (Table 3) incorrectly assigned one observation from the CON to the CAM group (96.7% of the samples were correctly assigned). When cross-validation was carried out, two samples were mislabeled: one CON observation was assigned to the CAM group and one CAM observation was assigned to the FIB group. As a result, cross-validation properly classified 93.3% of lamb meat samples. Finally, variance analysis showed that the odd/anteiso FA ratio clearly differed between the three treatments, while the iso/(anteiso+odd) FA ratio was different for FIB compared to CON and CAM treatments (Table 4).

Table 3. Fisher's classification functions.

	Treatments [1]		
	CON	**CAM**	**FIB**
Intersection	−97.266	−79.860	−75.070
Odd FA [2]	23.388	16.984	10.411
Iso FA	168.825	139.528	141.459
Anteiso FA	17.119	59.838	75.023

[1] CON: typical commercial concentrate rich in starch and based on cereals and soybean meal. CAM: typical commercial concentrate which replaces 50% of soybean meal with camelina meal. FIB: concentrate rich in neutral detergent fiber based on fibrous by-products and not including cereals or soybean meal. [2] FA: fatty acids.

Table 4. Mean separation analysis for the ratios of odd- and branched-chain fatty acids with discriminating ability in the discriminant functions (DF) 1 and 2.

	Treatments [1]			
	CON (0.31)	**CAM (0.40)**	**FIB (1.54)**	**SEM [3]**
DF1: Odd FA/Anteiso FA [2]	8.15 ^{a}	4.93 ^{b}	3.12 ^{c}	0.419
DF2: Iso FA/(Anteiso FA+Odd FA)	0.22 ^{b}	0.19 ^{b}	0.29 ^{a}	0.011

[1] CON: typical commercial concentrate rich in starch and based on cereals and soybean meal. CAM: typical commercial concentrate which replaces 50% of soybean meal with camelina meal. FIB: concentrate rich in neutral detergent fiber based on fibrous by-products and not including cereals or soybean meal. Values in parenthesis under each treatment are the neutral detergent fiber (NDF) to starch ratio of the respective concentrate. [2] FA: fatty acids. [3] SEM: Standard Error of the Mean. ^{a,b,c} Within a row, means without a common superscript letter are significantly different according to Tukey's test at $p < 0.05$.

4. Discussion

Comprehensive discussion regarding growth performance, carcass and meat quality traits as well as IMF composition of the lambs can be found elsewhere [21,22]. In summary, experimental animals were homogeneous, maintained under the same housing and management conditions, and the feeding trial was carried out simultaneously in all experimental groups. Since no differences were found in productive traits or intramuscular fat content between treatments, any differences in intramuscular OBCFA profile should only be ascribed to the type of diet fed to each group of lambs.

Our results support those of a recently published research on the validity of the FA profile of meat as a tool to identify the dietary background of lambs by means of discriminant analysis [5] but also indicate that it is not necessary to use an extensive set of FA in IMF for that purpose. The OBCFA contents in IMF suffice to discriminate the meat samples according to the composition of the diet consumed by the lambs during the fattening period. In DF1, odd and anteiso FA (right and left sides, respectively, Figure 1) showed the highest discriminant ability, whereas iso FA (upper-right side, Figure 1) had the highest discriminant ability in DF2. These results suggested that a high NDF to starch ratio of the concentrate, being the lowest for CON treatment and the highest for FIB treatment, would be negatively related to the odd/anteiso FA ratio and positively related to the iso/(anteiso+odd) FA ratio in IMF. This point was further confirmed by the results of the analysis of variance (Table 4). The negative and positive linear responses of the odd/anteiso and iso/(odd+anteiso) FA ratios, respectively, to the increase of the diet NDF to starch ratio that can be noted in Table 4 are in agreement with two papers published in the last few years where lambs were fed fattening diets with intended differences in their NDF/starch ratio [24,25] (Figure 2).

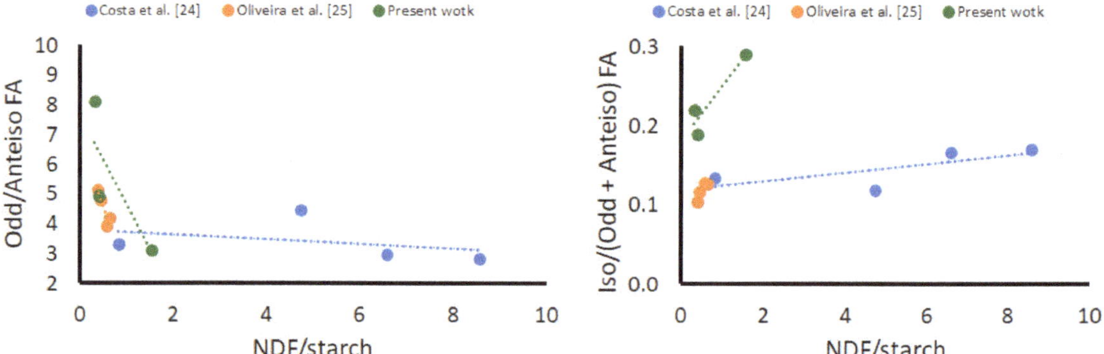

Figure 2. Responses of odd/anteiso and iso/(odd + anteiso) fatty acid (FA) ratios in the intramuscular fat of lambs to the neutral detergent (NDF) to starch ratio of the diet fed during the fattening period.

The effects of dietary treatments on ruminal bacteria populations might explain the validity of OBCFA as indicators of the NDF and starch proportions in the fattening diet fed to lambs. Diets rich in NDF raise the abundance of ruminal cellulolytic bacteria in growing lambs [17] without significant differences in meat OBCFA contents when forage is compared with fibrous by-products as an NDF source [26], and cellulolytic bacteria are usually enriched in iso FA, with some strains showing high levels of anteiso FA [12,13]. Moreover, it can be calculated that lambs classified postmortem as high-cellulolytic according to the number of main cellulolytic bacteria identified in the rumen contents had odd/anteiso and iso/(anteiso+odd) FA ratios in their IMF that were ~10% lower and ~17% higher, respectively, than that observed in the lambs classified as low-cellulolytic [27].

Our experimental concentrates did not include extra fat rich in polyunsaturated FA [21]. It is well-known that polyunsaturated FA in the diet had a negative impact on ruminal cellulolytic bacteria populations [28], which might change the OBCFA profile in IMF regardless of the NDF and starch proportions in the diet. In lambs fed the same diet either supplemented or not with linseed during the whole fattening period, the odd/anteiso FA ratio in IMF was ~ 30% lower in the linseed group compared with the control group [29]. Hence, the use of OBCFA profiles in IMF to separate lamb meat samples according to the NDF and starch proportions of the diet fed during the fattening period will not be appropriate when fat sources rich in polyunsaturated fatty acids are supplemented.

5. Conclusions

The present research shows that the quantification of OBCFA contents in IMF would suffice to separate meat samples according to the NDF to starch ratio of the diet fed to fattening lambs by means of LDA. The odd/anteiso and iso/(anteiso+odd) FA ratios showed negative and positive linear responses, respectively, to the NDF to starch ratio of the diet. A high odd/anteiso FA ratio along a low iso/(anteiso+odd) FA ratio in lamb meat would indicate a feeding regime rich in starch. Conversely, a low odd/anteiso FA ratio along a high iso/(anteiso+odd) FA ratio in lamb meat would indicate a feeding regime rich in NDF. Further research along these lines is recommended for ascertaining the validity of OBCFA contents in meat to authenticate the feeding background of lambs entering the meat chain supply.

Author Contributions: Conceptualization, A.L.M.M. and M.C.R.; methodology, A.L.M.M., P.G.-C. and F.R.D.; formal analysis, A.L.M.M. and P.G.-C.; investigation, M.Á.d.l.F., A.L.M.M., P.G.-C. and F.R.D.; data curation, M.C.R., A.L.M.M. and P.G.-C.; writing—original draft preparation, M.C.R., A.L.M.M. and A.S.; writing—review and editing, P.G.-C., A.L.M.M., A.S., F.R.D. and M.Á.d.l.F.; supervision, M.Á.d.l.F.; project administration, M.Á.d.l.F.; funding acquisition, M.Á.d.l.F. All authors have read and agreed to the published version of the manuscript.

Funding: This research was funded by the Spanish Ministry of Economy and Competitiveness (Project AGL2016-75159-C2-2-R).

Institutional Review Board Statement: Ethical review and approval were waived for this study, due to the fact that Spanish legislation and institutional requirements of the Universidad of Córdoba do not require it. This is because the assay was carried out under common zootechnical procedures and the animals did not suffer any intervention beyond those typical in such on-farm conditions.

Data Availability Statement: The data presented in this study are available in the article.

Conflicts of Interest: The authors declare no conflict of interest. The funders had no role in the design of the study; in the collection, analyses, or interpretation of data; in the writing of the manuscript, or in the decision to publish the results.

References

1. Sañudo, C.; Alfonso, M.; San Julián, R.; Thorkelsson, G.; Valdimarsdottir, T.; Zygoyiannis, D.; Stamataris, C.; Piasentier, E.; Mills, C.; Berge, P.; et al. Regional variation in the hedonic evaluation of lamb meat from diverse production systems by consumers in six European countries. *Meat Sci.* **2007**, *75*, 610–621. [CrossRef]
2. Watkins, P.J.; Frank, D.; Singh, T.K.; Young, O.A.; Warner, R.D. Sheepmeat flavor and the effect of different feeding systems: A review. *J. Agric. Food Chem.* **2013**, *61*, 3561–3579. [CrossRef] [PubMed]
3. Sepúlveda, W.S.; Maza, M.T.; Pardos, L. Aspects of quality related to the consumption and production of lamb meat. Consumers versus producers. *Meat Sci.* **2011**, *87*, 366–372. [CrossRef] [PubMed]
4. Monahan, F.J.; Schmidt, O.; Moloney, A.P. Meat provenance: Authentication of geographical origin and dietary background of meat. *Meat Sci.* **2018**, *144*, 2–14. [CrossRef] [PubMed]
5. Acciaro, M.; Decandia, M.; Sitzia, M.; Manca, C.; Giovanetti, V.; Cabiddu, A.; Addis, M.; Rassu, S.P.G.; Molle, G.; Dimauro, C. Discriminant analysis as a tool to identify bovine and ovine meat produced from pasture or stall-fed animals. *Ital. J. Anim. Sci.* **2020**, *19*, 1065–1070. [CrossRef]
6. Erasmus, S.W.; Muller, M.; Van Der Rijst, M.; Hoffman, L.C. Stable isotope ratio analysis: A potential analytical tool for the authentication of South African lamb meat. *Food Chem.* **2016**, *192*, 997–1005. [CrossRef] [PubMed]
7. Sun, S.; Guo, B.; Wei, Y.; Fan, M. Classification of geographical origins and prediction of δ13C and δ15N values of lamb meat by near infrared reflectance spectroscopy. *Food Chem.* **2012**, *135*, 508–514. [CrossRef]
8. Vasta, V.; Ratel, J.; Engel, E. Mass spectrometry analysis of volatile compounds in raw meat for the authentication of the feeding background of farm animals. *J. Agric. Food Chem.* **2007**, *55*, 4630–4639. [CrossRef]
9. Wang, J.; Xu, L.; Xu, Z.; Wang, Y.; Niu, C.; Yang, S. Liquid chromatography quadrupole time-of-flight mass spectrometry and rapid evaporative ionization mass spectrometry were used to develop a lamb authentication method: A preliminary study. *Foods* **2020**, *9*, 1723. [CrossRef]
10. Arshad, M.S.; Sohaib, M.; Ahmad, R.S.; Nadeem, M.T.; Imran, A.; Arshad, M.U.; Kwon, J.H.; Amjad, Z. Ruminant meat flavor influenced by different factors with special reference to fatty acids. *Lipids Health Dis.* **2018**, *17*, 223. [CrossRef]
11. Vahmani, P.; Ponnampalam, E.N.; Kraft, J.; Mapiye, C.; Bermingham, E.N.; Watkins, P.J.; Proctor, S.D.; Dugan, M.E. Bioactivity and health effects of ruminant meat lipids. Invited Review. *Meat Sci.* **2020**, *165*, 108114. [CrossRef] [PubMed]
12. Fievez, V.; Colman, E.; Castro-Montoya, J.M.; Stefanov, I.; Vlaeminck, B. Milk odd- and branched-chain fatty acids as biomarkers of rumen function-An update. *Anim. Feed Sci. Technol.* **2012**, *172*, 51–65. [CrossRef]
13. Vlaeminck, B.; Fievez, V.; Cabrita, A.R.J.; Fonseca, A.J.M.; Dewhurst, R.J. Factors affecting odd-and branched-chain fatty acids in milk: A review. *Anim. Feed Sci. Technol.* **2006**, *131*, 389–417. [CrossRef]
14. Gómez-Cortés, P.; Juárez, M.; de la Fuente, M.A. Milk fatty acids and potential health benefits: An updated vision. *Trends Food Sci. Technol.* **2018**, *81*, 1–9. [CrossRef]
15. Bessa, R.J.B.; Maia, M.R.G.; Jeronimo, E.; Belo, A.T.; Cabrita, A.R.J.; Dewhurst, R.J.; Fonseca, A.J.M. Using microbial fatty acids to improve understanding of the contribution of solid associated bacteria to microbial mass in the rumen. *Anim. Feed Sci. Technol.* **2009**, *150*, 197–206. [CrossRef]
16. Prado, L.A.; Ferlay, A.; Noziere, P.; Schmidely, P. Predicting duodenal flows and absorption of fatty acids from dietary characteristics in ovine and bovine species: A meta-analysis approach. *Animal* **2019**, *13*, 727–739. [CrossRef]
17. Ellison, M.J.; Conant, G.C.; Lamberson, W.R.; Cockrum, R.R.; Austin, K.J.; Rule, D.C.; Cammack, K.M. Diet and feed efficiency status affect rumen microbial profiles of sheep. *Small Rumin. Res.* **2017**, *156*, 12–19. [CrossRef]
18. Wang, Y.; Cao, P.; Wang, L.; Zhao, Z.; Chen, Y.; Yang, Y. Bacterial community diversity associated with different levels of dietary nutrition in the rumen of sheep. *Appl. Microbiol. Biotechnol.* **2017**, *101*, 3717–3728. [CrossRef]

19. Vlaeminck, B.; Fievez, V.; Demeyer, D.; Dewhurst, R.J. Effect of forage: Concentrate ratio on fatty acid composition of rumen bacteria isolated from ruminal and duodenal digesta. *J. Dairy Sci.* **2006**, *89*, 2668–2678. [CrossRef]
20. Zhang, Y.; Liu, K.; Hao, X.; Xin, H. The relationships between odd-and branched-chain fatty acids to ruminal fermentation parameters and bacterial populations with different dietary ratios of forage and concentrate. *J. Anim. Physiol. Anim. Nutr.* **2017**, *101*, 1103–1114. [CrossRef]
21. Avilés Ramirez, C.; Peña Blanco, F.; Horcada, A.; Nuñez Sánchez, N.; Requena Domenech, F.; Guzman Medina, P.; Martínez Marín, A.L. Effects of concentrates rich in by-products on growth performance, carcass characteristics and meat quality traits of light lambs. *Anim. Prod. Sci.* **2019**, *59*, 593–599. [CrossRef]
22. Gómez-Cortés, P.; Galisteo, O.O.; Avilés Ramirez, C.; Peña Blanco, F.; de la Fuente, M.A.; Núñez Sánchez, N.; Martínez Marín, A.L. Intramuscular fatty acid profile of feedlot lambs fed concentrates with alternative ingredients. *Anim. Prod. Sci.* **2019**, *59*, 914–920. [CrossRef]
23. Klecka, W.R. *Discriminant Analysis*; Sage Publications: Beverly Hills, CA, 1980.
24. Costa, M.; Alves, S.P.; Francisco, A.; Almeida, J.; Alfaia, C.M.; Martins, S.V.; Prates, J.A.M.; Santos-Silva, J.; Doran, O.; Bessa, R.J.B. The reduction of starch in finishing diets supplemented with oil does not prevent the accumulation of trans-10 18:1 in lamb meat. *J. Anim. Sci.* **2017**, *95*, 3745–3761. [CrossRef] [PubMed]
25. Oliveira, M.A.; Alves, S.P.; Santos-Silva, J.; Bessa, R.J.B. Effect of dietary starch level and its rumen degradability on lamb meat fatty acid composition. *Meat Sci.* **2017**, *123*, 166–172. [CrossRef]
26. Santos-Silva, J.; Francisco, A.; Alves, S.P.; Portugal, P.; Dentinho, T.; Almeida, J.; Soldado, D.; Jeronimo, E.; Bessa, R.J.B. Effect of dietary neutral detergent fibre source on lambs growth, meat quality and biohydrogenation intermediates. *Meat Sci.* **2019**, *147*, 28–36. [CrossRef]
27. Zhang, Z.; Niu, X.; Li, F.; Li, F.; Guo, L. Ruminal cellulolytic bacteria abundance leads to the variation in fatty acids in the rumen digesta and meat of fattening lambs. *J. Anim. Sci.* **2020**, *98*, skaa228. [CrossRef]
28. Huws, S.A.; Kim, E.J.; Cameron, S.J.S.; Girdwood, S.E.; Davies, L.; Tweed, J.; Vallin, H.; Scollan, N.D. Characterization of the rumen lipidome and microbiome of steers fed a diet supplemented with flax and echium oil. *Microb. Biotechnol.* **2015**, *8*, 331–341. [CrossRef]
29. Li, F.; Zhang, Z.; Li, X.; Zhu, B.; Guo, L.; Li, F.; Weng, X. Effect of duration of linseed diet supplementation before slaughter on the performances, meat fatty acid composition and rumen bacterial community of fattening lambs. *Anim. Feed Sci. Technol.* **2020**, *263*, 114457. [CrossRef]

Article

Characterization and Discrimination of Key Aroma Compounds in Pre- and Postrigor Roasted Mutton by GC-O-MS, GC E-Nose and Aroma Recombination Experiments

Huan Liu , Teng Hui, Fei Fang, Qianli Ma, Shaobo Li, Dequan Zhang and Zhenyu Wang *

Key Laboratory of Agro-Products Processing, Institute of Food Science and Technology, Chinese Academy of Agricultural Sciences, Ministry of Agriculture and Rural Affairs, Beijing 100193, China; sd_lh1990@126.com (H.L.); htengui@163.com (T.H.); fangfeixdf@163.com (F.F.); mql201228@163.com (Q.M.); lishaobo@caas.cn (S.L.); dequan_zhang0118@126.com (D.Z.)

* Correspondence: food2006wzy@163.com; Tel.: +86-(10)-6281-8740

Abstract: The key aroma compounds in the pre- and postrigor roasted mutton were studied in this study. The results showed that 33 and 30 odorants were detected in the pre- and postrigor roasted mutton, respectively. Eight aroma compounds, including 3-methylbutanal, pentanal, hexanal, heptanal, octanal, nonanal, 1-octen-3-ol, and 2-pentylfuran, were confirmed as key odorants by aroma recombination and omission experiments. The aroma profiles of pre- and postrigor roasted mutton both presented great fatty, roasty, meaty, grassy, and sweet odors. Particularly, the concentrations of hexanal, heptanal, octanal, nonanal, 1-octen-3-ol, and 2-pentylfuran in postrigor roasted mutton were significantly higher ($p < 0.05$) than those in the prerigor roasted mutton. The postrigor *back strap* was more suitable for roasting than the prerigor *back strap*. The pre- and postrigor roasted mutton could be obviously discriminated based on the aroma compounds by orthogonal partial least squares discrimination analysis (OPLS-DA) and principal component analysis (PCA). Hexanal and 1-octen-3-ol might potential markers for the discrimination of the pre- and postrigor roasted mutton.

Keywords: roasted mutton; pre- and postrigor; key aroma compounds; marker; recombination and omission experiments

Citation: Liu, H.; Hui, T.; Fang, F.; Ma, Q.; Li, S.; Zhang, D.; Wang, Z. Characterization and Discrimination of Key Aroma Compounds in Pre- and Postrigor Roasted Mutton by GC-O-MS, GC E-Nose and Aroma Recombination Experiments. *Foods* **2021**, *10*, 2387. https://doi.org/10.3390/foods10102387

Academic Editors: Benjamin W.B. Holman and Eric Nanthan Ponnampalam

Received: 8 September 2021
Accepted: 29 September 2021
Published: 8 October 2021

1. Introduction

The production of mutton was 4.88 million tons in China in 2019, with a growth rate of 2.6%. Roasted mutton is the most popular meat product due to its unique aroma. However, a few studies have been reported to characterize the aroma compounds of roasted mutton after chilling at 4 °C for 72 h (postrigor), among which hexanal, octanal and nonanal are the main odorants according to odor activity values (OAVs) and concentrations based on an internal standard [1,2]. The majority of consumers prefer roasted mutton without chilling (prerigor) rather than those from chilled carcasses in most areas in China, and they believe that roasted mutton without chilling (prerigor) is more flavory than any chilled sample. However, the assumption is only a traditional point, and no scientific data deny it. To date, only one study about aroma compounds in the roasted mutton was reported at different aging times by the universal steam oven, among which higher peak areas of total aroma compounds were found in cooked mutton aged for 3 days than those aged for 1 day. However, the study did not elucidate key aroma compounds and aroma profiles in mutton by the traditional charcoal roasting process and did not clarify the aroma differences among the samples [2]. Our previous results showed that the shear force of roasted mutton aged for 1–24 h first increased ($p < 0.05$) and significantly decreased ($p < 0.05$) when aged for 1–7 days. Roasted mutton aged for 1 day had the highest shear force value. Sheep muscles aged for 1–12 h, 1 day and 3–7 days were considered to be in the prerigor, rigor and postrigor phases based on shear force and pH values, respectively [2]. The determination of key aroma compounds can provide data support for the selection of raw meat, the slaughter process, and cooking method optimization [3]. However, the aroma compounds in the roasted mutton are unclear. In particular, the data on the differences in key aroma compounds in the pre- and postrigor roasted mutton are rather scarce.

Recently, the sensomics approach has been widely applied in the characterization of key aroma compounds in samples [4,5]. Key aroma compounds could not be determined by the concentrations and OAVs of odorants alone. The OAV of decanal was higher than 1 in Beijing Youji broth (19) and commercial Broiler broth (2), whereas this odorant did not significantly affect the aroma profile of chicken broth, indicating that it was not the key aroma compound [3]. In contrast, aroma recombination experiment has recently been applied to determine the key aroma compounds in various foods, such as wine [4]. GC-MS can accurately identify, quantitate and determine key aroma compounds in samples. However, the mass spectrometry of GC-MS cannot be first translated into sensory perception response and second visually present the difference of samples. Interestingly, "E-sensing" technologies can clarify the overall aroma difference by simulating the human sense of nose, including e-nose [6]. The flash GC e-nose was a combination of GC and e-nose, which could effectively separate compounds and identify differences, such as virgin olive oils [7]. In particular, the integration of e-nose and GC-MS could comprehensively elucidate the aroma difference in samples, such as roasted bread, heated oil and virgin olive oil [8–10].

This study aimed to confirm the key aroma compounds and their differences in the pre- and postrigor roasted mutton. (i) The key aroma compounds were accurately identified and quantitated by gas chromatography olfactometry mass spectrometry (GC-O-MS). Afterward, (ii) the key aroma compounds in samples were determined by OAVs, contribution rates, and recombination and omission experiments. Then, (iii) it was confirmed that the postrigor *back strap* was more suitable for roasting than the prerigor *back strap*. Finally, (iv) the potential markers discriminating the pre- and postrigor roasted mutton were determined by GC-O-MS, flash GC e-nose, orthogonal partial least squares discrimination analysis (OPLS-DA), and principal component analysis (PCA).

2. Materials and Methods

2.1. Chemicals and Reagents

Standards of most volatile compounds were obtained from Sigma-Aldrich (Shanghai, China): 1-octen-3-ol (98%), (*E*)-2-octen-1-ol (97%), 1-heptanol (98%), propanal (97%), pentanal (98%), hexanal (98%), heptanal (97%), octanal (99%), (*E*)-2-octenal (97%), nonanal (99.5%), (*E*)-2-nonenal (97%), benzaldehyde (99.5%), 2-pentylfuran (98%), and 2,3-pentanedione (97%). The 3-methylbutanal (98%) was supplied by Aladdin (Shanghai, China). The n-alkanes (C_7-C_{40}, $\geq$97%, external standard) was obtained from o2si Smart Solutions (Shanghai, China). The 2-methyl-3-heptanone (99%) was supplied by Dr. Ehrenstorfer (Beijing, China) as an internal standard.

2.2. Sample Preparation

All animal procedures performed in this study were approved by the Animal Care and Use Committee of the Institute of Food Science and Technology, Chinese Academy of Agricultural Sciences (Beijing, China). A total of 120 sheep (6-month-old, small-tail sheep × Mongolian sheep with 27.40 ± 2.64 kg carcass weight) were pastured together in Inner Mongolia Province in China. All sheep had the same genetic background and were fed the same diet. 12 Sheep were randomly selected from 120 sheep. The *back strap* was obtained according to the cutting technical specification of mutton, which was same with *backstrap 5101* in the seventh edition of Handbook of Australian Meat [11]. The pre- and postrigor muscles were applied in each carcass. The left carcass was treated with prerigor, and the right half was treated with postrigor in each carcass. The prerigor *back straps* were the muscles from 12 carcasses (pH: 6.42 ± 0.08), which were quickly frozen at −35 °C within 45 min after slaughter. The postrigor *back straps* were the muscles from the 12 carcasses (pH: 5.42 ± 0.27), which were kept at 4 °C for 72 h and thereafter frozen at −35 °C. All samples were wrapped with nylon/polyethylene, transported to our lab by cold-chain logistics and stored at −80 °C. The muscles were incubated (MIR-154-PC, Panasonic, Japan) at 4 ± 1 °C overnight thaw until the core temperatures dropped to the range of −3 and −5 °C. After being trimmed off connective tissue and surface fat, the samples were cut into cubes (3 × 1.5 × 1.5 cm^3). The samples were roasted for 10 min by traditional burning charcoal. The roasting process ended when the core temperature reached 77.6–79.9 °C in the samples (surface temperature: 85–97 °C).

2.3. GC-O-MS Analysis

Aroma compounds were extracted by the headspace solid-phase microextraction (HS-SPME) with a carboxen−polydimethylsiloxane fused silica (CAR/PDMS, 75 μm) coating fiber [12]. The minced sample and 2-methyl-3-heptanone (internal standard, 1.5 μL, 1.7 $\mu g \cdot \mu L^{-1}$) were put into a 20 mL vial sealed with a PTFE-silicon stopper. The vial was incubated at 55 °C for 10 min and the aroma compounds were extracted at 55 °C for 45 min. Immediately, the coating fiber was desorbed at 250 °C for 3 min. The analysis was prepared on an Agilent gas chromatograph (7890B) coupled with an olfactometry (ODP C200, Gerstal, Mulheim an der Ruhr, Germany) and 5977A mass selective detector. The aroma compounds were separated by a fused-silica capillary column (60 m × 250 μm × 0.25 μm, DB-Wax capillary column, Agilent Technologies, Santa Clara, CA, USA). The temperature program of the GC oven was 40 °C for 3 min, raised to 70 °C at 2 °C/min, increased to 130 °C at 3 °C/min, ramped to 230 °C at 10 °C/min and maintained for 10 min. The helium (99.99%) was prepared as a carrier gas with a flow rate of 1.4 mL/min. The injector temperature was kept at 250 °C with a splitless inlet. The electron ionization mode was positive ion (70 eV) with an acquisition range from 40 to 500 m/z in full-scan mode.

2.4. Identification Analysis of Aroma Compounds

Aroma compounds were identified by mass spectrometry library, linear retention indices (LRIs), odor qualities, and authentic flavor standards. LRIs were obtained according the retention time of n-alkanes (C_7–C_{40}) by linear interpolation. The aroma compounds were also determined by professional panelists using GC-O. Meanwhile, the authentic standards of aroma compounds were analyzed with the same detection procedure as that used for the samples. The aroma compounds were confirmed by retention times between authentic flavor standards and samples.

2.5. Quantitation Analysis of Aroma Compounds

Aroma compounds were quantitated by calibration curves of authentic flavor standards following semiquantitation of an internal standard. First, the concentrations of aroma compounds in the samples were evaluated according to the ratio of the concentration and peak area of the internal standard. In particular, the aroma compound concentrations with OAVs greater than 1 were calibrated by a 5-point standard curve of authentic flavor standards. Prior to quantitation analysis, the roasted mutton was prepared to obtain an artificial odorless matrix based on previous studies [12]. The calibration curves of aroma compounds in the roasted mutton were constructed by the above odorless matrix and authentic flavor standards with different concentrations. 2-Methyl-3-heptanone was put into the mixture to calibrate the peak area of aroma compounds. The odorless matrix without flavor standards was considered the control. Authentic flavor standards were analyzed by GC-SIM with the same detection procedure as that used for the samples. Authentic flavor standards, scanned ions (m/z) and calibration equations were obtained. The calibration curves of aroma compounds all have great linear correlations, where x is the ratio of the concentration of aroma compound to the internal standard and y is their peak ratio.

2.6. OAVs and Contribution Rate Analysis

The OAVs of aroma compounds were determined by dividing concentrations with their threshold [13]. The contribution rate was the OAV ratio of single aroma compound to total aroma compounds.

2.7. Aroma Recombination and Omission Experiments

The recombination and omission experiments were performed by a triangle test of sensory evaluation in a climate-controlled (26 ± 1 °C) sensory room [3]. A total of 50 sensory panelists aged 24–49 years old were screened and selected based on GBT 16291.1–2012. The panelists were trained for flavor recognition based on ISO 4121:2003. All panelists had been trained weekly and could describe and recognize odor qualities. Flavor profiles were determined using a scale from 0 to 5, which represented not detectable (0), very weak (1), weak (2), moderate (3), strong (4) and very strong (5) odors, respectively. The recombination model (model 1) was constructed by the above odorless matrix and authentic flavor standards with OAVs greater than 1. The sensory panelists evaluated the aroma similarity between model 1 and roasted mutton by a triangle test. Afterwards, the omission model (model 2) was prepared by omitting one aroma compound from model 1. The panelists estimated the aroma difference between model 1 and model 2. Finally, the recombination model (model 3) was prepared by an odorless matrix and aroma compounds that significantly affected the aroma profile of the samples. The panelists evaluated the aroma similarity between model 3 and the samples.

2.8. Flash GC E-Nose Analysis of Aroma Profile

A Heracles II e-nose (Alpha M.O.S., Toulouse, France) equipped with MXT-5 and MXT-1701 flame ionization detectors (FIDs) was used for the analysis of aroma compounds and aroma differences. The samples were treated as reported by Melucci and co-workers [7]. Briefly, the sample was heated at 50 °C for 30 min. Then, 3000 μL of headspace gas was injected into the GC port at a speed of 125 μL/s. The column temperature was 50 °C, rose to 250 °C at 2 °C/s and was maintained for 10 s. The temperatures of the GC port and FID were 200 °C and 260 °C, respectively. The aroma compounds were identified by retention indices from MXT-5 and MXT-1701 columns and determined by comparison with GC-MS data. The aroma differences in the pre- and postrigor roasted mutton were determined by PCA.

2.9. Statistical Analysis

All analyses were conducted in 12 measurements. Comparisons among roasted mutton of different aging times were performed using independent-samples t-test. The statistical analysis of aroma compounds in the roasted mutton were conducted at a level of $p < 0.05$ with SPSS 19.0 software (IBM Corporation, Armonk, NY, USA). Origin 2017 software and SIMCA 14.1 were used to perform plotting figures.

3. Results

3.1. Identification and Quantitation of Aroma Compounds in the Roasted Mutton

As presented in Tables 1–3, 33 aroma compounds were identified by GC-O-MS, among which 33 and 30 compounds were detected in the pre- and postrigor roasted mutton, respectively. Butanoic acid, pentanoic acid and 2,6-dimethylpyrazine were only found in prerigor samples. The aldehydes (10) and alcohols (7) with maximum types were the major odorants in the samples (Table 3). 3-Methylbutanal, pentanal, hexanal, heptanal, octanal, nonanal, and 1-octen-3-ol might be important odorants based on their high odor qualities (O) from GC-O-MS. The characteristic ion fragments of aroma compounds were obtained according to the identification of authentic flavor standards (Table 2).

Table 1. Aroma compounds, linear retention indices (LRIs), and identification methods in the pre- and postrigor roasted mutton.

Compounds [a]	LRIs		Identification [d]
	Literature [b]	Calculated [c]	
Pentane	– [e]	–	MS
Hexane	–	–	MS, S
1-Heptene	750	757	MS, LRI
Propanal	798	795	MS, LRI, O, S
Octane	–	–	MS, S
Acetone	814	816	MS, LRI
Methyl Ester Acetic Acid	827	827	MS, LRI
Ethyl Acetate	887	890	MS, LRI
2-Butanone	900	905	MS, LRI
3-Methylbutanal	915	918	MS, LRI, O, S
Pentanal	979	980	MS, LRI, O, S
2,3-Pentanedione	1060	1059	MS, LRI, O, S
Hexanal	1094	1088	MS, LRI, O, S
Heptanal	1188	1188	MS, LRI, O, S
2-Pentylfuran	1230	1234	MS, LRI, O, S
1-Pentanol	1261	1259	MS, LRI
Octanal	1291	1293	MS, LRI, O, S
2,5-Octanedione	–	1329	MS
2,6-Dimethylpyrazine	1338	1337	MS, LRI
1-Hexanol	1359	1362	MS, LRI
Nonanal	1396	1398	MS, LRI, O, S
(*E*)-2-Octenal	1434	1437	MS, LRI, O, S
1-Octen-3-Ol	1456	1458	MS, LRI, O, S
1-Heptanol	1462	1464	MS, LRI, O, S
2-Ethyl-1-Hexanol	1499	1497	MS, LRI
Benzaldehyde	1534	1537	MS, LRI, O, S
(*E*)-2-Nonenal	1549	1550	MS, LRI, O, S
1-Octanol	1573	1571	MS, LRI
2,3-Butanediol	1583	1589	MS, LRI
(*E*)-2-Octen-1-Ol	1622	1624	MS, LRI, O, S
Butanoic Acid	1644	1642	MS, LRI
Pentanoic Acid	1720	1724	MS, LRI
Hexanoic Acid	1854	1856	MS, LRI

[a] The aroma compounds in the pre- and postrigor roasted mutton. [b] Reported data in literatures. [c] Data calculated based on the retention time of n-alkanes (C_7–C_{40}) by linear interpolation. [d] Identified methods. MS, mass spectrometry library of GC-MS; LRI, linear retention indices; O, odor qualities; S, authentic flavor standards. [e] Not found or calculated.

To better understand the contributions of odorants to the aroma profile in samples, quantitation analysis was performed (Tables 2 and 3). The pre- and postrigor roasted mutton both contained 15 compounds (OAVs > 1), which were quantitated based on the standard calibration curves of 5 points (Table 2). The major aroma compounds in the samples were propanal (105.86–152.67 ng/g), pentanal (1398.14–1407.06 ng/g), 2,3-pentanedione (115.22–208.95 ng/g), hexanal (3218.71–4383.43 ng/g), heptanal (744.04–1294.82 ng/g), 1-pentanol (162.93–165.20 ng/g), octanal (264.68–506.82 ng/g), 2,5-octanedione (170.57–537.81 ng/g), nonanal (119.41–197.01 ng/g), and 1-octen-3-ol (219.01–498.46 ng/g). In particular, the concentration of only 3-methylbutanal of 15 aroma compounds (OAVs > 1) in the prerigor roasted mutton was significantly higher ($p < 0.05$) than that of postrigor mutton. The concentrations of the 13 key aroma compounds in the prerigor roasted mutton were significantly lower ($p < 0.05$) than postrigor mutton, except 3-methylbutanal and pentanal.

Table 2. Ion fragments and standard calibration curves of aroma compounds (OAVs >1) in the pre- and postrigor roasted mutton.

Compounds	Ion Fragments [a]	Standard Calibration Curves [b]	R^2
Propanal	27, 28, 29, 58	y = 0.0001x + 0.0022	0.995
3-Methylbutanal	41,43,44, 58	y = 0.00004x + 0.0009	0.990
Pentanal	29, 41, 44, 58	y = 0.0002x + 0.0016	0.990
2,3-Pentanedione	27, 29, 43, 57	y = 0.0001x − 0.0002	0.987
Hexanal	41, 44, 56, 57	y = 0.0008x + 0.1074	0.989
Heptanal	41, 43, 44, 70	y = 0.0003x + 0.0021	0.999
2-Pentylfuran	53, 81, 82, 138	y = 0.004x + 0.0012	0.998
Octanal	41,43, 56, 84	y = 0.0002x + 0.018	0.994
Nonanal	41, 43, 56, 57	y = 0.0011x + 0.0023	0.997
(*E*)-2-Octenal	29, 41, 55, 70	y = 0.0004x + 0.0007	0.988
1-Octen-3-Ol	43, 55, 57, 72	y = 0.0004x + 0.0086	0.999
1-Heptanol	41, 55, 56, 70	y = 0.0019x − 0.0071	0.997
Benzaldehyde	51, 77, 105, 106	y = 0.0067x − 0.0265	0.994
(*E*)-2-Nonenal	41, 43, 55, 70	y = 0.0105x − 0.0097	0.992
(*E*)-2-Octen-1-Ol	41, 43, 55, 57	y = 0.0011x − 0.0038	0.999

[a] Selected ion fragments based on the authentic flavor standards. [b] Equations of standard calibration curves, where x is the concentration ratio of authentic flavor standards to internal standard and y is the peak area ratio of authentic flavor standards to internal standard. The pre- and postrigor roasted mutton both contained the 15 aroma compounds.

Table 3. Concentrations, OAVs, and contribution rates of aroma compounds in the pre- and postrigor roasted mutton.

Compounds	Concentration (ng/g) [a]		OAVs [b]		Contribution Rates [c]	
	Pre-Rigor	Post-Rigor	Pre-Rigor	Post-Rigor	Pre-Rigor	Post-Rigor
Pentane	21.38 ± 1.68 [a]	10.13 ± 0.68 [b]	0	0	0	0
Hexane	13.78 ± 0.77 [a]	5.85 ± 0.38 [b]	0	0	0	0
1-Heptene	3.80 ± 0.15 [b]	4.42 ± 0.25 [a]	0	0	0	0
Propanal	105.86 ± 2.99 [b]	152.67 ± 10.72 [a]	11.14 ± 0.31 [b]	16.07 ± 1.13 [a]	0.49 ± 0.02	0.49 ± 0.03
Octane	14.69 ± 0.50 [b]	21.45 ± 0.89 [a]	0	0	0	0
Acetone	13.33 ± 0.71 [b]	16.00 ± 0.71[a]	0	0	0	0
Methyl Ester Acetic Acid	8.72 ± 0.58	7.62 ± 0.42	0	0	0	0
Ethyl Acetate	9.46 ± 0.60 [b]	15.32 ± 1.08 [a]	0.10 ± 0.01 [b]	0.15 ± 0.01 [a]	0	0.01 ± 0.00
2-Butanone	4.93 ± 0.26 [b]	8.71 ± 0.59 [a]	0	0	0	0
3-Methylbutanal	85.31 ± 1.94 [a]	48.79 ± 4.97 [b]	426.56 ± 9.71 [a]	243.96 ± 24.84 [b]	18.85 ± 0.63 [a]	7.40 ± 0.68 [b]
Pentanal	1398.14 ± 33.58	1407.06 ± 81.28	116.51 ± 2.80	117.26 ± 6.77	5.12 ± 0.08 [a]	3.58 ± 0.17 [b]
2,3-Pentanedione	115.22 ± 3.01 [b]	208.95 ± 11.61 [a]	5.76 ± 0.15 [b]	10.45 ± 0.58 [a]	0.26 ± 0.01 [b]	0.32 ± 0.02 [a]
Hexanal	3218.71 ± 75.44 [b]	4383.43 ± 114.32 [a]	715.27 ± 16.77 [b]	974.1 ± 25.40 [a]	31.62 ± 1.10	29.92 ± 0.66
Heptanal	744.04 ± 23.91 [b]	1294.82 ± 44.14 [a]	248.02 ± 7.97 [b]	431.61 ± 14.71 [a]	10.88 ± 0.17 [b]	13.21 ± 0.24 [a]
2-Pentylfuran	13.26 ± 0.36 [b]	20.91 ± 0.73 [a]	2.21 ± 0.06 [b]	3.49 ± 0.12 [a]	0.10 ± 0.00 [b]	0.11 ± 0.00 [a]
1-Pentanol	162.93 ± 6.09	165.20 ± 9.19	0.04 ± 0.00	0.04 ± 0.00	0	0
Octanal	264.68 ± 27.51 [b]	506.82 ± 28.84 [a]	378.12 ± 39.31 [b]	724.02 ± 41.20 [a]	16.31 ± 1.36 [b]	22.17 ± 1.09 [a]
2,5-Octanedione	170.57 ± 8.26 [b]	537.81 ± 9.87 [a]	0	0	0	0
2,6-Dimethylpyrazine	11.07 ± 0.45 [a]	0 [b]	0.03 ± 0.00 [a]	0 [b]	0	0
1-Hexanol	25.39 ± 1.79 [b]	48.87 ± 2.02 [a]	0.01 ± 0.00 [b]	0.02 ± 0.00 [a]	0	0
Nonanal	119.41 ± 4.51 [b]	197.01 ± 6.38 [a]	119.41 ± 4.51 [b]	197.01 ± 6.38 [a]	5.23 ± 0.13 [b]	6.05 ± 0.18 [a]
(*E*)-2-Octenal	8.86 ± 0.29 [b]	21.52 ± 0.69 [a]	2.96 ± 0.10 [b]	7.17 ± 0.23 [a]	0.13 ± 0.01 [b]	0.22 ± 0.01 [a]
1-Octen-3-Ol	219.01 ± 9.90 [b]	498.46 ± 10.96 [a]	219.01 ± 9.90 [b]	498.46 ± 10.96 [a]	9.57 ± 0.27 [b]	15.32 ± 0.31 [a]
1-Heptanol	15.26 ± 0.38 [b]	17.00 ± 0.33 [a]	5.09 ± 0.12 [b]	5.67 ± 0.0.11 [a]	0.22 ± 0.00 [a]	0.18 ± 0.00 [b]
2-Ethyl-1-Hexanol	1.76 ± 0.12 [b]	3.25 ± 0.26 [a]	0	0	0	0
Benzaldehyde	7.62 ± 0.04 [b]	8.88 ± 0.05 [a]	2.54 ± 0.01 [b]	2.96 ± 0.02 [a]	0.11 ± 0.00 [a]	0.09 ± 0.00 [b]
(*E*)-2-Nonenal	1.59 ± 0.00 [b]	1.80 ± 0.01 [a]	19.92 ± 0.04 [b]	22.58 ± 0.14 [a]	0.88 ± 0.02 [a]	0.70 ± 0.02 [b]
1-Octanol	12.24 ± 0.42 [b]	21.77 ± 0.64 [a]	0.11 ± 0.00 [b]	0.20 ± 0.01 [a]	0[b]	0.01 ± 0.00 [a]
2,3-Butanediol	4.31 ± 0.57	5.65 ± 0.59	0	0	0	0
(*E*)-2-Octen-1-Ol	14.94 ± 0.39 [b]	23.06 ± 0.93 [a]	4.98 ± 0.13 [b]	7.69 ± 0.31 [a]	0.22 ± 0.00	0.24 ± 0.01
Butanoic Acid	1.06 ± 0.10 [a]	0 [b]	0	0	0	0
Pentanoic Acid	1.63 ± 0.11 [a]	0 [b]	0	0	0	0
Hexanoic Acid	27.29 ± 1.69 [b]	50.54 ± 4.28 [a]	0.01 ± 0.00 [b]	0.02 ± 0.00 [a]	0	0

[a] Concentrations of aroma compounds were calculated according to standard calibration curves of 5 points. [b] OAVs were calculated by dividing concentrations with their threshold. [c] Contribution rates were the OAV rates of individual compound to all compounds.

3.2. Determination of Key Aroma Compounds in the Roasted Mutton

As shown in Table 3, the pre- and postrigor roasted mutton both contained 15 aroma compounds with OAVs greater than 1, including propanal, 3-methylbutanal, pentanal, 2,3-pentanedione, hexanal, heptanal, 2-pentylfuran, octanal, (E)-2-octenal, nonanal, 1-octen-3-ol, 1-heptanol, benzaldehyde, (E)-2-nonenal, and (E)-2-octen-1-ol. This result was also in agreement with the analysis of odor qualities. Among them, the highest OAVs were determined for hexanal (715.27-974.10), followed by 3-methylbutanal (243.96–426.56), octanal (378.12–724.02), 1-octen-3-ol (219.01–498.46), heptanal (248.02–431.61), and nonanal (119.41–197.01). The OAV of only 3-methylbutanal of the 15 aroma compounds in the prerigor roasted mutton was greater ($p < 0.05$) than that of postrigor mutton. The 13 aroma compounds had the reverse trends ($p < 0.05$). The changes in the contribution rates of aroma compounds were in accordance with those of OAVs, among which hexanal (29.92–31.62%) presented the highest contribution rate, followed by octanal (16.31–22.17%), 3-methylbutanal (7.40–18.85%), 1-octen-3-ol (9.57–15.32%), and heptanal (10.88–13.21%). These results preliminarily indicated that the 15 aroma compounds (OAVs > 1) with high contribution rates might be key odorants from the difference of aroma profiles in pre- and postrigor samples.

3.3. Confirmation of Key Aroma Compounds in the Roasted Mutton

The odorless matrix was constructed with 74.67% ultrapure water and authentic flavor standards (OAVs > 1) in the samples. The recombination model (model 1, 15 odorants) with all aroma compounds with OAVs greater than 1 revealed an extremely high similarity with the original roasted mutton in terms of the aroma profile by the triangle test. The results of the omission experiments (model 2, 14 odorants) indicated that 8 odorants significantly affected the overall aroma profile ($p < 0.05$) of the samples, including 3-methylbutanal, pentanal, hexanal, heptanal, 2-pentylfuran, octanal, nonanal, and 1-octen-3-ol. In particular, the model without hexanal and 1-octen-3-ol presented a noticeable difference ($p < 0.01$) compared to the aroma profile in model 1. Finally, the recombination model with the 8 aroma compounds mentioned above (model 3) showed a high similarity (4.51 out of 5 points) in comparison with roasted mutton, as illustrated in Figure 1. In particular, the pre- and postrigor roasted mutton both had fatty, roasty, meaty, grassy, and sweet odors. The intensity of the aroma profile in the postrigor roasted mutton was significantly greater ($p < 0.05$) than prerigor sample.

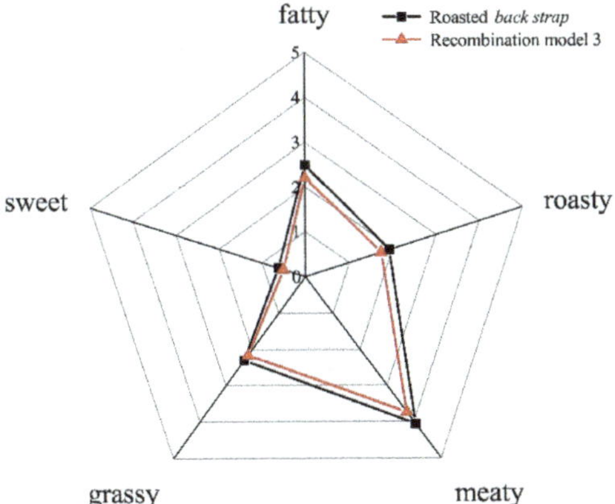

Figure 1. Aroma profiles of roasted mutton compared with the aroma recombination model 3.

3.4. Potential Markers Analysis for Discriminating the Pre- and Postrigor Roasted Mutton Based on Aroma Compounds

As presented in the score scatter plot of OPLS-DA ($R^2X = 0.92$, $R^2Y = 0.99$, $Q^2 = 0.99$) (Figure 2), the pre- and postrigor roasted mutton were obviously separated. R^2 and Q^2 revealed the fitness and predictive ability of the model, respectively. The prerigor roasted mutton was in the second and third quadrants, in which aldehydes, acids, esters, alkanes, and nitrogen-containing compounds were the predominant chemical families, such as 3-methylbutanal, pentanoic acid, and 2,6-dimethylpyrazine. The postrigor roasted mutton was located in the first and fourth quadrants of the model, among which alcohols, aldehydes, ketones, and furans had an important contribution, including 1-octen-3-ol, hexanal, 2,3-pentanedione, and 2-pentylfuran. The aroma compounds (variable importance for the projection ≥ 1) were generally considered as potential markers to discriminate samples. A total of 20 aroma compounds were identified to show differences between pre- and postrigor mutton (Figure 2c), such as 2,5-octanedione, 2,6-dimethylpyrazine, 1-octen-3-ol, and hexanal. The results also indicated that the postrigor roasted mutton had richer aroma compounds than the prerigor roasted mutton.

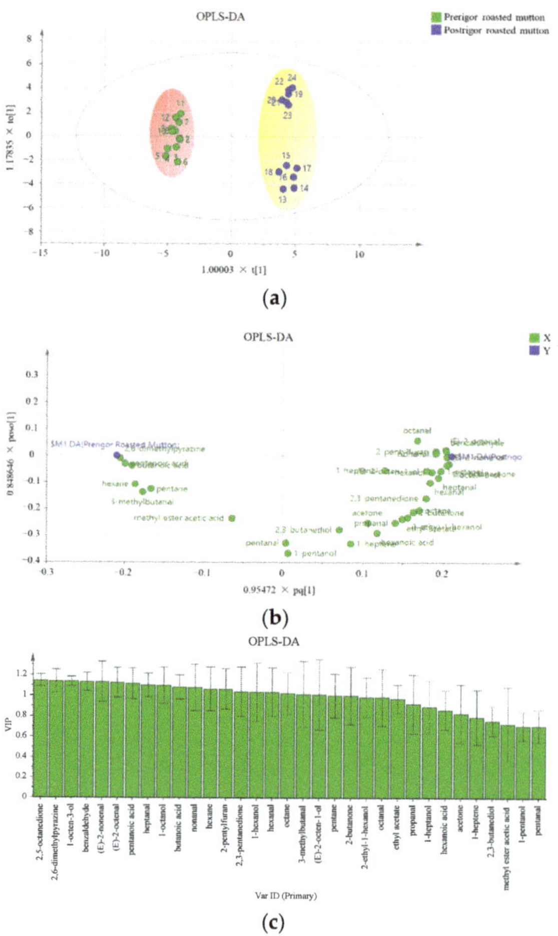

Figure 2. OPLS-DA of aroma compounds in the pre- and postrigor roasted mutton. (**a**) score scatter plot. (**b**) loading scatter plot. (**c**) VIP plot.

To further quickly determine the differential aroma compounds in the pre- and postrigor roasted mutton, flash GC e-nose and PCA were used. As illustrated in Table 4, 11 aroma compounds were identified in the two samples by flash GC e-nose. Among these, hexanal had the maximum peak area, followed by pentanal, 1-octen-3-ol, and heptanal. In particular, the peak areas of most aroma compounds, including hexanal and 1-octen-3-ol, in the postrigor samples was significantly greater ($p < 0.05$) than that in the prerigor samples, which was consistent with the GC-O-MS results. PCA of the flash GC e-nose was performed to determine the correlation pattern with individual composition variables in the discrimination between the two samples. As presented in Figure 3, the accumulative variance contribution rate of the first two PCs was 98.81% higher than 85% (PC1 of 97.68% and PC2 of 1.13%), which was sufficient to discriminate between these two samples. The general aroma feature could be well distinguished by a flash GC e-nose coupled with PCA. Based on above analysis, hexanal and 1-octen-3-ol might be potential markers to discriminate the pre- and postrigor roasted mutton.

Table 4. Peak areas of aroma compounds detected by flash GC e-nose in the pre- and postrigor roasted mutton.

Compounds (Peak Area)	Roasted Mutton	
	Pre-Rigor	Post-Rigor
Propanal	1039.59 ± 31.03	1036.75 ± 33.45
Hexane	489.23 ± 30.71	545.17 ± 49.23
3-Methylbutanal	251.64 ± 3.01 [b]	273.25 ± 7.56 [a]
Pentanal	7548.89 ± 108.17 [b]	8558.67 ± 189.74 [a]
2,3-Pentanedione	175.81 ± 19.61	177.83 ± 19.76
2,3-Butanediol	1875.48 ± 63.96	2063.67 ± 89.75
Hexanal	77261.81 ± 1382.88 [b]	87650.17 ± 1309.27 [a]
1-Hexanol	112.67 ± 2.40 [b]	135.58 ± 4.32 [a]
Heptanal	2079.92 ± 43.80 [b]	2627.92 ± 132.13 [a]
1-Octen-3-Ol	2498.33 ± 52.11 [b]	2934.42 ± 129.93 [a]
Octanal	614.85 ± 22.62 [b]	712.08 ± 12.94 [a]

Data with different superscripts ([a],[b]) within each row indicate significant difference ($p < 0.05$).

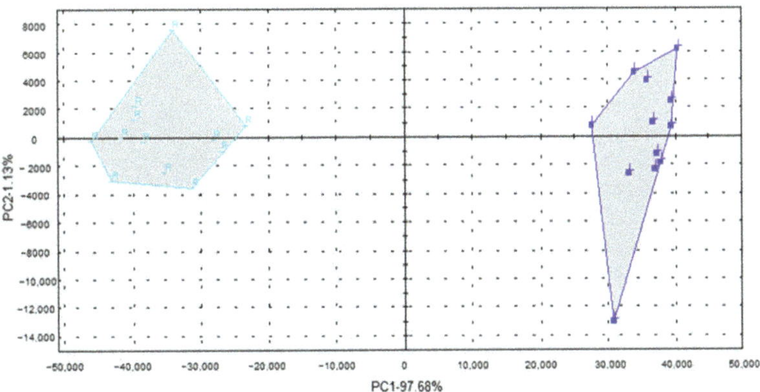

Figure 3. PCA of aroma compounds in the pre- and postrigor roasted mutton by flash GC e-nose. R and L represented the prerigor roasted mutton and postrigor roasted mutton, respectively.

4. Discussion

4.1. Aldehydes and Alcohols Were Key Aroma Compounds in the Pre- and Postrigor Roasted Mutton

It was reported that aldehydes and alcohols were the most important aroma compounds in roasted meat [14]. It was clearly observed that these compounds mainly contribute to the overall aroma of samples, such as hexanal (OAVs: 715.27–974.10) and 1-octen-3-ol (OAVs: 219.01–498.46). This result was in agreement with previous studies [1,2], which showed that hexanal had the most abundant concentration in roasted mutton, followed by 1-octen-3-ol, nonanal, and octanal. In particular, 8 of 15 odorants (OAVs > 1) comprising 6 aldehydes and 1 alcohol were confirmed as key odorants by the recombination and omission experiments. This result also corresponded to the studies, in which hexanal, heptanal, octanal, nonanal, and 1-octen-3-ol had the higher concentrations and OAVs in grilled goat meat [15]. The roasted mutton had strong roasty, fatty, grassy, meaty, and sweet odors, which were mainly caused by aldehydes and alcohols derived from the degradation of lipids and Strecker degradation of amino acids [16]. The phospholipids contained more unsaturated fatty acids than triacylglycerols, which caused the former's predominant contributions to the formation of fatty aldehydes and alcohols [17]. Pentanal, hexanal, heptanal, and 1-octen-3-ol could be generated from the oxidation of unsaturated fatty acids, which were responsible for the grassy note [3,18–20]. Aldehydes containing octanal and nonanal might predominantly contribute to fatty aromas [12]. 3-Methylbutanal, a Strecker aldehyde, was detected in the Maillard reaction with a seasoning-like odor [21]. In addition, the ketones and alkylfurans, including 2,3-pentanedione and 2-pentylfuran, could also be generated from the decomposition of lipids, which could generate roasty and meaty notes, respectively [22–24]. In particular, the aroma profile of roasted mutton was formed by the synergistic effect of key odorants rather than a single component [25]. Meanwhile, the concentrations of most key aroma compounds in the postrigor roasted mutton were significantly higher than those of the prerigor mutton, such as hexanal, heptanal, octanal, nonanal, 1-octen-3-ol, and 2-pentylfuran. This result indicated the postrigor *back strap* was more suitable for roasting than the prerigor *back strap*. This phenomenon was also consistent with the study reported by Coppock and Macleod, who clarified that the aging time generated more aroma compounds in the boiled beef [26]. Both thermal oxidation and autoxidation could produce the aldehydes and alcohols in meat and meat products. The richer aroma compounds in the postrigor roasted mutton could be explained by the autoxidation during aging [27,28].

4.2. Pre- and Postrigor Roasted Mutton were Discriminated Based on Key Aroma Compounds by GC-O-MS and GC E-Nose

In this study, GC-MS provided reliable and comprehensive diagnostic information for the detection of 8 key compounds, among which the concentration differences of 8 key odorants were responsible for the discrimination of the overall aroma profile of pre- and postrigor roasted mutton. In particular, hexanal and 1-octen-3-ol predominantly contributed to the aroma profile and caused the aroma difference of samples by using GC-O-MS. Meanwhile, the aroma profiles were obviously separated in the pre- and postrigor roasted mutton by using flash GC-O-MS, GC e-nose, OPLS-DA, and PCA, which was in agreement with aroma analysis of other food [29]. The characterization and discrimination of aroma compounds in the pre- and postrigor roasted mutton could also be successfully identified by GC-O-MS, among which hexanal and 1-octen-3-ol were key odorants and resulted in the difference of aroma profile in samples. The flash GC e-nose performance in the discrimination was consistent with respect to GC-O-MS, which was identical to previous studies [8,28]. These results indicated that hexanal and 1-octen-3-ol might be potential markers for discriminating the pre- and postrigor roasted mutton. This was in accordance with previous studies, among which hexanal and 1-octen-3-ol were indicators of oxidative stability and flavor acceptability in foods [30,31]. In addition, the combination of GC-MS with an e-nose could provide a comprehensive analysis for the characterization and discrimination of aroma compounds.

5. Conclusions

In this study, a total of 33 and 30 odorants were identified in the pre- and postrigor roasted mutton, among which they belonged to 8 chemical classes, such as aldehydes, ketones, alcohols, furans, acids, esters, and nitrogen-containing compounds. Eight odorants were confirmed to be the key aroma compounds in the roasted mutton, including hexanal, octanal, 1-octen-3-ol, nonanal, heptanal, pentanal, 3-methylbutanal, and 2-pentylfuran. The sensory evaluation of the recombination model including 8 key aroma compounds scored 4.51 out of 5 points. Only the concentration of 3-methylbutanal of 8 key aroma compounds in the prerigor roasted mutton was significantly higher than that of the postrigor mutton. Other 6 key aroma compounds, including hexanal, octanal, 1-octen-3-ol, nonanal, heptanal, and 2-pentylfuran, had the reverse trends. The pre- and postrigor roasted mutton could be discriminated based on the aroma compounds by GC-O-MS, flash GC e-nose, OPLS-DA, and PCA. Hexanal and 1-octen-3-ol might be potential markers to discriminate the pre- and postrigor roasted mutton. This study confirmed the key aroma compounds in the roasted mutton. Most importantly, this study provided the scientific data to clarify that the postrigor *back strap* was more suitable for roasting.

Author Contributions: Conceptualization, D.Z. and Z.W.; methodology, H.L.; software, T.H.; validation, H.L.; formal analysis, H.L.; investigation, H.L. and S.L.; resources, H.L., Q.M. and S.L.; data curation, H.L.; writing—original draft, H.L.; writing—review & editing, H.L., T.H., F.F., D.Z. and Z.W.; visualization, T.H. and F.F.; supervision, D.Z. and Z.W.; project administration, Z.W.; funding acquisition, D.Z. All authors have read and agreed to the published version of the manuscript.

Funding: This study has been financially supported by Agricultural Science and Technology Innovation Program (CAAS-ASTIP-2021-IFST-SN2021-04), National Agricultural Science and Technology Innovation Program (meat science and nutrition) (CAAS-ASTIP-2018-IAPPST-2), Meat Processing Key Laboratory of Sichuan Province Open Fund Project in 2021 (21-R-01).

Institutional Review Board Statement: Not applicable.

Informed Consent Statement: Not applicable.

Data Availability Statement: The data presented in this study are available on request from the corresponding author.

Conflicts of Interest: The authors declare no conflict of interest.

References

1. Liu, H.; Ma, J.; Pan, T.; Suleman, R.; Wang, Z.; Zhang, D. Effects of roasting by charcoal, electric, microwave and superheated steam methods on (non)volatile compounds in oyster cuts of roasted lamb. *Meat Sci.* **2021**, *172*, 108324. [CrossRef] [PubMed]
2. Xiao, X.; Hou, C.; Zhang, D.; Li, X.; Ren, C.; Ijaz, M.; Hussain, Z.; Liu, D. Effect of pre- and post-rigor on texture, flavor, heterocyclic aromatic amines and sensory evaluation of roasted lamb. *Meat Sci.* **2020**, *169*, 108220. [CrossRef]
3. Fan, M.; Xiao, Q.; Xie, J.; Cheng, J.; Sun, B.; Du, W.; Wang, Y.; Wang, T. Aroma Compounds in Chicken Broths of Beijing Youji and Commercial Broilers. *J. Agric. Food Chem.* **2018**, *66*, 10242–10251. [CrossRef]
4. Marcq, P.; Schieberle, P. Characterization of the key aroma compounds in a commercial Fino and a commercial Pedro Ximenez sherry wine by application of the sensomics approach. *J. Agric. Food Chem.* **2021**, *69*, 5125–5133. [CrossRef]
5. Jonas, M.; Schieberle, P. Characterization of the Key Aroma Compounds in Fresh Leaves of Garden Sage (*Salvia officinalis L.*) by Means of the Sensomics Approach: Influence of Drying and Storage and Comparison with Commercial Dried Sage. *J. Agric. Food Chem.* **2021**, *69*, 5113–5124. [CrossRef]
6. Rasekh, M.; Karami, H.; Wilson, A.; Gancarz, M. Classification and Identification of Essential Oils from Herbs and Fruits Based on a MOS Electronic-Nose Technology. *Chemosensors* **2021**, *9*, 142. [CrossRef]
7. Melucci, D.; Bendini, A.; Tesini, F.; Barbieri, S.; Zappi, A.; Vichi, S.; Conte, L.; Toschi, T.G. Rapid direct analysis to discriminate geographic origin of extra virgin olive oils by flash gas chromatography electronic nose and chemometrics. *Food Chem.* **2016**, *204*, 263–273. [CrossRef]
8. Rusinek, R.; Kmiecik, D.; Gawrysiak-Witulska, M.; Malaga-Toboła, U.; Tabor, S.; Findura, P.; Siger, A.; Gancarz, M. Identification of the olfactory profile of rapeseed oil as a function of heating time and ratio of volume and surface area of contact with oxygen using an electronic nose. *Sensors* **2021**, *21*, 303. [CrossRef]
9. Rusinek, R.; Gawrysiak-Witulska, M.; Siger, A.; Oniszczuk, A.; Ptaszy'nska, A.A.; Knaga, J.; Malaga-Toboła, U.; Gancarz, M. Effect of supplementation of flour with fruit fiber on the volatile compound profile in bread. *Sensors* **2021**, *21*, 2812. [CrossRef] [PubMed]
10. García-González, D.L.; Aparicio, R. Coupling MOS sensors and gas chromatography to interpret the sensor responses to complex food aroma: Application to virgin olive oil. *Food Chem.* **2010**, *120*, 572–579. [CrossRef]
11. AUS-MEAT. *Handbook of Australian Meat*, 7th ed.; AUS-MEAT Ltd.: South Brisbane, Australia, 2005.

12. Liu, H.; Wang, Z.; Zhang, D.; Shen, Q.; Pan, T.; Hui, T.; Ma, J. Characterization of key aroma compounds in Beijing roasted duck by gas chromatography-olfactometry-mass spectrometry, odor activity values and aroma-recombination experiments. *J. Agric. Food Chem.* **2019**, *67*, 5847–5856. [CrossRef] [PubMed]
13. Schieberle, P. New developments in methods for analysis of volatile flavor compounds and their precursors. In *Characterization of Food: Emerging Methods*; Gaonkar, A.G., Ed.; Elsevier Science BV: Amsterdam, The Netherlands, 1995; pp. 403–431.
14. Bueno, M.; Campo, M.M.; Cacho, J.; Ferreira, V.; Escudero, A. A model explaining and predicting lamb flavour from the aroma-active chemical compounds released upon grilling light lamb loins. *Meat Sci.* **2014**, *98*, 622–628. [CrossRef]
15. Elmore, J.S.; Cooper, S.L.; Enser, M.; Mottram, D.S.; Sinclair, L.A.; Wilkinson, R.G.; Wood, J.D. Dietary manipulation of fatty acid composition in lamb meat and its effect on the volatile aroma compounds of grilled lamb. *Meat Sci.* **2005**, *69*, 233–242. [CrossRef] [PubMed]
16. Amanpour, A.; Kelebek, H.; Selli, S. Characterization of aroma, aroma-active compounds and fatty acids profiles ofcv. Nizip Yaglik oils as affected by three maturity periods of olives. *J. Sci. Food Agric.* **2018**, *99*, 726–740. [CrossRef] [PubMed]
17. Fogerty, A.C.; Whitfield, F.B.; Svoronos, D.; Ford, G.L. The composition of the fatty acids and aldehydes of the ethanolamine and choline phospholipids of various meats. *Int. J. Food Sci. Technol.* **2007**, *26*, 363–371. [CrossRef]
18. Dach, A.; Schieberle, P. Characterization of the Key Aroma Compounds in a Freshly Prepared Oat (*Avena sativa L.*) Pastry by Application of the Sensomics Approach. *J. Agric. Food Chem.* **2021**, *69*, 1578–1588. [CrossRef]
19. Watanabe, A.; Kamada, G.; Imanari, M.; Shiba, N.; Yonai, M.; Muramoto, T. Effect of aging on volatile compounds in cooked beef. *Meat Sci.* **2015**, *107*, 12–19. [CrossRef]
20. Elmore, J.S.; Nisyrios, I.; Mottram, D.S. Analysis of the headspace aroma compounds of walnuts (*Juglans regia L.*). *Flavour Fragr. J.* **2005**, *20*, 501–506. [CrossRef]
21. Wang, W.; Zhang, L.; Wang, Z.; Wang, X.; Liu, Y. Physicochemical and sensory variables of Maillard reaction products obtained from Takifugu obscurus muscle hydrolysates. *Food Chem.* **2019**, *290*, 40–46. [CrossRef]
22. Jia, X.; Wang, L.; Zheng, C.; Yang, Y.; Wang, X.; Hui, J.; Zhou, Q. Key odorant differences in fragrant brassica napus and *Brassica juncea* oils revealed by gas chromatography-olfactometry, odor activity values, and aroma recombination. *J. Agric. Food Chem.* **2020**, *68*, 14950–14960. [CrossRef]
23. Yang, Y.; Sun, Y.; Pan, D.; Wang, Y.; Cao, J. Effects of high pressure treatment on lipolysis-oxidation and volatiles of marinated pork meat in soy sauce. *Meat Sci.* **2018**, *145*, 186–194. [CrossRef]
24. Elmore, J.S.; Mottram, D.S.; Dodson, A.T. Meat aroma analysis: Problems and solutions. In *Handbook of Flavor Characterization: Sensory Analysis, Chemistry, and Physiology*; Deibler, K.D., Delwiche, J., Eds.; Marcel Dekker, Inc.: New York, NY, USA, 2004; pp. 304–319.
25. Xiao, Z.; Xiang, P.; Zhu, J.; Zhu, Q.; Liu, Y.; Niu, Y. Evaluation of the perceptual interaction among sulfur compounds in mango by Feller's additive model, odor activity value, and vector model. *J. Agric. Food Chem.* **2019**, *67*, 8926–8937. [CrossRef] [PubMed]
26. Coppock, B.M.; MacLeod, G. The effect of ageing on the sensory and chemical properties of boiled beef aroma. *J. Sci. Food Agric.* **1977**, *28*, 206–214. [CrossRef]
27. Ma, Q.; Hamid, N.; Bekhit, A.E.D.; Robertson, J.; Law, T.F. Evaluation of pre-rigor injection of beef with proteases on cooked meat volatile profile after 1 day and 21 days post-mortem storage. *Meat Sci.* **2012**, *92*, 430–439. [CrossRef] [PubMed]
28. Tian, H.; Zhan, P.; Li, W.; Zhang, X.; He, X.; Ma, Y.; Guo, Z.; Zhang, D. Contribution to the aroma characteristics of mutton process flavor from oxidized suet evaluated by descriptive sensory analysis, gas chromatography, and electronic nose through partial least squares regression. *Eur. J. Lipid Sci. Technol.* **2014**, *116*, 1522–1533. [CrossRef]
29. Gancarz, M.; Malaga-Toboła, U.; Oniszczuk, A.; Tabor, S.; Oniszczuk, T.; Gawrysiak-Witulska, M.; Rusinek, R. Detection and measurement of aroma compounds with the electronic nose and a novel method for MOS sensor signal analysis during the wheat bread making process. *Food Bioprod. Process.* **2021**, *127*, 90–98. [CrossRef]
30. Asaduzzaman, M.; Biasioli, F.; Cosio, M.S.; Schampicchio, M. Hexanal as biomarker for milk oxidative stress induced by copper ions. *J. Dairy Sci.* **2017**, *100*, 1650–1656. [CrossRef]
31. Shahidi, F.; Yun, J.; Rubin, L.; Wood, D. The Hexanal Content as an Indicator of Oxidative Stability and Flavour Acceptability in Cooked Ground Pork. *Can. Inst. Food Sci. Technol. J.* **1987**, *20*, 104–106. [CrossRef]

Article

Effects of Scopoletin Supplementation and Stocking Density on Growth Performance, Antioxidant Activity, and Meat Quality of Korean Native Broiler Chickens

Sang Hun Ha [1,†], Hwan Ku Kang [2,†], Abdolreza Hosseindoust [1,3], Jun Young Mun [1], Joseph Moturi [1], Habeeb Tajudeen [1], Hwa Lee [4], Eun Ju Cheong [4,*] and Jin Soo Kim [1,3,*]

1 Department of Bio-Health Convergence, Kangwon National University, Chuncheon 24341, Korea; kayne7602@kangwon.ac.kr (S.H.H.); hosseindoust@kangwon.ac.kr (A.H.); 202016455@kangwon.ac.kr (J.Y.M.); motkondo@gmail.com (J.M.); habeebtop@gmail.com (H.T.)
2 Poultry Research Institute, National Institute of Animal Science, Pyeongchang 25342, Korea; magic100@korea.kr
3 Department of Animal Industry Convergence, Kangwon National University, Chuncheon 24341, Korea
4 Department of Forest Environmental System, Kangwon National University, Chuncheon 24341, Korea; dlghk018@kangwon.ac.kr
* Correspondence: ejcheong@kangwon.ac.kr (E.J.C.); kjs896@kangwon.ac.kr (J.S.K.)
† These authors contributed equally to this work.

Citation: Ha, S.H.; Kang, H.K.; Hosseindoust, A.; Mun, J.Y.; Moturi, J.; Tajudeen, H.; Lee, H.; Cheong, E.J.; Kim, J.S. Effects of Scopoletin Supplementation and Stocking Density on Growth Performance, Antioxidant Activity, and Meat Quality of Korean Native Broiler Chickens. *Foods* **2021**, *10*, 1505. https://doi.org/10.3390/foods10071505

Academic Editors: Benjamin W. B. Holman and Eric Nanthan Ponnampalam

Received: 8 May 2021
Accepted: 17 June 2021
Published: 29 June 2021

Publisher's Note: MDPI stays neutral with regard to jurisdictional clai-ms in published maps and institutio-nal affiliations.

Abstract: Stocking density stress is one of the most common management stressors in the poultry industry. The present study was designed to investigate the effect of dietary *Sophora koreensis* (SK; 0 and 20 mg/kg diet) and stocking density (SD; 14 and 16 chickens/m^2) on the antioxidant status, meat quality, and growth performance of native Korean chickens. There was a lower concentration of malondialdehyde (MDA) and a higher concentration of catalase, superoxide dismutase (SOD), and total antioxidant capacity in the serum and leg muscle with the supplementation of SK. The concentration of MDA was increased and concentrations of SOD were decreased in the leg muscle of chickens in low SD treatments. The SK-supplemented treatments showed an increased 3-ethylbenzothiazoline-6-sulfonate-reducing activity of leg muscles. The higher water holding capacity of breast muscle and a lower cooking loss and pH were shown in the SK-supplemented treatments. The addition of dietary SK resulted in a greater body weight gain and greater spleen and bursa Fabricius weight, as well as lower feed intake and abdominal fat. The low SD and supplementation of SK increased the concentrations of cholesterol. The concentration of glucose was increased in the low SD treatment. Corticosterone level was decreased in the SK-supplemented and low SD treatments. In conclusion, SK supplementation reduced the oxidative stress and increased meat quality and antioxidant status of chickens apart from the SD stress.

Keywords: broilers; stress; welfare; corticosterone; productivity

1. Introduction

Stocking density (SD) exceeding the comfort zone causes stress in farm animals [1]. Presently, the SD ratio is markedly increasing worldwide to minimize costs. The continuous increase in the SD of broiler chickens in the poultry industry to decrease production costs increases health and welfare issues [2]. In addition to low body weight gain, production of broiler chickens in a high-SD situation decreases meat quality including water-holding capacity and meat tenderness by increasing the oxidative reactions [1,3]. As the traditional broiler's meat is costly because of higher meat quality compared with modern broilers, the low meat quality can compromise the marketability. Moreover, the high stress level affects polyunsaturated fatty acids of meat, which increases the vulnerability to oxidative deterioration [3–5]. The control of lipid oxidation associated with SD entails the supplementation of antioxidant factors to block the production of free radicals [6]. Several important macromolecules or enzymes are under the influence of reactive oxygen species (ROS) and free radicals, which have the potential of increasing lipid peroxidation in organs [7–9]. An oxidative stress condition in animals refers to the progressive loss of anti-oxidative status caused by various internal stressors or environmental factors. Normally, the natural antioxidant defense system is able to prevent cells from oxidative injuries by enzymatic control of free radicals [10]. Superoxide dismutase (SOD), glutathione peroxidase (GPx), and catalase are the major enzymes to counteract free radicals and diminish the lipid peroxidation rates [11,12]. In addition, the lower antioxidant defense in the body results in the excessive generation of ROS leading to oxidative injuries [6]. Therefore, improvement of oxidative status may alleviate the detrimental effects of high SD on growth performance and meat quality.

Sophora koreensis (SK), from the Fabaceae family, is a perennial herb in the mountainous area of the Korean Peninsula [13]. The root and flower of SK species contain a large quantity of flavonoids [14], isoflavonoids [15], and scopoletin [16,17] having anti-oxidative capacity [18]. Scopoletin is a kind of phenolic coumarin with promising anti-inflammatory effects [19] that can protect the body from microbial attack and environmental stress, including mechanical injury [20]. The antioxidant properties of scopoletin were proved by scavenging superoxide anion, which may prevent stressful conditions related to oxidative damage [16]. In our dose-dependent pre-study, the antioxidant properties of SK in chickens had been proven (unpublished). Therefore, we hypothesized that the effects of SK can be better highlighted when used during a stressful condition. To our knowledge, there is a lack of reports regarding the antioxidant effects of scopoletin-rich diets, not only on the growth performance but also on the meat quality of broiler chickens during high SD. This experiment was thus designed to determine the effects of a supplementary scopoletin-rich feed additive on meat quality and antioxidant status in broilers under high SD stress.

2. Materials and Methods

2.1. Experimental Design, Chickens, and Diets

The experiment was approved by the Institutional Animal Care and Use Committee, Kangwon National University (KW-170519-1). Three hundred and twelve Korean native chickens (Hanhyup 3, 914.3 ± 26.3 g, 49 days old) were fed for 35 days. A scopoletin-rich product extracted from SK was added to the experimental diets from the first day of study. Four treatments included SK supplementation levels (0 and 20 ppm) and stocking densities (14 and 16 bird/m^2). Each treatment consisted of 6 replicates with 13 birds per replicate (floor pen, w2350 × d1500 × h850 mm), where the rice hull was used as litter. The temperature and humidity of the broiler house were controlled by an automatic ventilation system to have an average temperature and humidity of 22.1 °C and 46%, respectively. The diet contained 18% crude protein, 3100 kcal/kg metabolizable energy, 0.86% calcium, 0.33% available phosphorus, 0.21% sodium, 1.01% available lysine, 3.5% crude fiber, and 6.86% ether extract. White light was provided (25 lux at bird-head level), with a light schedule of 19L:5D (lights off from 2200 to 0300 h), and water and feed were maintained ad libitum. The mash-type diet was prepared to meet the nutrient requirements of Korean native chickens. The SK supplement contained 2090.5 mg scopoletin/kg (Table 1).

2.2. Sample Collection

At day 35, a total of 72 birds, 18 chickens per treatment (3 chickens/replicate), were used for carcass characteristics analysis, meat quality, relative organ weight, antioxidant status, and plasma metabolites evaluation. Birds in the same range of bodyweight (BW) based on the average BW of a treatment were applied. Blood samples were taken from the wing vein by using a syringe. Collected blood was moved into non-treated vacuum tubes, and was immediately sent for plasma separation, then centrifuged at 2500× g for 10 min. Serum was aspirated and located in a 2.5 mL centrifuge tube then stored at −20 °C before analysis for malondialdehyde (MDA), catalase, SOD, and total antioxidant capacity (TAC). All selected birds were decapitated at the first cervical vertebrae. After defeathering and removal of organs and feet, the carcass, breast muscles (both sides), drumsticks (both sides), and abdominal fat were weighed and then stored at −20 °C. The meat quality, antioxidant status, and weights of liver, spleen, and bursa of Fabricius were measured to calculate the relative carcass weight and internal organ weight, and then these parts were stored at −80 °C.

Table 1. Analyzed composition of *Sophora koreensis*.

Item	*Sophora koreensis*
Scopoletin (mg/kg)	2090.5
Dry matter %	93.12
Crude protein %	12.82
Ether extract %	1.72
Crude fiber %	29.69
Ash %	3.15
Calcium %	1.12
Phosphorus %	0.19
Amino Acids %	
Arg	3.21
His	0.61
Ile	9.58
Leu	23.03
Lys	11.16
Met	5.57
Phe	4.97
Thr	3.03
Trp	1.61
Val	1.06
Fatty acids %	
Palmitic acid	0.43
Oleic acid	1.06
Linoleic acid	0.35
Linolenic acid	0.59
Arachidonic acid	5.18

2.3. Antioxidant Status in Serum and Muscle

In order to assay the antioxidant factors activity in muscle and serum, the samples were pre-treated and measured as explained by Hosseindoust et al. [5] by using a Cayman kit manual (Enzyme activity assay, Cayman Chemical, Ann Arbor, MI, USA). The harvested samples from the muscle and serum were subjected to the evaluation of the concentration of MDA (Cat #10006438, Cayman Chemical, Ann Arbor, MI, USA), catalase (Cat #707002, Cayman Chemical, Ann Arbor, MI, USA), SOD (Cat #706002, Cayman Chemical, Ann Arbor, MI, USA), and TAC (Cat #709001, Cayman Chemical, Ann Arbor, MI, USA). To evaluate the absorption detection, a microplate reader (Power Wave XS, BIoTeK, Winooski, VT, USA) was applied according to the Cayman kit's manufacturer's manual.

2.4. Radical Scavenging Capacity

The evaluation of 3-ethylbenzothiazoline-6-sulfonate (ABTS)-reducing activity was performed using the supernatant collected from thigh meat according to Hosseindoust [5] and Blois [21]. In brief, 200 µL of supernatant was placed in a 5 mL centrifuge tube and added to 800 µL deionized distilled water. The determination of ABTS-reducing activity and the preparation of the ABTS solution were conducted following the method described by Erel [22]. The prepared ABTS solution was diluted with ethanol for adjusting absorbance approximately 0.70 at 734 by using a UV spectrophotometer (Optizen 3220UV, MECASYS, Daejeon, Korea). The diluted ABTS solution (3 mL) was mixed with 20 µL of supernatant and the absorbance was measured by a UV spectrophotometer at 734 nm. The ethanol was used as a blank. The percentage inhibition was obtained by the following equation: ABTS-reducing activity (%) = ((absorbance of the control − the absorbance of the sample)/absorbance of the control) × 100.

2.5. Meat Quality

Leg muscle (with bone, and without skin, tendon, and fat) meat color was examined using a Chroma Meter CR-400 instrument (Minolta Co., Osaka, Japan) according to International Commission on Illumination (CIE) L * (lightness), a * (redness), and b * (yellowness). Water holding capacity (WHC), cooking loss (%), shear force (%), pH, and MDA were examined using muscle pectoralis major according to Hosseindoust et al. [5]. The WHC was measured by placing a 0.5 g meat sample on a round plastic plate in a tube (Millipore Ultrafree-MC; Millipore, Bedford, MA). The samples were heated for 20 min (80 °C) in a water bath, then cooled (23 ± 1 °C), and centrifuged (2000× g) for 10 min (4 °C) to evaluate WHC as follows: WHC = (moisture content−water loss)/moisture content × 100. To measure the cooking loss, 3 g of leg muscle meat was placed in a plastic bag and heated in a water bath (85 °C) for 20 min, then cooled at room temperature to calculate cooking loss according to this equation: (sample weight before cooking/sample weight after cooking)/sample weight before cooking × 100 [5]. For shear force determination, a texture analyzer (TA-XT2i, Stable Microsystems, Surrey, UK) equipped with a Warner-Bratzler shear blade, a 25 kg load cell, and a test speed setting at 2.0 mm/s was used with the maximum force (kg) [23]. The pH of meat was evaluated as explained by Hosseindoust et al. [5]. In brief, a 5 g sample of meat was homogenized in distilled water (45 mL) by a homogenizer (DIAX 900, Heidolph, Kelheim, Germany) for 15 s, and then the pH was determined by using a pH meter (Orion 230A Thermo Fisher Scientific, Waltham, MA, USA). After performing the pH process, a Watman no. 2 (Hillsboro, OH, USA) was used to filter homogenized samples.

2.6. Blood Metabolites

The mixed blood samples with K_2 EDTA were used for determination of total cholesterol, total protein, triglyceride, glucose, glutamic pyruvic transaminase (GPT), glutamate oxaloacetate transaminase (GOT), albumin, phosphate, and calcium (Hemavet 950, Drew Scientific, Miami Lakes, FL, USA). The serum corticosterone was analyzed using the ELISA kit (Corticosterone ELISA kit, Enzo life Sciences, Farmingdale, NY, USA).

2.7. Statistical Analysis

The experimental values were analyzed by GLM procedure of SAS® 9.3 software (SAS Inst. Inc., Cary, NC, USA). The pen was used as the experimental unit for the analysis of growth performance, and an individual chicken was considered as the experimental unit for measuring the blood, meat quality, and carcass trait samplings. The difference of means was tested by Tukey test. The effect of dietary SK supplementation and stocking densities and their interactions were determined. A significant difference was expressed either $p < 0.01$ or $p < 0.05$.

3. Results

3.1. Antioxidant Factors

There was a significantly lower ($p < 0.01$) concentration of MDA in the serum with the supplementation of SK (Table 2). The serum concentrations of catalase ($p < 0.05$) and SOD ($p < 0.05$) were increased by supplementation of SK in the diet, while there was no difference between the SD treatments. The addition of dietary SK resulted in a greater ($p < 0.01$) TAC in the serum; however, the SD treatments did not change the TAC capacity. The low SD ($p < 0.05$) and supplementation of SK ($p < 0.01$) decreased the content of MDA in the leg muscle. The concentrations of catalase ($p < 0.01$) and SOD ($p < 0.05$) were increased by supplementation of SK in the diet, while there was a decrease in the concentration of SOD in the high SD treatment. There was an increased ($p < 0.01$) TAC in the leg muscle of chickens with increased supplementation of SK; however, there were no significant effects of SD on TAC in the leg muscle.

Table 2. Effect of dietary SK and SD on antioxidant activity of serum and leg muscle in Korean native chicken.

Stocking Density (n/m^2)	14		16		SEM	p-Values		
Sophora koreensis (ppm)	0	20	0	20		SD	SK	SD × SK
Serum								
MDA (nmol/mL)	10.55	5.87	11.41	6.22	0.55	0.584	<0.001	0.818
Catalase (nmol/min/mL)	0.23	0.31	0.21	0.31	0.02	0.844	0.027	0.903
SOD (U/mL)	44.73	61.63	49.33	63.38	0.77	0.053	<0.001	0.369
TAC (mM)	0.15	0.32	0.14	0.30	0.02	0.675	<0.001	0.957
Leg muscle								
MDA (nmol/mg)	0.59	0.51	0.66	0.57	0.01	0.029	0.004	0.794
Catalase (nmol/min/mg)	0.21	0.43	0.21	0.39	0.02	0.685	<0.001	0.591
SOD (U/mg)	39.97	68.78	32.30	50.85	1.53	<0.001	<0.001	0.109
TAC (mM)	0.13	0.37	0.11	0.33	0.01	0.184	<0.001	0.466

SEM, standard error of means; SD, stocking density effect; SK, *Sophora koreensis* supplementation effect; SD × SK, stocking density × *Sophora koreensis* supplementation effect interaction; MDA, malondialdehyde; SOD, Superoxide dismutase; TAC, total antioxidant capacity.

3.2. ABTS-Reducing Activity

The antioxidant capacity result of serum indicated that ABTS-reducing activity was enhanced in the SK-supplemented treatments (Figure 1). No difference in ABTS-reducing activity was detected between the SD treatments. The SK-supplemented treatments showed an increased ($p < 0.05$) ABTS-reducing activity of leg muscles compared with the non-SK-supplemented treatments; however, there was a greater ABTS-reducing activity of leg muscle in the SK-supplemented treatments.

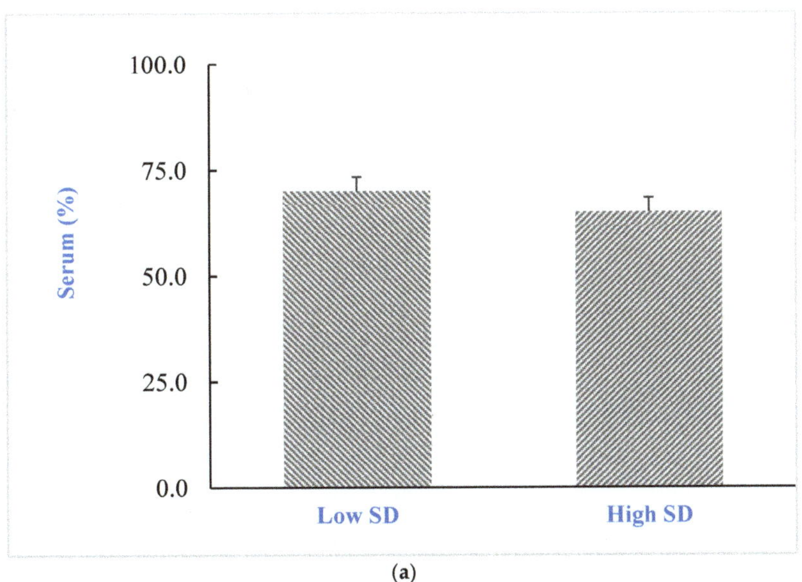

(a)

Figure 1. *Cont.*

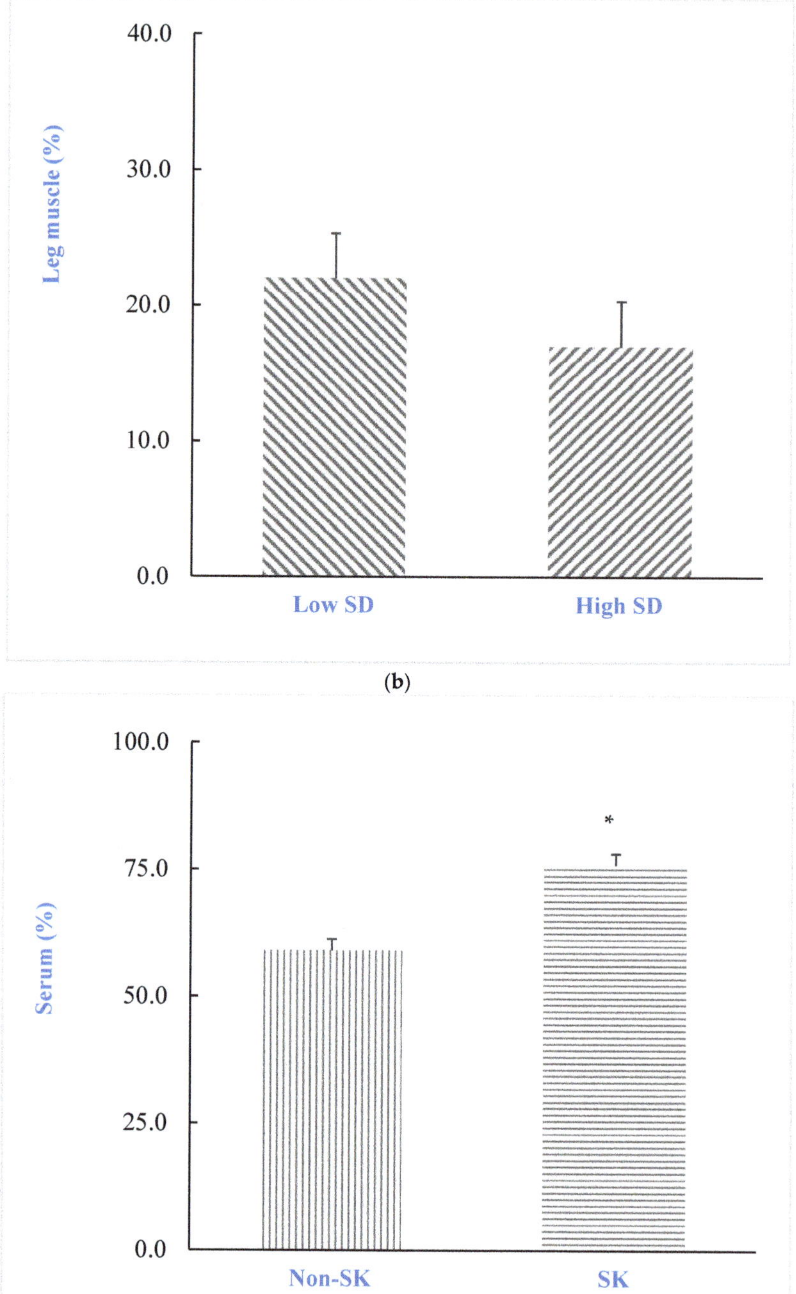

Figure 1. *Cont.*

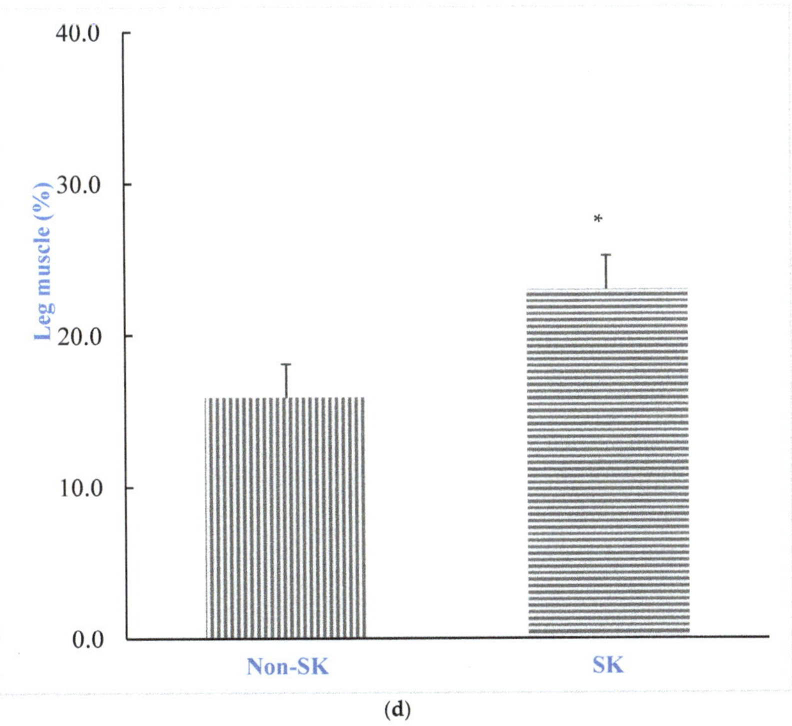

(d)

Figure 1. ABTS radical scavenging capacity (%) of different stocking density (SD) on serum (**a**) and leg muscle (**b**), and *Sophora koreensis* (SK) supplementation on serum (**c**) and leg muscle (**d**) of Korean native chicken. Non-SK, basal diet; SK, basal diet + 20 ppm SK; Asterisks (*) indicate statistical significance ($p < 0.05$).

3.3. Meat Color and Meat Quality

The effect of diets and SD on breast meat color and quality of chickens was shown in Table 3. There were no breast meat redness, lightness, and yellowness responses to supplementation of SK in the diet and rearing in different stocking densities. The higher water holding capacity of breast muscle was shown for the SK-supplemented treatments ($p < 0.01$); however, there was no change among the SD treatments. A lower cooking loss and breast muscle pH was reported in the SK-supplemented treatments ($p < 0.01$), although there were no changes in cooking loss and pH of breast muscle between the SD treatments. The shear force was not affected by treatments.

3.4. Growth Response, Carcass Traits, Immune Organ Ratio

Table 4 shows the influences of diets and SD on growth performance, carcass traits, and relative organ weight. The effect of SK supplementation on improving final BW, BW gain, and feed intake of chickens was significant ($p < 0.01$). There was no difference in feed conversion ratio between the SK treatments. There was no difference in final BW and BW gain between the SD treatments; however, the feed intake and feed conversion ratio of chickens were adversely affected by the high SD. The carcass yield, breast meat, and drumsticks showed no dietary SK effects; however, a lower ($p < 0.01$) abdominal fat was shown in the SK-supplemented treatments. The carcass yield, breast meat, drumstick percentage, and abdominal fat were decreased in the high SD treatments. There was no change in the relative weight of the liver. There were significant interactions between the SD and SK in increasing the relative weight of the spleen and bursa Fabricius. The main effects shown were that the relative weight of the spleen and bursa Fabricius were increased by supplementation of SK, but decreased by increasing the SD. There were no differences in the percentage of the spleen to bursa Fabricius and relative weight of thyroid between the treatments.

Table 3. Effect of dietary SK and SD on breast meat color and quality in Korean native chicken.

Stocking Density (n/m^2)	**14**		**16**		**SEM**	***p*-Values**		
Sophora koreensis (ppm)	0	20	0	20		SD	SK	SD × SK
Meat color								
Lightness (L *)	53.03	52.92	52.81	52.88	0.32	0.837	0.977	0.882
Redness (a*)	4.14	4.56	4.20	4.45	0.17	0.943	0.350	0.812
Yellowness (b*)	8.18	8.49	8.37	8.61	0.21	0.727	0.525	0.937
Meat quality								
Water holding capacity (%)	44.95	48.27	40.91	47.75	0.62	0.081	0.001	0.172
Cooking loss (%)	30.09	26.81	29.82	25.26	0.61	0.461	0.004	0.602
Shear force (n/cm^2)	23.63	24.91	22.95	24.12	0.39	0.360	0.110	0.965
pH	5.68	5.80	5.81	5.58	0.02	0.132	<0.001	0.089

SEM, standard error of means; SD, stocking density effect; SK, *Sophora koreensis* supplementation effect; SD × SK, stocking density × *Sophora koreensis* supplementation effect interaction.

Table 4. Effect of dietary SK and SD on growth performance, carcass traits, and relative weights of organs in Korean native chicken.

Stocking Density (n/m^2)	**14**		**16**		**SEM**	***p*-Values**		
Sophora koreensis (ppm)	0	20	0	20		SD	SK	SD × SK
Growth Performance								
Final BW (g/bird)	2319	2313	2149	2191	8.12	0.273	<0.001	0.153
BW gain (g/bird)	1408	1399	1231	1276	10.01	0.373	<0.001	0.199
FI (g/bird)	3949	3849	3458	3390	18.29	0.032	<0.001	0.667
FCR (g/bird)	2.81	2.75	2.82	2.66	0.02	0.037	0.356	0.276
Carcass traits (%)								
Carcass yield	71.05	71.56	69.41	69.82	0.20	<0.001	0.269	0.907
Breast meat	19.72	19.96	17.93	18.32	0.13	<0.001	0.228	0.770
Drumsticks	14.17	14.34	13.21	13.34	0.15	0.003	0.622	0.939
Abdominal fat	1.84	1.42	1.66	1.35	0.02	0.009	<0.001	0.211
Relative weights of organs (%)								
Liver	2.67	2.56	2.43	2.41	0.07	0.147	0.623	0.732
Spleen	0.088	0.114	0.079	0.083	0.01	<0.001	<0.001	0.002
Bursa of Fabricius	0.092	0.125	0.087	0.095	0.01	<0.001	<0.001	<0.001
Spleen/bursa	0.923	0.872	0.878	0.847	0.06	0.257	0.184	0.730
Thyroid	0.629	0.655	0.728	0.503	0.03	0.697	0.153	0.078

SEM, standard error of means; SD, stocking density effect; SK, *Sophora koreensis* supplementation effect; SD × SK, stocking density × *Sophora koreensis* supplementation effect interaction; BW, bodyweight; FI, feed intake; FCR, feed conversion ratio.

3.5. Blood Profile

The effects of dietary SK on blood profile are shown in Table 5. Results indicated that total cholesterol and glucose levels were decreased in the high SD treatment but that the concentration of blood total cholesterol was increased in the SK-supplemented treatments. There was no change in concentration of total protein, triglyceride, GPT, GOT, albumin, phosphate, and calcium among the treatments. The blood corticosterone level was significantly higher ($p < 0.05$) in the high SD treatment; however, corticosterone level was decreased in the SK-supplemented treatments.

Table 5. Effect of dietary SK and SD on blood profile in Korean native chicken.

Stocking Density (n/m^2)	**14**		**16**		**SEM**	***p*-Values**		
Sophora koreensis (ppm)	0	20	0	20		SD	SK	SD × SK
Total cholesterol (mg/dL)	106.3	115.6	102.7	105.8	1.49	0.036	0.050	0.302
Total protein (mg/dL)	2.88	2.74	2.73	2.70	0.03	0.144	0.206	0.373
Triglyceride (mg/dL)	55.89	54.09	51.37	54.20	1.62	0.503	0.875	0.482
Glucose (mg/dL)	256.1	254.0	243.1	239.5	2.82	0.024	0.619	0.894
GPT (U/L)	2.07	2.12	2.15	2.19	0.08	0.636	0.778	0.996
GOT (U/L)	214.3	225.7	210.9	217.8	3.22	0.388	0.172	0.728
Albumin (mg/dL)	1.12	1.03	1.10	1.09	0.01	0.568	0.097	0.150
Phosphate (µM/L)	10.14	10.18	10.15	10.02	0.12	0.758	0.850	0.738
Calcium (µM/L)	9.09	9.07	9.10	9.23	0.10	0.686	0.785	0.717
Corticosterone (ng/mL)	53.37	48.42	61.51	51.06	1.05	0.019	0.002	0.206

SEM, standard error of means; SD, stocking density effect; SK, *Sophora koreensis* supplementation effect; SD × SK, stocking density × *Sophora koreensis* supplementation effect interaction; GPT, glutamic pyruvic transaminase; GOT, glutamate oxaloacetate transaminase.

4. Discussion

Sophora koreensis has been known as a traditional herb to treat rheumatoid issues in Korea because of its antioxidant capacity [13,24]. The antioxidant enzymes including SOD, GPx, and catalase are the first factors against antioxidant reactions [11,12]. Antioxidant effects of scopoletin were shown earlier [16,17]. An increase in the aforementioned enzyme's production capacity can improve the antioxidant system by controlling the production of ROS. Superoxide dismutase is an important enzyme in the protection of cells from adverse effects of ROS [25]. In the current study, decreased concentration of MDA and increased TAC in the serum of chickens treated with SK may be related to the increased concentrations of catalase and SOD, which reduce the formation of peroxides and hydroperoxides in fat tissues [7]. The increased activities of catalase and SOD in the serum of broiler chickens fed SK diets show its potential to scavenge free radicals during the stressful condition of high SD. It has been reported that the stressful condition decreases the SOD and catalase production, which in turn increases MDA production [26]. There is a positive correlation between high MDA concentrations and lipid peroxidation [7,8]. Meanwhile, the reduction of lipid peroxidation in the thigh muscle may be reflected in the decrease of MDA concentration in the SK treatments. Scopoletin also showed high inhibitory activities against anti-inflammatory cytokines by decreasing TNF-α, IL-1β, and IL-6 secretions [17,19,27]. It has been reported that the anti-inflammatory influences of scopoletin are associated with a decrease in free radicals production. The production of MDA, as an important indicator of lipid peroxidation, is due to the exposure of free radicals to the plasma membrane [7,8,28]. Lee et al. [29] reported that the MDA levels were significantly increased in mice under alcoholic food stress, but that the supplementation of scopoletin prevented the increase in MDA concentration compared with the control treatment. Several challenge experiments with high inflammatory condition confirmed that the decrease in inflammation mediates cell damage, lipid peroxidation, and increases the inactivation of antioxidant enzymes [9,10,28]. Several researchers have reported that high SD could increase the stress level and decrease the antioxidant status by increasing MDA and decreasing SOD concentrations in serum [1,2,30]. Although increased catalase and TAC seem to be consistent in the plasma and leg muscle in the SK-supplemented treatments, this trend is missing for chickens in the SD treatments. However, chickens in the SD treatments showed a higher MDA and lower SOD content in the leg muscle. Although the decrease in the SOD concentration was in line with the increased MDA in the leg muscle, the concentration of the aforementioned parameters in the serum was not in agreement with the leg muscle results.

Although scopoletin increases the antioxidant status [17], literature did not study the effects of supplemental scopoletin on meat quality related to SD stress in chickens focusing on the antioxidant capacity. Therefore, this study aimed to test the influences of scopoletin on the antioxidant capacity of meat, plasma, and the possible interactions with the quality and color of meat. Although the scavenging capacity of scopoletin against ABTS in chicken meat has not been studied, an in vitro study on the antioxidant role of phenolic compounds reported that the α-diphenyl-β-picrylhydrazyl radical-scavenging activity of scopoletin was around 11,800 times higher than vitamin C, making it a potent antioxidant compound [31]. The increased scavenging capacity of ABTS of meat due to scopoletin was in line with an enhanced TAC in the leg muscle, indicating the reduction of oxidative damage in muscle tissues with the presence of scopoletin. Furthermore, an increased SOD, as well as decreased MDA, content in the serum and meat confirm the capability of scopoletin in decreasing oxidative stress. Therefore, the result of the current study showed that the supplementation of 20 ppm scopoletin adequately increased the ABTS-reducing activity in chickens.

In the present study, the abdominal fat percentage of chicken was shown to be lower in the SK-supplemented group than in the non-supplemented group, which was in line with the results reported by Rajaei et al. [32], who reported that the addition of 5 mL/L of noni juice, as a rich source of scopoletin, in drinking water had a significant effect in decreasing abdominal fat content in broiler chickens. Scopoletin is known as a stimulator of fatty acid oxidative genes including PPARa, Acsl1, CPT, Acox, and Acaa1a, and an inhibitor of lipogenic genes such as sterol regulatory element-binding protein-1c and fatty acid synthase in the white adipose tissue and liver in rat [29,33]. The reduced abdominal fat may be due to the supplementation of scopoletin as a bioactive component that reduces oxidative stress and improves carbohydrate and fat metabolism. Serum corticosterone can reflect the welfare status of chickens with the environment [34], and several production and behavioral parameters can be under the influence of corticosterone hormone. Hosseindoust et al. [5] reported that stress hormone is a decisive index for deposition of protein and carcass percentage in chickens. Supplementation with *Morinda citrifolia* L. as a rich source of scopoletin increased the absorption of amino acids, which subsequently increased the carcass rate of broiler chickens [32]. In the current study, the decrease of abdominal fat in the SK-supplemented treatments may not be due to the serum corticosterone because of the insignificant difference in breast meat and drumstick. However, chickens in the SD treatments showed a lower breast meat and drumstick percentage as well. The lower protein deposition may decrease the percentage of muscle to fat and be responsible for the higher relative abdominal fat. There is a positive correlation between corticosterone concentration and abdominal, thigh, or cervical adipose tissue's fat contents [35]; however, the degradation of skeletal muscle can be increased when the serum corticosterone concentration increases [36]. Therefore, the lower protein deposition in muscular organs including breast muscle and drumstick may be responsible for the higher relative abdominal fat.

The health status and BW of animals are under the adverse effect of SD stress, which may cause economic loss [1,2]. Corticosterone is one of the most common end-products of stress and will be secreted to the blood during stressful environments [5,34,37]. Long-time exposure to restraints disrupts the hypothalamic-pituitary-adrenal axis and increases the concentrations of plasma corticosterone [38]. Consistently, the current study showed that the supplementation of SK in the diet during SD stress led to the decrease of corticosterone in the serum. The excessive production of ROS compromises cell growth by degrading cytoskeletal proteins, as well as causing lipids peroxidation during stressful conditions [9]. Our result is in line with those from other researchers who have employed SD stress in poultry [34].

Our study showed that dietary SK supplementation improved the meat quality of chickens by increasing water holding capacity and decreasing cooking loss. The water holding capacity is known as a determinantal factor of meat quality [3]. The significantly lower cooking loss in SK-supplemented treatments may be due to the lower serum corticosterone level. It was reported that excessive levels of serum corticosterone not only reduced the BW gain through reducing anabolism and increasing catabolism processes [39], but also induced lipid peroxidation [40,41], which may influence meat quality. The protective roles of antioxidant enzymes and controlling the effects of free radical damage seem essential to reduce the adverse influences of stocking density stress. Antioxidant properties of scopoletin may be reflected in the MDA concentration of meat. The lower concentration of serum corticosterone may indicate that the chickens fed SK diets had relatively lower stress regardless of SD because the SD did not decrease cooking loss of breast meat. There is a positive relationship between corticosterone production and the secretion of inflammatory cytokines including IL-1b, IL-6, IL-10, IL-12-a, and IL-18 [19,37], which can adversely affect meat quality. Corticosterone secretion has been shown to be a factor to decrease meat quality by degrading protein in muscle and decreasing the fatty acid transport protein expression [35]. Song et al. [42] stated that the concentration of uric acid, as a factor to show protein catabolism in tissues, increased in blood when the concentration of corticosterone increased in the blood. In addition, the production of ROS can be increased by corticosteroid hormones [43]. The high antioxidant status leads to lower exposure to hydroxyl and peroxyl radicals and possibly the protection of lipid tissues from oxidation through chelating free radicals [4,6]. The result of radical scavenging capacity in this study shows that scopoletin is a potent antioxidant factor to protect fatty acids oxidation and increase meat quality.

The BW gain of chickens in the high SD treatments was 11.9% lower than the low SD treatments. In addition, feed intake was 13.9% lower in the high SD treatments. The reduction of weight gain and feed intake may show that the SD stress adversely affected the performance of broiler chickens. There are important biomarkers such as corticosterone that decrease growth performance during a stressful period [34,38]. The greater BWG in SK-supplemented groups may be because of the antioxidant effects of scopoletin, which may stimulate protein synthesis.

5. Conclusions

In conclusion, to improve radical scavenging capacity and controlling lipids peroxidation, the antioxidant capacity of broiler chickens can be improved by SK supplementation during high SD stress. Dietary SK improved meat quality through increasing ABTS radical scavenging capacity in the serum and leg muscle. In addition, lower abdominal fat and higher immune organs weight were shown in chickens fed SK. Therefore, our study suggests that SK is a practically useful feed additive to improve the meat quality and weight gain of chickens regardless of SD stress.

Author Contributions: Conceptualization and writing—original draft preparation, A.H., S.H.H., H.K.K. and J.S.K.; methodology, A.H. and S.H.H.; software, H.K.K.; validation, S.H.H.; formal analysis, S.H.H.; investigation, S.H.H., H.K.K. and H.T.; resources, A.H.; data/table curation, H.K.K., H.T. and H.L.; writing—review and editing, J.Y.M., E.J.C. and J.M. All authors have read and agreed to the published version of the manuscript.

Funding: This work was carried out with the support of "Cooperative Research Program for Agriculture Science and Technology Development (Project No. PJ012821012021)" Rural Development Administration, Republic of Korea.

Institutional Review Board Statement: The experiment was approved by the Institutional Animal Care and Use Committee, Kangwon National University (KW-170519-1).

Informed Consent Statement: Informed consent was obtained from all subjects involved in the study.

Conflicts of Interest: The authors declare no conflict of interest.

References

1. Goo, D.; Kim, J.H.; Park, G.H.; Delos Reyes, J.B.; Kil, D.Y. Effect of heat stress and stocking density on growth performance, breast meat quality, and intestinal barrier function in broiler chickens. *Animals* **2019**, *9*, 107. [CrossRef]
2. Dawkins, M.S.; Donnelly, C.A.; Jones, T.A. Chicken welfare is influenced more by housing conditions than by stocking density. *Nature* **2004**, *427*, 342–344. [CrossRef]

3. Kim, T.K.; Yong, H.I.; Jung, S.; Kim, H.W.; Choi, Y.S. Technologies for the Production of Meat Products with a Low Sodium Chloride Content and Improved Quality Characteristics—A Review. *Foods* **2021**, *10*, 957. [CrossRef]
4. Shakeri, M.; Cottrell, J.J.; Wilkinson, S.; Le, H.H.; Suleria, H.A.; Warner, R.D.; Dunshea, F.R. Growth performance and characterization of meat quality of broiler chickens supplemented with betaine and antioxidants under cyclic heat stress. *Antioxidants* **2019**, *8*, 336. [CrossRef]
5. Hosseindoust, A.; Oh, S.M.; Ko, H.S.; Jeon, S.M.; Ha, S.H.; Jang, A.; Son, J.S.; Kim, G.Y.; Kang, H.K.; Kim, J.S. Muscle Antioxidant Activity and Meat Quality Are Altered by Supplementation of Astaxanthin in Broilers Exposed to High Temperature. *Antioxidants* **2020**, *9*, 1032. [CrossRef] [PubMed]
6. Xiao, Z.; He, L.; Hou, X.; Wei, J.; Ma, X.; Gao, Z.; Yuan, Y.; Xiao, J.; Li, P.; Yue, T. Relationships between Structure and Antioxidant Capacity and Activity of Glycosylated Flavonols. *Foods* **2021**, *10*, 849. [CrossRef] [PubMed]
7. Ore, A.; Akinloye, O.A. Oxidative stress and antioxidant biomarkers in clinical and experimental models of non-alcoholic fatty liver disease. *Medicina* **2019**, *55*, 26. [CrossRef] [PubMed]
8. Bao, Y.; Zhu, Y.; Ren, X.; Zhang, Y.; Peng, Z.; Zhou, G. Formation and inhibition of lipid alkyl radicals in roasted meat. *Foods* **2020**, *9*, 572. [CrossRef]
9. El-Bahr, S.M.; Shousha, S.; Alfattah, M.A.; Al-Sultan, S.; Khattab, W.; Sabeq, I.I.; Ahmed-Farid, O.; El-Garhy, O.; Albusadah, K.A.; Alhojaily, S.; et al. Enrichment of Broiler Chickens' Meat with Dietary Linseed Oil and Lysine Mixtures: Influence on Nutritional Value, Carcass Characteristics and Oxidative Stress Biomarkers. *Foods* **2021**, *10*, 618. [CrossRef]
10. Arnold, M.; Rajagukguk, Y.V.; Gramza-Michałowska, A. Functional Food for Elderly High in Antioxidant and Chicken Eggshell Calcium to Reduce the Risk of Osteoporosis—A Narrative Review. *Foods* **2021**, *10*, 656. [CrossRef] [PubMed]
11. Kopec, W.; Jamroz, D.; Wiliczkiewicz, A.; Biazik, E.; Pudlo, A.; Korzeniowska, M.; Hikawczuk, T.; Skiba, T. Antioxidative Characteristics of Chicken Breast Meat and Blood after Diet Supplementation with Carnosine, L-histidine, and β-alanine. *Antioxidants* **2020**, *9*, 1093. [CrossRef] [PubMed]
12. Lee, J.; Hosseindoust, A.; Kim, M.; Kim, K.; Choi, Y.; Lee, S.; Lee, S.; Cho, H.; Kang, W.S.; Chae, B. Biological evaluation of hot-melt extruded nano-selenium and the role of selenium on the expression profiles of selenium-dependent antioxidant enzymes in chickens. *Biol. Trace. Elem. Res.* **2020**, *194*, 536–544. [CrossRef]
13. Lee, W.K.; Tokuoka, T.; Heo, K. Molecular evidence for the inclusion of the Korean endemic genus "Echinosophora" in Sophora (Fabaceae), and embryological features of the genus. *J. Plant. Res.* **2004**, *117*, 209–219. [CrossRef]
14. Choi, E.J.; Kwon, H.C.; Sohn, Y.C.; Nam, C.W.; Park, H.B.; Kim, C.Y.; Yang, H.O. Four flavonoids from Echinosophora koreensis and their effects on alcohol metabolizing enzymes. *Arch. Pharm. Res.* **2009**, *32*, 851–855. [CrossRef]
15. Iinuma, M.; Ohyama, M.; Tanaka, T.; Mizuno, M.; Hong, S.K. An isoflavanone from roots of Echinosophora koreensis. *Phytochemistry* **1991**, *30*, 3153–3154. [CrossRef]
16. Shaw, C.Y.; Chen, C.H.; Hsu, C.C.; Chen, C.C.; Tsai, Y.C. Antioxidant properties of scopoletin isolated from Sinomonium acutum. *Phytother. Res.* **2003**, *17*, 823–825. [CrossRef]
17. Bibi Sadeer, N.; Montesano, D.; Albrizio, S.; Zengin, G.; Mahomoodally, M.F. The versatility of antioxidant assays in food science and safety—Chemistry, applications, strengths, and limitations. *Antioxidants* **2020**, *9*, 709. [CrossRef]
18. Sohn, H.Y.; Son, K.H.; Kwon, C.S.; Kwon, G.S.; Kang, S.S. Antimicrobial and cytotoxic activity of 18 prenylated flavonoids isolated from medicinal plants: *Morus alba* L., Morus mongolica Schneider, *Broussnetia papyrifera* (L.) Vent, Sophora flavescens Ait and Echinosophora koreensis Nakai. *Phytomedicine* **2004**, *11*, 666–672. [CrossRef]
19. Sakthivel, K.M.; Vishnupriya, S.; Priya Dharshini, L.C.; Rasmi, R.R.; Ramesh, B. Modulation of multiple cellular signalling pathways as targets for anti-inflammatory and anti-tumorigenesis action of Scopoletin. *J. Pharm. Pharmacol.* **2021**, rgab047. [CrossRef]
20. Tanaka, Y.; Data, E.S.; Hirose, E.S.; Taniguchi, T.; Uritani, L. Biochemical changes in secondary metabolites in wounded and deteriorated cassava roots. *Agric. Biol. Chem.* **1983**, *47*, 693–700.
21. Blois, M.S. Antioxidant determinations by the use of a stable free radical. *Nature* **1958**, *181*, 1199–1200. [CrossRef]
22. Erel, O. A novel automated direct measurement method for total antioxidant capacity using a new generation, more stable ABTS radical cation. *Clin. Biochem.* **2004**, *37*, 277–285. [CrossRef] [PubMed]
23. Shim, Y.H.; Kim, J.S.; Hosseindoust, A.; Choi, Y.H.; Kim, M.J.; Oh, S.M.; Ham, H.B.; Kumar, A.; Kim, K.Y.; Jang, A. Investigating meat quality of broiler chickens fed on heat processed diets containing corn distillers dried grains with solubles. *Korean J. Food Sci. Anim. Res.* **2018**, *38*, 629–635.
24. Chen, S. Natural products triggering biological targets-a review of the anti-inflammatory phytochemicals targeting the arachidonic acid pathway in allergy asthma and rheumatoid arthritis. *Curr. Drug Targets* **2011**, *12*, 288–301. [CrossRef] [PubMed]
25. Lee, S.Y.; Hur, S.J. Effect of Treatment with Peptide Extract from Beef Myofibrillar Protein on Oxidative Stress in the Brains of Spontaneously Hypertensive Rats. *Foods* **2019**, *8*, 455. [CrossRef]
26. Xiong, Y.; Yin, Q.; Li, J.; He, S. Oxidative Stress and Endoplasmic Reticulum Stress Are Involved in the Protective Effect of Alpha Lipoic Acid Against Heat Damage in Chicken Testes. *Animals* **2020**, *10*, 384. [CrossRef]
27. Kim, H.J.; Jang, S.I.; Kim, Y.J.; Chung, H.T.; Yun, Y.G.; Kang, T.H.; Jeong, O.S.; Kim, Y.C. Scopoletin suppresses pro-inflammatory cytokines and PGE2 from LPS-stimulated cell line, RAW 264.7 cells. *Fitoterapia* **2014**, *75*, 61–266. [CrossRef]

28. Liu, S.L.; Deng, J.S.; Chiu, C.S.; Hou, W.C.; Huang, S.S.; Lin, W.C.; Liao, J.C.; Huang, G.J. Involvement of heme oxygenase-1 participates in anti-inflammatory and analgesic effects of aqueous extract of Hibiscus taiwanensis. *Evid. Based Complement. Altern. Med.* **2012**, *2012*, 132859. [CrossRef]
29. Lee, H.I.; Yun, K.W.; Seo, K.I.; Kim, M.J.; Lee, M.K. Scopoletin prevents alcohol-induced hepatic lipid accumulation by modulating the AMPK–SREBP pathway in diet-induced obese mice. *Metabolism* **2014**, *63*, 593–601. [CrossRef] [PubMed]
30. Ma, H.; Xu, B.; Li, W.; Wei, F.; Kim, W.K.; Chen, C.; Sun, Q.; Fu, C.; Wang, G.; Li, S. Effects of alpha-lipoic acid on the behavior, serum indicators, and bone quality of broilers under stocking density stress. *Poult. Sci.* **2020**, *99*, 4653–4661. [CrossRef]
31. Yi, B.; Hu, L.; Mei, W.; Zhou, K.; Wang, H.; Luo, Y.; Wei, X.; Dai, H. Antioxidant phenolic compounds of cassava (Manihot esculenta) from Hainan. *Molecules* **2011**, *16*, 157. [CrossRef] [PubMed]
32. Rajaei-Sharifabadi, H.; Ellestad, L.; Porter, T.; Donoghue, A.; Bottje, W.G.; Dridi, S. Noni (Morinda citrifolia) modulates the hypothalamic expression of stress-and metabolic-related genes in broilers exposed to acute heat stress. *Front. Genet.* **2017**, *8*, 192. [CrossRef] [PubMed]
33. Lee, H.I.; Lee, M.K. Coordinated regulation of scopoletin at adipose tissue–liver axis improved alcohol-induced lipid dysmetabolism and inflammation in rats. *Toxicol. Lett.* **2015**, *237*, 210–218. [CrossRef]
34. Von Eugen, K.; Nordquist, R.E.; Zeinstra, E.; van der Staay, F.J. Stocking density affects stress and anxious behavior in the laying hen chick during rearing. *Animals* **2019**, *9*, 53. [CrossRef] [PubMed]
35. Wang, X.J.; Wei, D.I.; Song, Z.; Jiao, H.C.; Lin, H. Effects of fatty acid treatments on the dexamethasone-induced intramuscular lipid accumulation in chickens. *PLoS ONE* **2012**, *7*, e36663. [CrossRef]
36. Scanes, C.G. Biology of stress in poultry with emphasis on glucocorticoids and the heterophil to lymphocyte ratio. *Poult. Sci.* **2016**, *95*, 2208–2215. [CrossRef]
37. Shini, S.; Shini, A.; Kaiser, P. Cytokine and chemokine gene expression profiles in heterophils from chickens treated with corticosterone. *Stress.* **2010**, *13*, 185–194. [CrossRef]
38. Mohammed, A.; Mahmoud, M.; Murugesan, R.; Cheng, H.W. Effect of a Synbiotic Supplement on Fear Response and Memory Assessment of Broiler Chickens Subjected to Heat Stress. *Animals* **2021**, *11*, 427. [CrossRef]
39. Dong, H.; Lin, H.; Jiao, H.C.; Song, Z.G.; Zhao, J.P.; Jiang, K.J. Altered development and protein metabolism in skeletal muscles of broiler chickens (Gallus gallus domesticus) by corticosterone. *Comp. Biochem. Physiol. Part A Mol. Integr. Physiol.* **2007**, *147*, 189–195. [CrossRef]
40. Lin, H.; Decuypere, E.; Buyse, J. Oxidative stress induced by corticosterone administration in broiler chickens (Gallus gallus domesticus): 2. Short-term effect. *Comp. Biochem. Physiol. Part B Mol. Biol.* **2004**, *139*, 745–751. [CrossRef]
41. Lin, H.; Gao, J.; Song, Z.G.; Jiao, H.C. Corticosterone administration induces oxidative injury in skeletal muscle of broiler chickens. *Poult. Sci.* **2009**, *88*, 1044–1051. [CrossRef] [PubMed]
42. Song, D.; King, A. Effects of heat stress on broiler meat quality. *World Poult. Sci. J.* **2015**, *71*, 701–709. [CrossRef]
43. Sato, H.; Takahashi, T.; Sumitani, K.; Takatsu, H.; Urano, S. Glucocorticoid generates ROS to induce oxidative injury in the hippocampus, leading to impairment of cognitive function of rats. *J. Clin. Biochem. Nutr.* **2010**, *47*, 224–232. [CrossRef]

Article

Evaluation of Greek Cattle Carcass Characteristics (Carcass Weight and Age of Slaughter) Based on SEUROP Classification System

Kostoula Nikolaou [1], Panagiota Koutsouli [2,*] and Iosif Bizelis [2]

1 Department of Bovine Sector and Equity, Directorate General of Agriculture, Directorate of Animal Husbandry Systems, Hellenic Ministry of Rural Development and Food, Veranzerou 46, 10176 Athens, Greece; knikolaou@minagric.gr

2 Laboratory of Animal Breeding and Husbandry, Department of Animal Science, Faculty of Animal Biosciences, Agricultural University of Athens, 11855 Athens, Greece; jmpiz@aua.gr

* Correspondence: panagiota@aua.gr; Tel.: +30-210-5292-4440

Received: 21 October 2020; Accepted: 26 November 2020; Published: 28 November 2020

Abstract: In Greece, all cattle carcasses produced from a variety of breed types are classified according to the SEUROP system. The objective of this study was to evaluate Greek carcass characteristics such as carcass weight and age of slaughter based on SEUROP classification system (muscle conformation and fat deposit classes) and to describe the effect of main factors such as breed, gender, year of slaughter, farm's geographical region and month of slaughter on these carcass parameters. It is the first study that evaluates local breeds, revealing the wide diversity of the Greek cattle breeding conditions. The analyzed records consisted of 323,046 carcasses from 2011 to 2017. All the examined factors significantly affected the mean carcass weight (298.9 ± 0.2 kg) and the mean slaughter age (559.1 ± 0.3 days). Carcasses from beef meat breeds had on average higher mean carcass weight while the local breeds had lower. The mean slaughter age and carcass weight were higher in winter than in summer. The local and the dairy breeds were classified in similar muscle conformation classes. Finally, Greek cattle carcasses from almost all regions were satisfactory for their quality carcass traits with good muscle conformation (R, O and U class) and low-fat deposit (class 1 to 3).

Keywords: beef; local breeds; carcass weight; age of slaughter; SEUROP system

---

1. Introduction

Beef is the third most widely consumed meat in the world and is considered to be a highly nutritious and valued food [1]. Intended for culinary and meat-processing purposes beef meat must meet certain qualitative requirements in terms of its sensory characteristics such as a suitable color, a desirable flavor features, an appropriate texture and a high level of tenderness [2]. The characteristics of beef carcass have a significant effect on meat quality and play a decisive role in determining its value. The value of the animal carcass and the cost of producing that carcass determine the profitability in cattle production systems [3]. Furthermore, major factors that affect the value of the carcasses and the cost of meat production include the animal's genotype, nutritional and management practices applied on the farm. The quality classification systems of carcasses are widely used as tool in the beef industry, making the business transactions easier while at the same time support the primary sector by providing it with useful information. The term "classification" defined as a set of descriptive terms describes the features of the carcass, which are useful to those involved in the trading of carcasses [4]. In the European Union (EU), the adoption of the SEUROP classification system within the member states established in 1981 is obligatory to record, monitor and collect data according to EU legislation [5] that concerns the carcass weight, the gender and the age of slaughter, the muscular conformation and the state of fattening of the carcass. Therefore, these data are measurable indicators that determine the quality characteristics and define the economic value of the carcass.

The general view by all sectors involved (slaughterhouses, producers, services) for the beef carcass classification system in the EU is that it operates well and provides, mainly for producers, a reliable basis for the deadweight sale of finished cattle [6]. Although in a recent study [7] is considered that, the SEUROP grid may be based on global indicators but it does not consider the carcass as a complex and heterogeneous entity, so, in the same SEUROP classification, it could include different muscles with higher or lower commercial value. The lack of a strong and clear link between sensory scores and European carcass classification standards shows that the European beef industry can not only rely on them but also needs to integrate quality into the carcass value [8]. Therefore, a study proposed alternative measures to be included in the SEUROP system in order to enable meat quality and to deliver consistent beef quality to the consumers [8].

While the SEUROP carcass grading system can be appropriate today, the ongoing changes in the production and marketing of cattle internationally could require in the future providing additional meat quality characteristics. A significant number of research studies tried to investigate the relationship between meat quality characteristics and the carcass parameters of SEUROP system mainly muscle conformation and fat deposit. Regarding marbling, a recent study [9] indicated that European classification scores explain only a slight proportion of the variance in marbling score (32%, 46%, 34% and 21% for the entire cattle group, young bulls, females and steers, respectively). Moreover, a significant correlation was observed among carcass yield and SEUROP conformation and fatness scores with intramuscular fat, slaughter body weight and hot carcass weight [10]. As a first step in developing a new way to assess the overall quality of beef carcasses in Europe it was proposed [11] a set of 5 indicators to include in the SEUROP system [hind quarter weight, meat color, retail-cut yield, rib-eye area and marbling score].

Beef production in the EU is approximately stable around 600,000 tons per month and holds the 3rd position after the United States of America (USA) and the Federative Republic of Brazil. In 2018, almost 7930 million tons of bovine meat (calve, young cattle, heifer, cow, bull and bullock) were produced in the EU from 87 million bovine animals. The highest production of European beef meat came from France (19%), Germany (15%) and the United Kingdom (12%), while almost half of the veal production in EU came from Spain (23%) and the Netherlands (23%) [12]. Considering that, the primary production of beef in EU consists of almost two thirds of dairy cows it is obvious that milk production is the main objective for most European cattle farms and only a small part of their income comes from beef production.

The average carcass weight in EU increased by about 24 kg/head from 2000 to 2015 [13], despite the fact that the EU beef consumption corresponds to 10.9 kg/capita/year with large fluctuations between its member states [14].However, beef consumption in the developed world has been declining for the past 20 years, with rates falling to 12% in the EU, 19% in the USA and 20% in Australia [15]. Many studies have evaluated the causes of this declining trend that could be attributed to the negative criticism received by beef meat on issues related to the environment, public health, safety and authenticity, including the lack of consistency in the quality of beef meat [15–18]. Since 2003, the World Health Organization (WHO) has developed specific guidelines that pointed out the relationship between dietary fat and incidence of lifestyle diseases [19]. Supporting not only sensory and nutritional quality is therefore a priority issue for the beef meat industry in order to overcome the decline trend in consumption [7]. However, consumers increasingly appear to prefer high-quality meat cuts that are, characterized by consistently high levels of eating quality [18].

Numerous studies evaluating the endogenous factors that affect the quality characteristics of carcasses, pointed out the effect of genotype and gender of the cattle [20–28]. Regarding the exogenous factors, regional differences due to climate and geographical morphology heterogeneity, affect the calving season, weaning weight, reproductive efficiency, feed costs and the animal's growth gain, configuring the final quantity and quality characteristics of the slaughtered cattle [29,30].

The beef sector in Greece has a great interest for study because it presents a lot of peculiarities. Being the southernmost country in Europe, it differs significantly, not only for the climatic conditions in contrast to the northern European countries, but also for the diversity in the breeding conditions of bovine animals. In addition, there is a large variety of cattle breeds that are bred and slaughtered in Greece, because local breeds do not meet the Greek beef meat's demand. Specifically, the Greek local breeds have not evaluated in the past according their carcass characteristics either compared with other European beef breeds.

This study aims:(i) to describe the effect of main factors (breed, gender, year of slaughter, farm's geographical region and month of slaughter) on the carcass weight and age of slaughter for various types of cattle carcasses (calve, young cattle, heifer and young bull); (ii) to evaluate the beef carcasses produced in Greece based on the SEUROP classification system. It is hypothesized that the information concerning the effect of the farm's geographical region on the carcass characteristics will give more insight on the development of the sector. In addition, for the first time it will be presented the carcass characteristics from four Greek meat breed carcasses (Greek Red, Greek Blonde, Vrachiceratiki or Greek Brachyceros and Local cattle) as well as from the Greek Buffalo (Bubalus bubalis).

2. Materials and Methods

Field data (n = 979,806) were collected from the Integrated Veterinary Information System (IVIS) and the online application "ARTEMIS" of the Hellenic Agricultural Organization "ELGO-DIMITRA" from 132 approved slaughterhouses, geographically distributed in all 13 regions of the country from years 2011 to 2017. The registration of the data is obligatory based on the national legislation. The data included the gender, the breed and the geographical region of the farm, the date of birth and the date of slaughter, the carcass weight and the SEUROP classification categories.

The EU definition of carcass is "the whole body of a slaughtered animal as presented after bleeding, evisceration and skinning". According to European legislation [31], the beef carcass is weighed as soon as possible after slaughter and not later than 60 min after the animal has been stuck and the presentation of the beef carcass should be (a) without the head and the feet; the head shall be separated from the carcass at the atloido-occipital joint and the feet shall be severed at the carpametacarpal or tarsometatarsal joints; (b) without the organs contained in the thoracic and abdominal cavities with or without the kidneys, the kidney fat and the pelvic fat; (c)without the sexual organs and the attached muscles and without the udder or the mammary fat.

The EU classification system classified bovine carcasses according to their gender and age into 6 categories using the letters A, B, C, D, E and Z. The definition of each letter is A: carcasses of uncastrated male animals aged from 12 months to less than 24 months; B: carcasses of uncastrated male animals aged from 24 months; C: carcasses of castrated male animals aged from 12 months; D: carcasses of female animals that have calved; E: carcasses of other female animals aged from 12 months and Z: carcasses of animals aged from eight months to less than 12 months. In addition to the latter categories, in European legislation [31] there is one more with the letter V for the carcasses of animals aged less than eight months. The beef carcasses in category V were not obliged to be classified according to SEUROP system. In our study, the category C was not used because there were no carcasses slaughtered in Greece in this category.

The SEUROP system defines six classes in order to classify carcasses according to their muscle conformation. The S class is "superior"; the E class is "excellent"; the U class is "very good"; the R class is "good"; the O is "fair" and the P class is "poor". Regarding to the fat deposit, the EU system classified bovine carcasses into five classes using the numbers 1–5. Specifically, class 1 is low; class 2 is slight; class 3 is average; class 4 is high; class 5 is very high.

The final selected dataset for analysis excluded the crossbred animals and consisted from 323,046 carcasses derived from 24 purebred cattle breeds including all animals with age of slaughter from 210 to 975 days with a sufficient number of observations (>100).

For the statistical processing of carcass weight and age of slaughter data, the analysis of variance was used (one-way ANOVA) in order to detect significant differences between the relative means for breed, gender, slaughter year, slaughter month, farm's geographical region and categories of carcass classification, muscle conformation and fattening. For multiple comparisons, the Bonferroni criterion used was set at significance level of $p \leq 0.05$. All statistical analyses were performed with the statistical program SPSS Statistics for Windows (IBM SPSS statistics Version 22.0, 2020).

3. Results

Data showed that a high percentage of carcasses (n = 503,000) resulted from random and unidentified crossbreeding (51.3%). The statistical data processing showed that the carcass weight and the age of slaughter averaged 298.9 ± 0.2 kg and 559.1 ± 0.3 days (about 1.5 years), respectively.

3.1. The Effect of Breed on Carcass Characteristics

The total number of carcasses in Table 1 was 321,381. The breeds with the largest number of carcasses were Limousin (28.8%) and Holstein (21.7%). Additionally, 12.9% of beef carcasses slaughtered were from the local breed of Greek Red. It is worth noting that a remarkable number of carcasses were Metis (9.5%) and Baltata Romameasca (6.7%), breeds originated mainly from Romania, a favorable destination to buy cattle for fattening due to its short distance from Greece.

Table 1. Number of carcasses per breed (N), means ± std. error for the age at slaughter (days, d) and the carcass weight (kg) from 24 cattle breeds (>100 observations) reared in Greece.

Breed Type	Breed Name	N	Age at Slaughter (d)	Carcass Weight (kg)
dairy	Holstein	69,861	578.0 ^{cghk} ± 0.6	251.1 ^{c} ± 0.3
dairy	Red and White	596	529.1 ^{ad} ± 6.6	269.0 ^{gko} ± 4.6
dairy	Baltata Neagra	562	572.6 ^{bdefg} ± 7.4	274.2 ^{dglp} ± 3.5
dual	Braunvieh	1952	582.7 ^{cfh} ± 3.4	262.1 ^{kl} ± 1.7
dual	Bruna	683	574.1 ^{behijl} ± 6.2	275.3 ^{dlno} ± 2.8
dual	Fleckvieh	2381	604.2 ^{ck} ± 2.7	292.3 ^{pr} ± 1.6
dual	Simmental	6778	591.6 ^{cjk} ± 1.6	293.3 ^{pt} ± 1.1
dual	Baltata Romameasca	21,461	566.2 ^{bel}± 1.0	303.7 ^{fh} ± 0.5
dual	Bruna de Maramures	801	565.0 ^{bdeh} ± 5.4	304.8 ^{dhrt} ± 3.0
dual	Salers	530	581.3 ^{behkl} ± 5.1	360.4 ^{jq} ± 3.2
beef	Limousin	92,560	568.3 ^{b} ± 0.4	328.9 ^{s} ± 0.3
beef	Montbelliard	513	620.3 ^{c} ± 7.0	343.2 ^{q} ± 3.7
beef	Aubrac	6851	565.3 ^{bel} ± 1.3	373.6 ^{ejnr} ± 1.1
beef	Blanc Bleu	1701	574.2 ^{behl}± 2.8	378.9 ^{e} ± 2.3
beef	Charolais	13,326	599.6 ^{ck}± 1.1	388.7 ± 0.8
beef	Blonde d' Aquitaine	7898	528.0 ^{a} ± 1.1	404.9 ^{i} ± 1.0
beef	Parthenaise	743	593.3 ^{cgjk} ± 3.6	425.6 ^{m} ± 3.1
crossed	Metis	30,517	563.8 ^{dl} ± 0.9	293.4 ^{dp} ± 0.5
crossed	Groase	8993	554.4 ^{i} ± 1.1	405.4 ^{i} ± 0.8
local	Vrachiceratiki	1488	528.1 ^{ad}± 4.6	171.3 ^{b}± 2.3
local	Greek Buffalo	2493	694.4 ± 3.3	200.1 ^{a} ± 1.1
local	Local	6344	567.5 ^{cd}± 2.1	206.8 ^{a} ± 1.2
local	Greek Red	41,358	471.9 ^{c} ± 0.7	251.5 ^{c} ± 0.4
local	Greek Blonde	991	547.2 ^{di} ± 4.5	290.3 ^{dp} ±3.4

Means within the same column followed by different superscript for each variable ($^{a, b, c, d, e, f, g, h, i, j, k, l, m, n, o, p, q, r, s, t}$) among breeds differ significantly $p \leq 0.05$.

Mean carcass weight ranged from 171.3 ± 2.3 kg (Vrachiceratiki) to 425.6 ± 3.1 kg (Parthenaise). Table 1 showed that an average carcass weight over 400 kg was observed for meat beef breeds as Parthenaise (425.6 ± 3.1 kg) and Blonde d'Aquitaine (404.9 ± 1.0 kg) and the crossed type Groase (405.4 ± 0.8 kg). Lower mean carcass weight was found in carcasses from local cattle breeds with small body conformation as Vrachiceratiki (171.3 ± 2.3 kg), Greek Buffalo (200.1 ± 1.1 kg) and Local (206.8 ± 1.2 kg). In contrast carcasses from Greek Red (251.5 ± 0.4 kg) and Greek Blonde (290.3 ± 3.4 kg) had higher mean carcass weight among the local breeds and good body conformation because the animals were upgraded crossbred with Limousin and Charolais respectively. On the other hand, a relatively low mean carcass weight had the carcass from Holsteins (251.1 ± 0.4 kg).

Figure 1 shows the distribution of frequencies of classes for muscle conformation (a1–a4) and fat deposit (b1–b4) scores given as explanatory spider web charts in grouped breed types of dairy (a1,b1), dual purpose (a2,b2), beef (a3,b3) and local (a4,b4) cattle breeds, respectively.

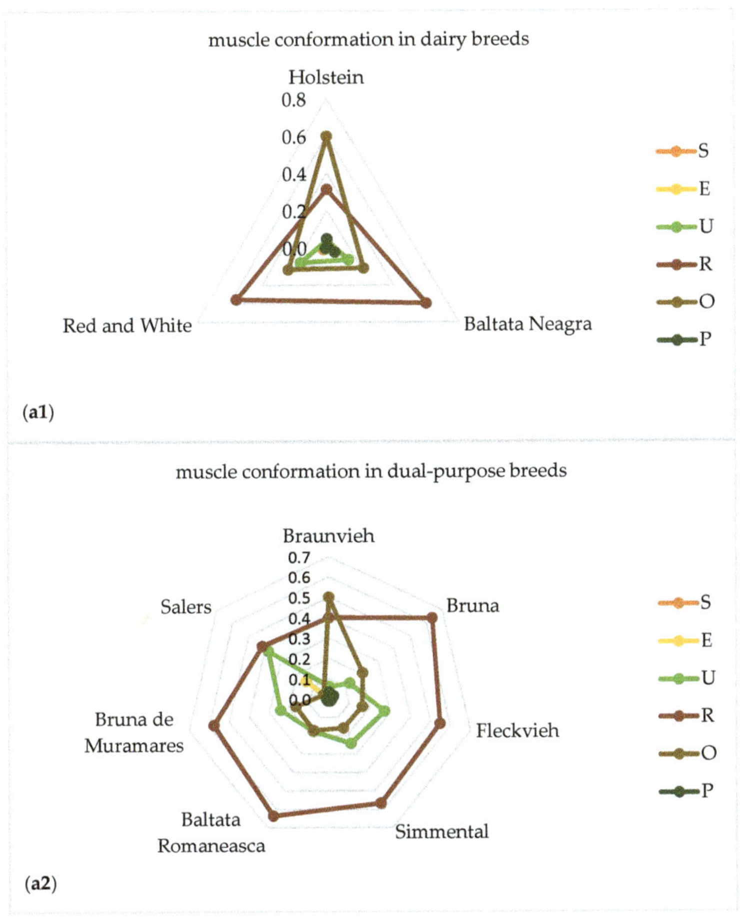

Figure 1. *Cont.*

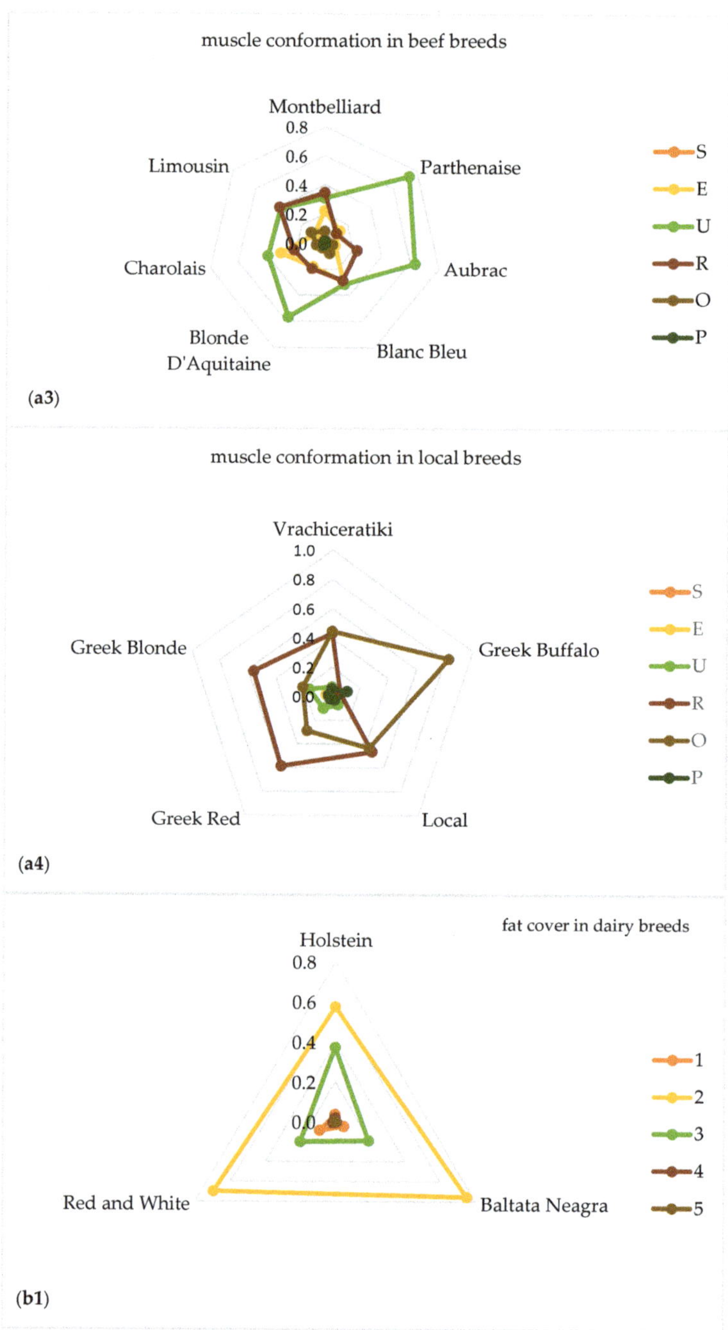

Figure 1. *Cont.*

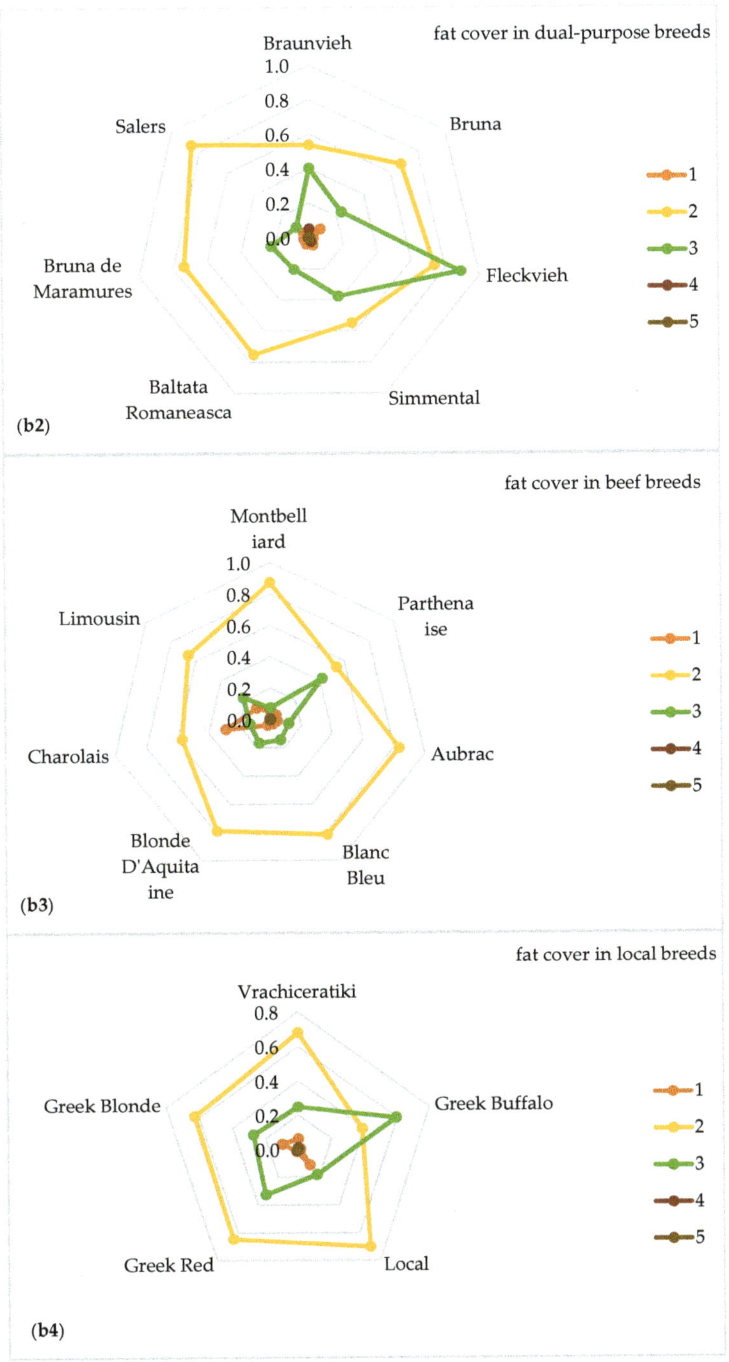

Figure 1. Proportions of muscle conformation (**a1–a4**) and fat cover (**b1–b4**) scores in grouped breed types of dairy (**a1**,**b1**), dual-purpose (**a2**,**b2**), beef (**a3**,**b3**) and local (**a4**,**b4**) cattle breeds according to SEUROP classification system.

According to the muscle conformation and the breed types of cattle it is obvious from the Figure 1 that the beef breeds had the highest value in conformation classes. The beef breeds that distinguished for their very good muscular conformation (class U) were Parthenaise, Blonde d' Aquitaine and Charolais. From the dual-purpose breeds, only Salers had classified with SEUROP grid in class E. The local and the dairy breeds had similar muscle classification classes. Within class O, the classified carcasses were Holstein and Greek Buffalo. It is worth noting that for the fat deposit, the majority of carcasses in all breed types classified in class 2. The breeds that classified in class 3 were Holstein, Greek Buffalo, Parthenaise and Fleckvieh. Greek Buffalo carcasses, although they had the second lowest mean carcass weight (200.1 ± 1.1 kg) from all breed types, it was showed that they had higher fat deposit similar to Parthenaise (425.6 ± 3.1 kg) that had the heaviest mean carcass weight among all breed types.

Mean slaughter age more than 600 d was observed in carcasses of Greek Buffalo (694.4 ± 3.3 d), Montbelliard (620.3 ± 7.0 d) and Fleckvieh (604.2 ± 2.7 d). These were mainly breeds reared in semi-extensive and dual-purpose farms, suitable for milk and meat production or very resilient cattle, not only able to produce plenty of milk but also to withstand environmental difficult conditions. On the contrary, lower mean age of slaughter was found in carcasses of the Greek Red (471.9 ± 0.7 d).

3.2. The Effect of Gender on Carcass Characteristics

The carcass weight of male carcasses (n = 268,463) was found significantly heavier (316.0 ± 0.2 kg) than females' (n = 54,583) which was observed to be 214.3 ± 0.3 kg ($p \leq 0.001$). The mean age of slaughter for male animals was 564.4 ± 0.3 d, while for females was 532.8 ± 0.8 d ($p \leq 0.001$). It is worth noting the fact that female carcasses of this study came from a large percentage of heifers, intended primarily, for slaughter and not for replacement of older females. As common reasons to remove females under 2 years of age from the breeding herd, referred the low daily gains before weaning, questionable inheritance, poor performance of dam and/or sire, undesirable conformation, or failure to exhibit a normal oestrus cycle [32]. The age of slaughter was shorter for females because they were destined to be bred only for fattening. Females slaughtered at older age and heavier carcass weight had increased fat composition rather than increased muscular conformation. The opposite effect would be for male carcasses.

3.3. The Effect Of Year Of Slaughter On Carcass Characteristics

The year of slaughter affected the mean carcass weight and mean age of slaughter significantly ($p \leq 0.001$) as it is shown in Table 2. Regarding the distribution of muscle conformation and fat deposit classes of carcasses during the seven years (2011–2017) as indicated by the SEUROP classification system the findings are presented in Figure 2, given as explanatory spider web charts.

Table 2. Number of carcasses, means ± std. error for the age of slaughter (days, d) and carcass weight (kg) during 2011 to 2017.

Year	N	Age of Slaughter (d)	Carcass Weight (kg)
2011	58,652	552.5 ^{a} ± 0.6	294.8 ^{a} ± 0.4
2012	52,634	560.4 ^{b} ± 0.7	296.2 ^{a} ± 0.4
2013	45,887	565.0 ^{c} ± 0.7	299.7 ^{b} ± 0.5
2014	42,514	563.6 ^{c} ± 0.7	299.5 ^{b} ± 0.5
2015	41,625	564.2 ^{c} ± 0.7	305.3 ^{c} ± 0.5
2016	40,229	557.5 ^{b} ± 0.8	303.5 ^{c} ± 0.5
2017	41,505	552.1 ^{a} ± 0.7	295.2 ^{a} ± 0.5
total	323,046	559.1 ± 0.3	298.8 ± 0.2

Means within the same column followed by different superscript for each variable ($^{a, b, c}$) among years differ significantly ($p \leq 0.05$).

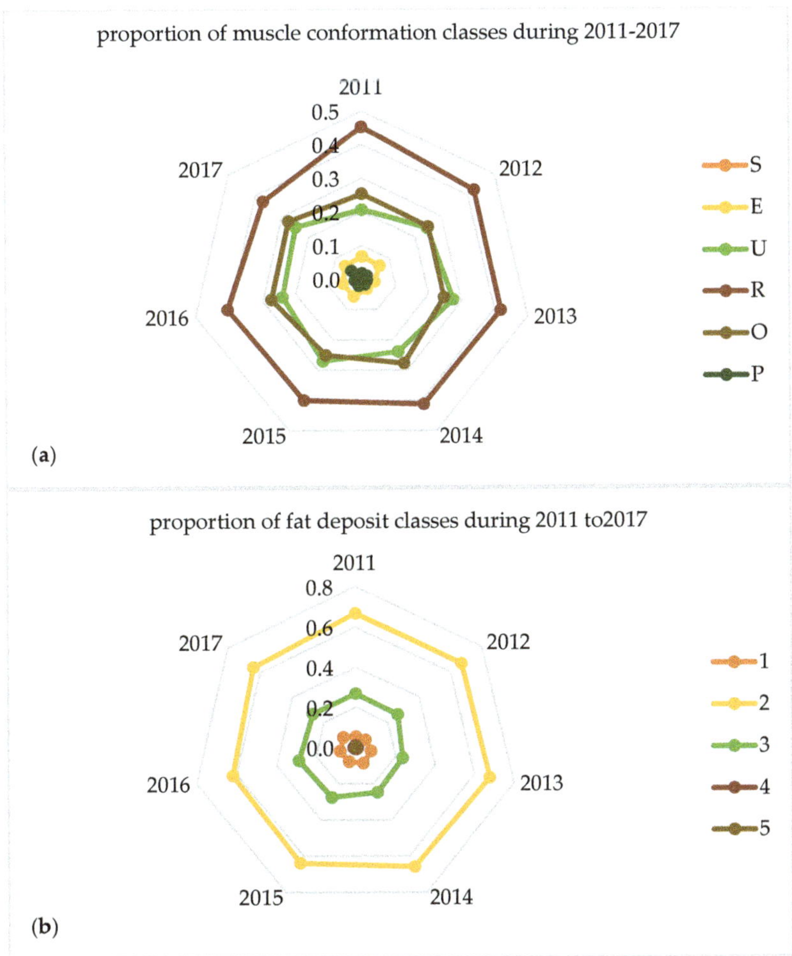

Figure 2. Proportion of muscle conformation (**a**) and fat deposit (**b**) classes for the carcasses during 2011–2017.

Over the course of seven years, fluctuations in both mean carcass weight and age of slaughter regardless of the gender were observed. The total number of cattle slaughtered was decreased over the years. In fact, the highest decrease was observed in 2016 and 2017, and reached the reduction rate of approximately 31.4% and 29.2%, respectively, compared to 2011, where it reached the highest value.

In Figure 2 the majority of carcasses were classified according to muscle conformation in classes R, O and U, while for the fat deposit in classes 2 and 3.

3.4. The Effect of Geographical Region of Farms on Carcass Characteristics

The largest number of cattle farms in the country was located in Northern Greece and specifically in the regions of Eastern Macedonia and Thrace and Central Macedonia and the lowest number was located in the region of Attica [33]. Table 3 shows that from the records of 322,609 cattle slaughtered in the 13 regions, the lightest carcass weight was found for carcasses derived from the region of Epirus (247.2 ± 0.8 kg). This observation is in accordance with the relatively small number of cattle farms in this area that cover only 7.5% of the total number of farms in the country. Taking into account the geographical criteria, the region of Epirus is a mountainous area in the northwestern part of the country where traditionally bred sheep and goats. Furthermore, the largest percentage of carcasses (32.2%) in the region of Epirus originated from Holstein breed, while 21.6% of them belonged to Greek breeds (Table S1) that showed low mean carcass weight as mentioned in Table 1.

Table 3. Number of cattle farms and beef carcasses (N), means ± std. error for the age at slaughter (days) and carcass weight (kg) distributed in the 13 geographical regions.

Geographical Region	N (Farms)	N (Carcasses)	Age at Slaughter (d)	Carcass Weight (kg)
ATTICA				
1. Attica	53	3684	522.2^{a} ± 3.2	268.2^{a} ± 1.3
CENTRAL GREECE				
2. Thessaly	2	53,994	498.5^{b} ± 0.6	270.0^{a} ± 0.4
3. Central Greece (Sterea)	573	15,578	527.1^{a} ± 1.1	297.7^{b} ± 0.8
4. Peloponnese	920	8914	569.5^{c} ± 1.5	268.0^{a} ± 1.3
5. Western Greece	1591	26,924	608.3^{di} ± 0.9	369.9^{e} ± 0.7
6. Ionian Islands	274	4152	623.0^{ei}± 2.1	344.4^{f} ± 1.5
AEGEAN ISLANDS & CRETE				
7. North Aegean	551	9200	616.3^{f}± 1.4	316.6^{g} ± 0.9
8. South Aegean	1183	7035	567.0^{dg} ± 1.8	279.0^{dh} ± 1.0
9. Crete	200	8459	572.0^{h} ± 1.7	259.8^{i} ± 0.9
NORTHERN GREECE				
10. Epirus	1103	12,058	544.8^{dei} ± 1.5	247.2^{c} ± 0.8
11. Western Macedonia	1099	14,185	561.5^{e} ± 1.2	275.5^{d} ± 0.8
12. Eastern Macedonia & Thrace	2946	39,078	565.3^{i} ± 0.8	268.9^{a} ±0,4
13. Central Macedonia	2642	119,348	570.6^{e} ± 0.4	318.0^{gj} ± 0.3
total	14,699	322,609	559.0 ± 0.3	298.8 ± 0.2

Means within the same column followed by different superscript for each variable ($^{a, b, c, d, e, f, g, h, i, j}$) among geographical regions differ significantly ($p \leq 0.05$).

Regarding the mean age of slaughter, it is interesting to note that the lowest slaughter age observed in the region of Thessaly, could be attributed to the number of cattle raised for fattening in this region. The total percentage reached 16.74% of the total number of carcasses, as well as the fact that 39% of these carcasses originated from Limousin breed and 38.4% from the local breed Greek Red (Table S1).

Furthermore, in Table 3, we observed that only in three regions (Western Greece, Ionian Islands and North Aegean) occurred the highest value of the mean age of slaughter (over 600 d) and the mean carcass weight (over 300 kg). In addition, the cattle breed that slaughtered in these regions was mainly Limousin, a pure meat breed (data in Supplementary Table S1).

Figure 3 given as explanatory spider web charts depicts the proportion of muscle conformation (a) and fat deposit (b) classes in the 13 regions of the country.

As it shown in Figure 3, the major proposition of muscle conformation classes of the Greek carcasses in all over the 13 regions was R. In the region of Grete, a wide proportion of carcasses is classified in conformation class U and in the Eastern Macedonia & Thrace, a large proportion of carcasses classified in class O. The class E appeared mainly in carcasses slaughtered in the North Aegean. According to fat deposit the class 2 appeared in all 13 regions and only in regions of Epirus and Grete there was a large proportion that classified their carcasses in classes 3 and 1, respectively.

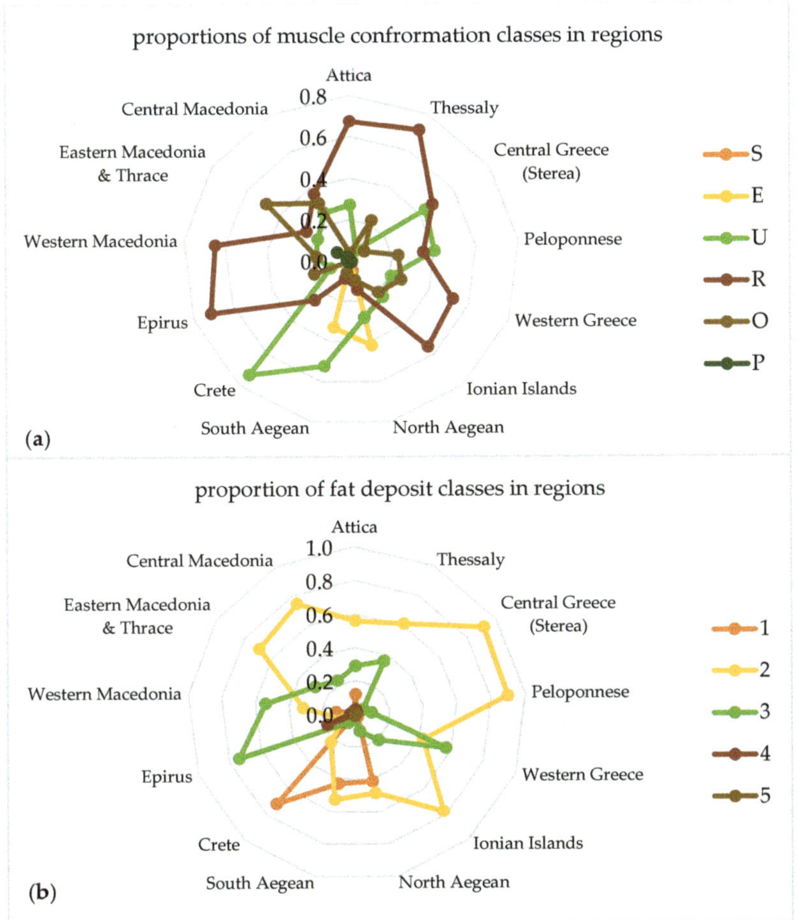

Figure 3. Proportions of muscle conformation (**a**) and fat deposit (**b**) classes distributed across 13 regions.

3.5. The Effect of Month of Slaughter on Carcass Characteristics

Regarding the total number of carcasses had significant differences between the first and the second half of the year (Table 4). It is worth noting that a remarkable number of carcasses (over 30,000) were slaughtered in the second half of the year especially on the 7th, 8th, 11th and 12th month. On the contrary the lowest number of carcasses showed on the 1st month of the year (n = 8997).

Comparisons among the months of slaughter showed that both the mean carcass weight and the mean age of slaughter differentiated significantly ($p \leq 0.001$). The lowest value for the mean carcass weight was in the 3rd month of the year (293.2 ± 0.7 kg) while the highest value was in the 11th month (301.4 ± 0.6 kg). The difference between the two values of the mean carcass weight was 8.2 kg.

As for the age of slaughter the lower value appeared in 5th and 6th month of the year (555.3 ± 0.9 d and 555.4 ± 0.9 d respectively) and the highest in the 1st month (568.1 ± 1.6 d).

In Figure 4 given as explanatory spider web charts, the distribution of conformation classes is homogeneous across the months and the highest proportion of muscle conformation in class R is widespread all over in the 12 months. Similarly, regarding the fat cover the highest proportion of fat deposit is classified in class 2.

Table 4. Number of beef carcasses per month, means ± std. error for the age at slaughter (days, d) and carcass weight (kg).

Month of Slaughter	N	Age at Slaughter (d)	Carcass Weight (kg)
1	8997	568.1 [a] ± 1.6	302.7 [a] ± 1.1
2	25,758	564.6 [ad] ± 0.9	298.5 [bde] ± 0.6
3	20,951	557.2 [b] ± 1.1	293.2 [c] ± 0.7
4	25,492	561.9 [cde] ± 0.9	298.6 [bde] ± 0.6
5	28,682	555.3 [b] ± 0.9	298.3 [bde] ± 0.6
6	27,820	555.4 [b] ± 0.9	297.3 [d] ± 0.6
7	30,423	554.8 [b] ± 0.8	297.1 [df] ± 0.6
8	30,280	558.0 [be] ± 0.8	299.4 [abd] ± 0.5
9	29,495	557.4 [b] ± 0.8	300.7 [ae] ± 0.6
10	29,447	557.0 [b] ± 0.8	299.3 [abd] ± 0.6
11	30,879	563.1 [ad] ± 0.8	301.4 [a] ± 0.6
12	32,567	561.8 [cde] ± 0.8	300.1 [ab] ± 0.6
total	320,791	559.0 ± 0.3	298.8 ± 0.2

Means within the same column followed by different superscript for each variable ([a, b, c, d, e,f]) among month of slaughter differ significantly ($p \leq 0.05$).

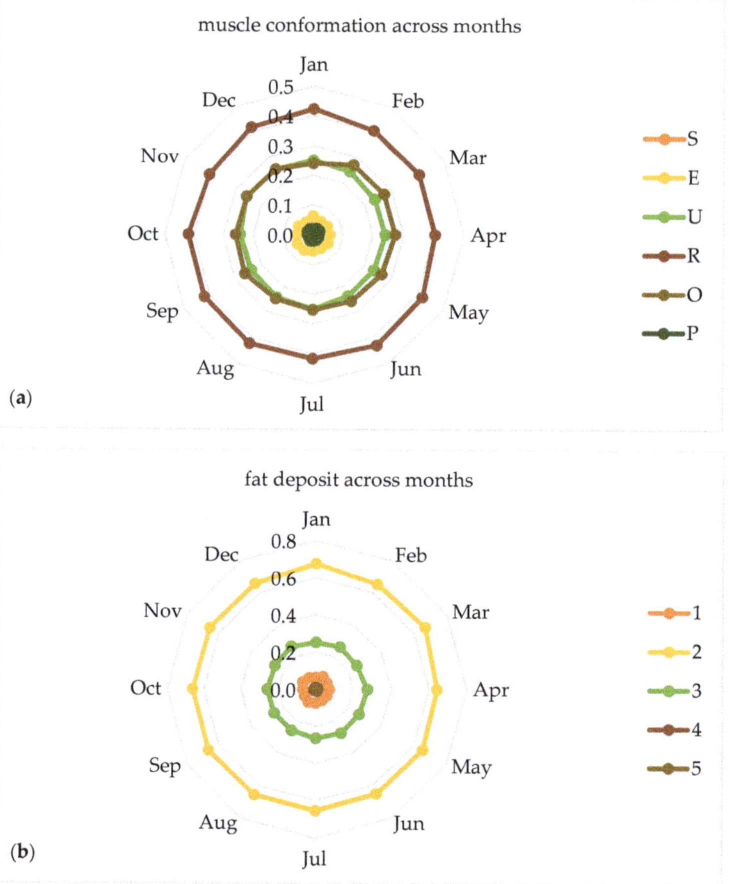

Figure 4. Proportions of muscle conformation (**a**) and fat deposit (**b**) classes distributed across 12 months.

3.6. Evaluation of Carcass Quality Characteristics Based on the SEUROP Classification System

Regarding the classification of carcasses into 6 categories (A,B,D,E,Z,V) based on gender and age at slaughter (Table 5), the heaviest mean carcass weight was recorded in the male carcasses of category B (329.2 ± 0.6 kg) and A (320.7 ± 0.2 kg).As for the female carcasses (categories D and E) the mean carcass weight ranged from 239.1 ± 0.8 kg and 219.4 ± 0.4 kg, respectively. Finally, in categories Z and V, it was appeared a lower mean carcass weight (218.1 ± 0.5 kg and 187.4 ± 1.6 kg, respectively) due to their younger age at slaughter.

Table 5. Number of beef carcasses (N), means ± std. error for the age at slaughter (days, d) and the carcass weight (kg) according to SEUROP classification scale [1] (category, muscle conformation and fat deposit).

SEUROP Classification Scale	N	Age at Slaughter (d)	Carcass Weight(kg)
category *			
A	218,732	549.5 ^{a} ± 0.2	320.7 ^{a} ± 0.2
B	31,506	813.3 ^{b} ± 0.4	329.2 ^{b} ± 0.6
D	7597	835.4 ^{c} ± 1.1	239.1 ^{c} ± 0.8
E	35,465	540.0 ^{d} ± 0.7	219.4 ^{d} ± 0.4
Z	27,107	318.2 ^{e} ± 0.2	218.1 ^{d} ± 0.5
V	2535	226.2 ^{f} ± 0.2	187.4 ^{e} ± 1.6
muscle conformation **			
S	1389	730.5 ^{a} ± 3.8	446.7 ^{a} ± 1.8
E	17,360	590.7 ^{b} ± 0.8	422.2 ^{b} ± 0.5
U	76,802	580.3 ^{c} ± 0.4	372.6 ^{c} ± 0.3
R	130,302	546.8 ^{d} ± 0.4	286.7 ^{d} ± 0.2
O	81,670	561.0 ^{e} ± 0.6	235.0 ^{e} ± 0.3
P	6515	600.6 ^{f} ± 2.5	209.4 ^{f} ± 1.0
fat deposit***			
1	21,911	602.2 ^{a} ± 1.0	348.8 ^{a} ± 0.8
2	199,138	553.6 ^{b} ± 0.3	301.8 ^{b} ± 0.2
3	79,909	574.6 ^{c} ± 0.5	289.7 ^{c} ± 0.3
4	2647	583.8 ^{d} ± 3.7	257.7 ^{d} ± 1.8
5	196	579.2 ^{abcd} ± 11.0	273.3 ^{cd} ± 5.9

[1] each class/category includes subclasses (+) & (−); *A: carcasses of uncastrated male animals aged from 12 months to less than 24 months; B: carcasses of uncastrated male animals aged from 24 months; D: carcasses of female animals that have calved; E: carcasses of other female animals aged from 12 months; Z: carcasses of animals aged from 8 months to less than 12 months; V: carcasses of animals aged less than 8 months; ** S: superior; E: excellent; U: very good; R: good; O: fair; P: poor; *** class 1 = low; class 2 = slight; class 3 = average; class 4 = high; class 5 = very high; means within the same column followed by different superscript for each variable (a,b,c,d,e,f) among; classes/categories differ significantly ($p \leq 0.05$).

According to Table 5 the largest percentage of carcasses (67.7%) belonged to category A (n = 218,732). The main reason was that the cattle imported to the country with the main purpose of breeding was fattening, they were preferred to be male, due to higher growth rate, better carcass performance and muscle conformation. Category V (animal carcasses under eight months old) had the lowest number of carcasses (n = 2,535) depicting the great need of Greek beef farmers to buy calves for breeding and fattening from abroad, as long as the demand was not met by the local market.

Regarding the classification of carcasses by category of muscle conformation according to SEUROP (Table 5), it is observed that the largest percentage of them were classified in class R (n = 130,302) as "good", followed by the other classes (O, U, E, P & S). These findings are explained by the fact that the majority of carcasses were male animals that were mainly classified in category R. Additionally, in category P where the number of carcasses was relatively low (n = 6515), they were classified carcasses of poor muscle conformation. In addition, mean carcass weight followed by a normal distribution with the highest value appearing in category S (446.7 ± 1.8 kg) and the lowest in category P (209.4 ± 1.0 kg). The mean carcass weight of category R was 286.7 ± 0.2 kg.

As for the mean age at slaughter, it is noticed that in category P there was a relatively extended number of days (600.6 ± 2.5 d) which could be explained by the fact that in this category mainly female carcasses were classified. Female cattle from this category removed from the livestock farm to the slaughterhouse due to low milk yields or possible accidents in the farm.

Regarding to the fattening state, most carcasses (n = 199,138) classified to class 2 followed by classes 3, 1, 4 and final class 5. The majority of carcasses in Greece (65.5%) had low fat deposit (class 1) while very large fat deposit (class 5) was found in a very small number of carcasses (n = 196).

4. Discussion

Our study based on a collection of seven-year records is a first attempt to give insight into the beef carcasses characteristics that produced in Greece and it will try to highlight trends that emerge into the European beef sector. Nearly the half of carcasses derived from crossbred animals of unknown genotype and this is a fact that arises from the need to supply Greek cattle farms with animals from other EU member states or from third countries, where the purchase of such crossbred animals is achieved at a lower price than that of purebred beef meat breeds. Additionally, many crossbred carcasses are offspring of dairy cows mated with bulls of meat breeds, which are also fattened in order to obtain the desired carcass weight.

Analyzing the purebred carcasses all investigated factors significantly affected the mean slaughter age and carcass weight. The mean carcass weight (298.9 ± 0.2 kg) compared with the average carcass weight in the EU-27 a decade ago, shows a trend to increase through this seven-year's period [34]. According to the above source [34], the average carcass weight has increased continuously since 2002 in the EU. In comparison with the beef sector in Ireland [35], the average carcass weight that failed to achieve a desired conformation score, was 301 kg; hence, huge prospect exists to improve this parameter in Greek carcasses too. The explanation could be the same for Irish beef carcasses. The carcasses that failed to achieve the desired fat or weight specification, on the one hand, could be attributed to the inability of producers to determine whether an animal is suitable for slaughter and on the other hand, could be the inability of cattle to reach a desirable carcass [35].

In our study, the mean age at slaughter of females was 532.8 ± 0.8 d, because the older female carcasses usually classified in higher age of slaughter, were excluded from our analysis. The observations in this study were consistent with previous results about the effect of breed on carcass characteristics. In a study considering 15 European cattle breeds [36], the dairy and local cattle breeds produced lighter carcasses as opposed to predominantly cattle breeds. Breed-specific differences in growth rate of local breeds could explain their relatively lower carcass weights as it was pointed out in another study [37]. Similarly, in our study lower mean carcass weight was found mainly in carcasses of local cattle breeds with moderate to poor body conformation. Additionally, we found that higher value of slaughter age generated heavier carcass weight which is in line with the observations of [38]. Regarding the use of dairy Holstein calves for beef production it is a common practice, which represents a significant portion of the meat consumed worldwide. As it is mentioned [39], Holstein calves finished in feedlot had higher fat content in carcass than those finished on pasture and they are excellent producers of lean meat, with good smoothness, flavor and juiciness. In this study, the descendants of Holstein dairy cows were sold to light live weight because it was not economically advantageous to be fattened into heavy ones.

The breed type was also reflected in muscle conformation and fat cover classes. The highest conformation classes such as E and U, was found in the beef meat breeds and the lowest classes such as O and P in dairy breeds. These results are in line with the results of [36] where the highest conformation score was in the double muscled Piedmontese and the lowest in Jersey. On the contrary, all the breed types in our study for the fatness classes were ranged in a similar way, mainly in the class 2. Similar results have reported in [40] that bulls showed greater muscle development, less fat deposition and were more efficient in producing leaner carcasses than steers which may be mainly attributed to the effects of male hormones on muscle protein anabolism. The class E in muscular conformation, which classified Salers carcasses in our study could explained according to the results of [41] where between Holstein and Salers breeds were observed that Salers cows had more muscle in carcass and Holstein cows were fatter than Salers cows.

Furthermore, the gender as a factor had a significant influence on mean carcass weight and on mean age at slaughter ($p \leq 0.001$). Similar to our findings it is referred by other studies [42,43] that bull carcasses are characterized by higher meat content with simultaneous lower content of fat compared to heifer carcasses. In addition, in a study with the double-muscled Belgian Blue bulls and cows, most of the carcass quality parameters were more favorable for males than for females [44]. In heifer's life, rearing factors applied during both pre-weaning and fattening periods influenced carcass and meat quality [45]. The relationship between tenderness and gender has evaluated by many studies [40,46] which found that meat from young bulls was significantly less tender than that of heifers. The male carcasses in this study came almost exclusively from cattle, which, whether imported from other countries or born in the country and they were bred for fattening and slaughtered when they gained the desired live weight. Considering that the local market system is based on carcass weight, the heaviest young bulls have a significant economic advantage over heifers in commercial scale. Therefore, this result explains further the dominance of young bulls in the local slaughter of beef meat. In addition, the encouraged to produce heavier carcasses due to favor slaughter pricing of heavier carcasses is a common practice in many countries according to a study in South Africa [47].

The effect of year of slaughter on carcass traits reflects to a large extent the fluctuations of environmental factors on the cattle farms from year to year and the ability of beef industry to adapt and respond. In a study of Slovenian cattle [48], the carcass weight of young bulls, heifers and cows varied among different years, but no trend could be noticed. Additionally, in Slovenia, in another study [49] within a decade from January 2005 to December 2015, the carcass weight of young bulls significantly increased from 345 to 354 kg in the first three years and then to 359 kg in 2013. The decline in the total number of beef carcasses was due to the outbreak of bovine nodular dermatitis in the country during 2016–2017 which affected mainly the areas of Northern Greece where the largest number of cattle farms exist. In addition, the decline trend for the number of male carcasses reaching 32.88% from 2011 to 2017, could be attributed to the same reason mentioned for the total decreased number of carcasses, since male carcasses made up a 83% of the total number of cattle slaughtered in the country.

Comparisons of the classification categories within the geographical regions of the cattle farms, it was observed that the best performance in terms of the carcass muscle conformation, number of carcasses and breed types was located in the Northern and Central regions of Greece. There is an obvious heterogeneity of the environment affecting the productive management of cattle farms in our country. Similar to our findings, cattle carcasses from the northern regions of Mexico had a higher marbling score than those in the southern regions and performed better overall [50]. The carcasses classification according to muscle conformation, focusing in the region of Central Macedonia showed that 37% of the total carcasses were slaughtered in the above region. Within this region a percentage of 26.4% of the total number of carcasses were classified in category U, 3.51% in category E and 36.38% in category R, while only 2.24% in category P. In addition to that, in the region of Central Macedonia, the mostly high-yielding cattle breeds were reared. More specifically, 75.25% of the total carcasses of the Blanc Blue breed and 77.3% of the total carcasses of the Blonde d' Aquitaine breed were bred and slaughtered in this region (Table S1). It is reasonable to consider that the cattle farmers of the above region seem to be more professionals regarding the management of their livestock and presented a business profile that focuses on their economic performance.

Significant differences were found among months of slaughter ($p \leq 0.001$). According to several studies on this factor, the seasonal changes in temperature affect the level of glycogen after slaughter and the ultimate pH and therefore the quality of meat [51,52]. In addition, another study [53] showed that the quality classification grades of the carcass were higher during January, February and March compared with May through November. The above results [53] are in line with ours. It is worth noting that in another study [54] the annual trends typically reach the lightest Hot Carcass Weight (HCW) for the year in May and seasonal differences in HCW could be a result of the type of cattle marketed at this period. Hence, similarly, in our study the low mean carcass weight in March could be due to a lesser availability of high nutrition value feeds during late autumn and winter seasons or market issues. Furthermore, our results were consistent with the results of [38], that heavier carcasses were observed for slaughter in autumn and winter. These results confirmed by another study [55], where animals that slaughtered in spring recorded lower carcass weights. It is known that cattle imported for fattening during the summer months are slaughtered during the winter. The average fattening period is about five months. Summer season, due to the extreme weather conditions (high temperature, high humidity) prevailed stresses and disrupts the growth rate of the animals. As it is known [56], cattle are considered more sensitive to hot than to cool temperatures. As a result, those cattle have not gained sufficient live weight. On the contrary, in December the largest number of carcasses (n = 32,567) was observed due to the efforts to satisfy the high demand during the Christmas period, while the mean carcass weight was increased (300.1 ± 0.6 kg). Hence, the lowest number of carcasses slaughtered in the first month of the year reflects the decrease of meat consumption after holiday's period. It is also worth noting that the mean age at slaughter over 562 days was higher mainly during the winter months, from November to February, to allow animals acquire the desired carcass weight to cover the high consumption observed during this period.

The EU classification system presents differences on beef carcass quality among the member states. The variations in cattle delivered to a European slaughterhouse in terms of age, breed, weight and feeding production systems are large and make it very difficult or even impossible for the slaughtering industry to produce European beef of a standardized quality [57]. Hence, the comparison between beef carcasses that classified under SEUROP classification system could lead to useful information about the beef sector in EU. The results in the present study showed that the majority of carcasses were classified in the category R and in class 2 of fat cover, i.e., carcasses with good muscle conformation and low amount of fat. Fat cover is a more reliable indicator of meat quality than carcass conformation [58]. On the other hand, carcass conformation classes are a factor that influences purchase prices. It is notable that in another study in Poland in all cattle categories, the better the conformation class, the higher the purchase price [59]. In accordance with that the Spanish beef market demands young bullock cattle with superior muscling that will yield a higher percentage of lean, and therefore, carcass conformation is the key factor for carcass economic value [60]. The results of the latter study for low fat carcasses are in accordance with a study conducted in Finland [61] where consumers favor in low fat products. The above studies have motivated beef industry to suggest that two thirds of the carcasses would have a EUROP fat score of 2 and one third a EUROP fat score of 3 and also to give penalties for carcasses less than 320 kg with fat scores 3–5 and for carcasses over 320 kg with fat scores 4–5. In France [18] although conformation has been a more important component, French consumers prefer beef with less visual fat at the retail level. In contrast, this trend for low fat carcasses if it is compared with other European but not EU member states, the results did not converge. For example, in Serbia [62], beef carcasses were evaluated as having conformation R in 59% of cases but the carcass fat tissue coverage degree was rated as 4 for 87% of carcasses.

5. Conclusions

In Greece, the carcasses are produced from a variety of cattle breed types. In our study, beef breeds classified in highest muscle conformation classes such as E and U, while in lowest classes such as O and P classified mainly dairy and local cattle breeds. From the dual-purpose breeds only Salers had a large proportion of classified carcasses in class E. Local breeds and Holstein cattle had lower mean carcass weight and in comparison with other EU countries, the lower value of the mean carcass weight in main beef breeds that produced in Greece it is due to different breeding and diet conditions. Mean carcass weight and mean age at slaughter were significantly differed among the relative levels of each factor examined. Male carcasses were 83% of the total number of cattle slaughtered in Greece, which reflects the dominance of young bulls in the local market system. There was a decreasing trend in the total number of cattle reared for meat during the studied years. Northern and central regions of Greece produced carcasses with the best performance in terms of carcass muscle conformation, number of carcasses and breed types, so the development of beef sector in Greece is based mainly on these regions. Higher values of mean carcass weight and mean age at slaughter were observed in winter than in the summer months. According to the SEUROP classification system, Greek carcasses had good muscle conformation (class R) and low amount of fat (class 2), which could reveal an EU trend for low fat deposit in beef meat.

Supplementary Materials: The following are available online at http://www.mdpi.com/2304-8158/9/12/1764/s1, Table S1: Number of carcasses (counts) classified for their muscle conformation in each breed and region and the relative percentages.

Author Contributions: Conceptualization, I.B.; methodology, I.B.; software, P.K.; validation, I.B., K.N.; investigation, K.N.; data curation, K.N., P.K.; writing—original draft preparation, K.N.; writing—review and editing, I.B., P.K., K.N. All authors have read and agreed to the published version of the manuscript.

Funding: This research received no external funding.

Acknowledgments: The authors would like to acknowledge the Ministry of Rural Development and Food and Hellenic Agricultural Organisation "ELGO-DIMITRA" for the courtesy of dataset of bovine carcasses used for analysis.

Conflicts of Interest: The authors declare no conflict of interest.

References

1. Scollan, N.D.; Hocquette, J.-F.; Nuernberg, K.; Dannenberger, D.; Richardson, I.; Moloney, A. Innovations in beef production systems that enhance the nutritional and health value of beef lipids and their relationship with meat quality. *Meat Sci.* **2006**, *74*, 17–33. [CrossRef] [PubMed]
2. Sadowska, A.; Świderski, F.; Kostyra, E.; Rachtan-Janicka, J.; Najman, K. Effect of ageing time on quality characteristics of different bovine muscles. *Int. J. Food Sci. Technol.* **2020**, *55*, 1189–1198. [CrossRef]
3. Crosson, P.; Finneran, E.; McGee, M. Drivers of Profit for Beef Production Systems. Beef Manual Section 2, Chapter 6. 2016. Available online: https://www.teagasc.ie (accessed on 14 July 2020).
4. Polkinghorne, R.; Thompson, J. Meat standards and grading. *Meat Sci.* **2010**, *86*, 227–235. [CrossRef] [PubMed]
5. Regulation (EU) No 1308/2013 of the European Parliament and of the Council of 17 December 2013 establishing a common organisation of the markets in agricultural products and repealing Council Regulations (EEC) No 922/72, (EEC) No 234/79, (EC) No 1037/2001 and (EC) No 1234/2007. *Off. J. Eur. Union* **2013**, *347*, 671. Available online: http://data.europa.eu/eli/reg/2013/1308/2019-01-01 (accessed on 15 July 2020).
6. Agriculture and Horticulture Development Board (AHDB). Review of the EU Carcass Classification System for Beef and Sheep (EPES 0708/01). A Report for DEFRA, AHDB Industry Consulting, November 2008., Kenilworth, Warwickshire, 1-97. Available online: https://webarchive.nationalarchives.gov.uk/20120312125104/http://archive.defra.gov.uk/evidence/economics/foodfarm/reports/carcaseclassification/Full%20Version.pdf (accessed on 24 November 2020).
7. Ellies-Oury, M.-P.; Hocquette, J.-F.; Chriki, S.; Conanec, A.; Farmer, L.J.; Chavent, M.; Saracco, J. Various Statistical Approaches to Assess and Predict Carcass and Meat Quality Traits. *Foods* **2020**, *9*, 525. [CrossRef]
8. Bonny, S.P.F.; Pethick, D.W.; Legrand, I.; Wierzbicki, J.; Allen, P.; Farmer, L.J.; Polkinghorne, R.J.; Hocquette, J.-F.; Gardner, G.E. European conformation and fat scores have no relationship with eating quality. *Animal* **2016**, *10*, 996–1006. [CrossRef]
9. Liu, J.; Chriki, S.; Ellies-Oury, M.P.; Legrand, I.; Pogorzelski, G.; Wierzbicki, J.; Farmer, L.; Troy, D.; Polkinghorne, R.; Hocquette, J.F. European conformation and fat scores of bovine carcasses are not good indicators of marbling. *Meat Sci.* **2020**, *170*, 108233. [CrossRef]
10. Giaretta, E.; Mordenti, A.L.; Canestrari, G.; Brogna, N.; Palmonari, A.; Formigoni, A. Assessment of muscle Longissimus thoracis et lumborum marbling by image analysis and relationships between meat quality parameters. *PLoS ONE* **2018**, *13*, e0202535. [CrossRef]

11. Monteils, V.; Sibra, C.; Ellies, M.-P.; Botreau, R.; De La Torre, A.; Laurent, C. A set of indicators to better characterize beef carcasses at the slaughterhouse level in addition to the EUROP system. *Livest. Sci.* **2017**, *202*, 44–51. [CrossRef]
12. Eurostat. Available online: https://ec.europa.eu/eurostat/en/web/products-eurostat-news/-/DDN-20170201-1 (accessed on 10 June 2020).
13. Hocquette, J.-F.; Ellies-Oury, M.-P.; Lherm, M.; Pineau, C.; Deblitz, C.; Farmer, L.J. Current situation and future prospects for beef production in Europe—A review. *Asian Australas. J. Anim. Sci.* **2018**, *31*, 1017–1035. [CrossRef]
14. OECD/FAO, OECD Food and Agriculture Organization of the United Nations. *OECD-FAO Agricultural Outlook 2019–2028*; OECD Publishing: Paris, France; Food Agriculture Organization of the United Nations: Rome, Italy, 2019. [CrossRef]
15. Farmer, L.J.; Farrell, D.T. Review: Beef-eating quality: A European journey. *Animal* **2018**, *12*, 2424–2433. [CrossRef] [PubMed]
16. Henchion, M.; McCarthy, M.; Resconi, V.C.; Troy, D.J. Meat consumption: Trends and quality matters. *Meat Sci.* **2014**, *98*, 561–568. [CrossRef] [PubMed]
17. Hocquette, J.-F. Is in vitro meat the solution for the future? *Meat Sci.* **2016**, *120*, 167–176. [CrossRef] [PubMed]
18. Ellies-Oury, M.-P.; Lee, A.; Jacob, H.; Hocquette, J.-F. Meat consumption–What French consumers feel about the quality of beef? *Ital. J. Anim. Sci.* **2019**, *18*, 646–656. [CrossRef]
19. World Health Organization (WHO). Diet, Nutrition and the Prevention of Chronic Diseases. Report of a Joint Who/Fao Expert Consulation, Who Technical Report Series 916, 2003, Geneva. Available online: https://www.who.int/dietphysicalactivity/publications/trs916/download/en/ (accessed on 2 November 2020).
20. Geay, Y. Dressing percentage in relation to weight, sex and breed. In *Patterns of Growth and Development in Cattle*; De Boer, H., Martin, J., Eds.; Martinus Nijhoff: London, UK, 1978; Volume 2, pp. 35–46.
21. Pyatt, N.A.; Berger, L.L.; Faulkner, D.B.; Walker, P.M.; Rodriguez-Zas, S.L. Factors affecting carcass value and profitability in early-weaned Simmental steers: I. Five-year average pricing. *J. Anim. Sci.* **2005**, *83*, 2918–2925. [CrossRef]
22. Albertí, P.; Ripoll, G.; Goyache, F.; Lahoz, F.; Olleta, J.; Panea, B.; Sañudo, C. Carcass characterisation of seven Spanish beef breeds slaughtered at two commercial weights. *Meat Sci.* **2005**, *71*, 514–521. [CrossRef]
23. McPhee, M.J.; Oltjen, J.W.; Famula, T.R.; Sainz, R.M. Meta-analysis of factors affecting carcass characteristics of feedlot steers. *J. Anim. Sci.* **2006**, *84*, 3143–3154. [CrossRef]
24. Du Plessis, I.; Hoffman, L. Effect of slaughter age and breed on the carcass traits and meat quality of beef steers finished on natural pastures in the arid subtropics of South Africa. *S. Afr. J. Anim. Sci.* **2007**, *37*, 143–153. [CrossRef]
25. Lucero-Borja, J.; Pouzo, L.; De La Torre, M.; Langman, L.; Carduza, F.; Corva, P.; Santini, F.; Pavan, E. Slaughter weight, sex and age effects on beef shear force and tenderness. *Livest. Sci.* **2014**, *163*, 140–149. [CrossRef]
26. Bonny, S.P.F.; Hocquette, J.-F.; Pethick, D.W.; Farmer, L.J.; Legrand, I.; Wierzbicki, J.; Allen, P.K.; Polkinghorne, R.J.; Gardner, G.E. The variation in the eating quality of beef from different sexes and breed classes cannot be completely explained by carcass measurements. *Animal* **2016**, *10*, 987–995. [CrossRef]
27. Motoyama, M.; Sasaki, K.; Watanabe, A. Wagyu and the factors contributing to its beef quality: A Japanese industry overview. *Meat Sci.* **2016**, *120*, 10–18. [CrossRef] [PubMed]
28. Eriksson, S.; Ask-Gullstrand, P.; Fikse, W.F.; Jonsson, E.; Eriksson, J.-Å.; Stålhammar, H.; Wallenbeck, A.; Hessle, A. Different beef breed sires used for crossbreeding with Swedish dairy cows—Effects on calving performance and carcass traits. *Livest. Sci.* **2020**, *232*, 232. [CrossRef]
29. Sartwelle, J.D.; Outlaw, J.L.; Richardson, J.W. Financial Impacts of Regional Differences in Beef Cattle Operations. Selected Paper prepared for presentation. In Proceedings of the Southern Agricultural Economics Association Annual Meetings, Orlando, FL, USA, 5–8 February 2006.
30. Asem-Hiablie, S.; Rotz, C.A.; Stout, R.; Stackhouse-Lawson, K. Management characteristics of beef cattle production in the Northern Plains and Midwest regions of the United States. *Prof. Anim. Sci.* **2016**, *32*, 736–749. [CrossRef]

31. Commission Delegated Regulation (EU) 2017/1182 of 20 April 2017 supplementing Regulation (EU) No 1308/2013 of the European Parliament and of the Council as regards the Union scales for the classification of beef, pig and sheep carcasses and as regards the reporting of market prices of certain categories of carcasses and live animals. *Off. J. Eur. Union* **2017**, *171*, 74–99. Available online: http://data.europa.eu/eli/reg_del/2017/1182/oj (accessed on 1 November 2020).
32. Macintire, W.H. Résum of Fluoride Research at the University of Tennessee Agricultural Experiment Station, 1920-1954. *J. AOAC Int.* **1955**, *38*, 913–931. [CrossRef]
33. Statistics. Available online: https://www.statistics.gr/el/statistics?p_p_id=documents_WAR_publicationsportlet_INSTANCE_VBZOni0vs5VJ&p_p_lifecycle=2&p_p_state=normal&p_p_mode=view&p_p_cacheability=cacheLevelPage&p_p_col_id=column-2&p_p_col_count=4&p_p_col_pos=3&_documents_WAR_publicationsportlet_INSTANCE_VBZOni0vs5VJ_javax.faces.resource=document&_documents_WAR_publicationsportlet_INSTANCE_VBZOni0vs5VJ_ln=downloadResources&_documents_WAR_publicationsportlet_INSTANCE_VBZOni0vs5VJ_documentID=331385&_documents_WAR_publicationsportlet_INSTANCE_VBZOni0vs5VJ_locale=el (accessed on 30 September 2020).
34. Ataide Dias, R.; Mahon, G.; Dore, G. *EU Cattle Population in December 2007 and Production Forecasts for 2008*; The Publications Office of the European Union: Luxembourg, 2008; Volume 49, pp. 1–12. Available online: https://op.europa.eu/en/publication-detail/-/publication/aaa06517-347f-476c-85cc-60119f06fb0e (accessed on 14 September 2020).
35. Kenny, D.; Murphy, C.P.; Sleator, R.D.; Judge, M.M.; Evans, R.D.; Berry, D.P. Animal-level factors associated with the achievement of desirable specifications in Irish beef carcasses graded using the EUROP classification system. *J. Anim. Sci.* **2020**, *98*. [CrossRef]
36. Albertí, P.; Panea, B.; Sañudo, C.; Olleta, J.; Ripoll, G.; Ertbjerg, P.; Christensen, M.; Gigli, S.; Failla, S.; Concetti, S.; et al. Live weight, body size and carcass characteristics of young bulls of fifteen European breeds. *Livest. Sci.* **2008**, *114*, 19–30. [CrossRef]
37. Xie, X.; Meng, Q.; Ren, L.; Shi, F.; Zhou, B. Effect of cattle breed on finishing performance, carcass characteristics and economic benefits under typical beef production system in China. *Ital. J. Anim. Sci.* **2012**, *11*, 58. [CrossRef]
38. Monteils, V.; Sibra, C. Rearing practices in each life period of beef heifers can be used to influence the carcass characteristics. *Ital. J. Anim. Sci.* **2019**, *18*, 734–745. [CrossRef]
39. Dos Santos, P.V.; Paris, W.; De Menezes, L.F.G.; Vonz, D.; Da Silveira, M.F.; Tubin, J. Carcass physical composition and meat quality of holstein calves, terminated in different finishing systems and slaughter weights. *Ciência Agrotecnologia* **2013**, *37*, 443–450. [CrossRef]
40. Chriki, S.; Picard, B.; Faulconnier, Y.; Micol, D.; Brun, J.-P.; Reichstadt, M.; Jurie, C.; Durand, D.; Renand, G.; Journaux, L.; et al. A Data Warehouse of Muscle Characteristics and Beef Quality in France and A Demonstration of Potential Applications. *Ital. J. Anim. Sci.* **2013**, *12*, e41. [CrossRef]
41. Jurie, C.; Picard, B.; Hocquette, J.-F.; Dransfield, E.; Micol, D.; Listrat, A. Muscle and meat quality characteristics of Holstein and Salers cull cows. *Meat Sci.* **2007**, *77*, 459–466. [CrossRef] [PubMed]
42. Węglarz, A. Meat quality defined based on pH and colour depending on cattle category and slaughter season. colour and pH as determinants of meat quality dependent on cattle category and slaughter season. *Czech. J. Anim. Sci.* **2010**, *55*, 548–556. [CrossRef]
43. Blanco, M.; Ripoll, G.; Delavaud, C.; Casasús, I. Performance, carcass and meat quality of young bulls, steers and heifers slaughtered at a common body weight. *Livest. Sci.* **2020**, *240*, 104156. [CrossRef]
44. Fiems, L.O.; De Campeneere, S.; Van Caelenbergh, W.; De Boever, J.L.; Vanacker, J.M. Carcass and meat quality in double-muscled Belgian Blue bulls and cows. *Meat Sci.* **2003**, *63*, 345–352. [CrossRef]
45. Soulat, J.; Picard, B.; Léger, S.; Monteils, V. Prediction of beef carcass and meat quality traits from factors characterising the rearing management system applied during the whole life of heifers. *Meat Sci.* **2018**, *140*, 88–100. [CrossRef] [PubMed]
46. Hanzelková, Š.; Simeonovová, J.; Hampel, D.; Dufek, A.; Šubrt, J. The effect of breed, sex and aging time on tenderness of beef meat. *Acta Veter- Brno* **2011**, *80*, 191–196. [CrossRef]
47. Agbeniga, B.; Webb, E.C. Influence of carcass weight on meat quality of commercial feedlot steers with similar feedlot, slaughter and post-mortem management. *Food Res. Int.* **2018**, *105*, 793–800. [CrossRef]
48. Petrič, N.; Čepon, M.; Žgur, S. Results of beef carcass grading in Slovenia from 1997 to 2006. *Poljoprivreda* **2007**, *13*, 94–98.

49. Zgur, S.; Cepon, M. The effect of transition from EUROP 5-point scale to 15-point scale beef carcass classification on carcass distribution of young slaughtered bulls in Slovenia. *Poljoprivreda* **2015**, *21*, 207–211. [CrossRef]
50. Méndez, R.D.; Meza, C.O.; Berruecos, J.M.; Garcés, P.; Delgado, E.J.; Rubio, M.S. A survey of beef carcass quality and quantity attributes in Mexico1. *J. Anim. Sci.* **2009**, *87*, 3782–3790. [CrossRef] [PubMed]
51. Kadim, I.; Mahgoub, O.; Al-Ajmi, D.; Al-Maqbaly, R.; Al-Mugheiry, S.; Bartolome, D. The influence of season on quality characteristics of hot-boned beef m. longissimus thoracis. *Meat Sci.* **2004**, *66*, 831–836. [CrossRef] [PubMed]
52. Florek, M.; Litwińczuk, Z.; Skałeck, P. Influence of slaughter season of calves and ageing time on meat quality. *Pol. J. food Nutr. Sci.* **2009**, *59*, 309–314.
53. Kreikemeier, K.K.; Unruh, J.A.; Eck, T.P. Factors affecting the occurrence of dark-cutting beef and selected carcass traits in finished beef cattle. *J. Anim. Sci.* **1998**, *76*, 388–395. [CrossRef] [PubMed]
54. Gray, G.D.; Moore, M.C.; Hale, D.S.; Kerth, C.R.; Griffin, D.B.; Savell, J.W.; Raines, C.R.; Lawrence, T.E.; Belk, K.E.; Woerner, D.R.; et al. National Beef Quality Audit–2011: Survey of instrument grading assessments of beef carcass characteristics1,2. *J. Anim. Sci.* **2012**, *90*, 5152–5158. [CrossRef] [PubMed]
55. Taylor, J.; Ladds, P.; Goddard, M. Carcass weights of cattle in relation to breed, sex, age, region of origin and season. *Aust. J. Exp. Agric.* **1991**, *31*, 745. [CrossRef]
56. Baker, F.H.; Beck, A.M.; Binkerd, E.F.; Blosser, T.H.; Brown, K.I.; Corah, L.R.; Curtis, S.E. *Scientific Aspects of the Welfare of Food Animals.ed.*; Council for Agricultural Science and Technology: Ames, IA, USA, 1981; p. 91.
57. Nielsen, N.A.; Jeppesen, L.F. The Beef Market in the European Union. Available online: https://pure.au.dk/ws/files/114/wp75.pdf (accessed on 14 July 2020).
58. Nogalski, Z.; Pogorzelska-Przybyłek, P.; Sobczuk-Szul, M.; Purwin, C. The effect of carcase conformation and fat cover scores (EUROP system) on the quality of meat from young bulls. *Ital. J. Anim. Sci.* **2019**, *18*, 615–620. [CrossRef]
59. Janiszewski, P.; Borzuta, K.; Lisiak, D.; Powałowski, K.; Samardakiewicz, Ł. Effect of carcass conformation and fatness on beef pH and characterization of the purchase structure of domestic beef cattle. *Sci. Ann. Pol. Soc. Animal Prod.* **2015**, *11*, 65–74.
60. Beriain, M.J.; Indurain, G.; Carr, T.R.; Insausti, K.; Sarries, V.; Purroy, A. Contrasting appraisals of quality and value of beef carcasses in Spain and the United States. *Revue Méd. Vét.* **2013**, *16*, 337–342.
61. Huuskonen, A.K.; Pesonen, M.; Kämäräinen, H.; Kauppinen, R. A comparison of purebred Holstein-Friesian and Holstein-Friesian × beef breed bulls for beef production and carcass traits. *Agric. Food Sci.* **2013**, *22*, 262–271. [CrossRef]
62. Petrovic, M.Z.; Djokovic, R.; Vasilev, D.; Djordjevic, V.Z.; Dimitrijevic, M.; Stajkovic, S.; Karabasil, N. Analysis of Beef Meat Quality in a Slaughterhouse in Raska District. *Meat Technol.* **2018**, *59*, 23–27. [CrossRef]

Publisher's Note: MDPI stays neutral with regard to jurisdictional claims in published maps and institutional affiliations.

 *foods*

Article

The Meat Quality Characteristics of Holstein Calves: The Story of Israeli 'Dairy Beef'

Ariel Shabtay [1], Einav Shor-Shimoni [1], Ala Orlov [1], Rotem Agmon [1], Olena Trofimyuk [1], Ofir Tal [2] and Miri Cohen-Zinder [1,*]

1 Beef Cattle Section, Newe-Ya'ar Research Center, Agricultural Research Organization, P.O. Box 1021, Ramat Yishay 30095, Israel; shabtay@volcani.agri.gov.il (A.S.); einav@volcani.agri.gov.il (E.S.-S.); Lab-IASA@excellence.org.il (A.O.); rotema@volcani.agri.gov.il (R.A.); Tropymuk@gmail.com (O.T.)

2 Institute of Plant Sciences, Newe-Ya'ar Research Center, Agricultural Research Organization, P.O. Box 1021, Ramat Yishay 30095, Israel; ofirt@volcani.agri.gov.il

* Correspondence: mirico@volcani.agri.gov.il

Abstract: Global animal production systems are often criticized for their lack of sustainability and insufficient resilience to ensure food security. The 'farm-to-fork' approach aims at orienting food systems towards the creation of a positive environmental impact, nutritious, healthy, safe and sufficient foods, and fairer economic returns for primary producers. Many countries rely on an imported supply of live animals to fulfill their needs for fresh meat. In Israel, ~60% of the sources of fresh beef come from the import of live animals. In order to encourage sustainable beef production in Israel, the proportion of local beef should be raised at the expense of imported animals. However, for this to be achieved, the superior performance of local beef should be justified. The current study was conducted to compare between the meat quality characteristics of local (Israeli Holstein; N = 205) vs. imported (Australian; N = 169) animals. Generally, while the imported calves presented a higher dressing percentage ($p < 0.0001$), the local animals were characterized by tenderer meat ($p < 0.0001$), longer sarcomeres ($p < 0.0001$), higher a* color attributes and pH ($p < 0.001$), superior cooking ($p = 0.002$) and thawing loss ($p < 0.0001$), higher intra-muscular fat (IMF) content, and a higher PUFA proportion ($p < 0.01$ and $p < 0.0001$, respectively) and PUFA:SFA ratio. The findings shown herein may provide sound arguments for stakeholders and policy makers to facilitate sustainable local beef production in Israel.

Keywords: Holstein; beef; imported animals; local breeds; sustainability; meat quality

Citation: Shabtay, A.; Shor-Shimoni, E.; Orlov, A.; Agmon, R.; Trofimyuk, O.; Tal, O.; Cohen-Zinder, M. The Meat Quality Characteristics of Holstein Calves: The Story of Israeli 'Dairy Beef'. *Foods* **2021**, *10*, 2308. https://doi.org/10.3390/foods10102308

Academic Editors: Benjamin W. B. Holman, Eric Nanthan Ponnampalam and Thierry Astruc

Received: 8 August 2021
Accepted: 23 September 2021
Published: 29 September 2021

1. Introduction

It has become evident that the sustainability of food systems is critical to their resilience to the recurrence of natural disasters and health crises. This principle is being demonstrated by the current COVID-19 outbreak, which triggered disruptions to import/export activities, in parallel with intensifying calls for shorter supply chains and increased local production [1].

The 'farm-to-fork' approach is a comprehensive strategy, seeking to address the challenges of sustainable food systems by orienting food systems towards the creation of a positive environmental impact, nutritious, healthy, safe and sufficient foods, and fairer economic returns, particularly for primary producers. Unique to animal production systems, the 'farm-to-fork' initiative aims at avoiding carbon leakage through animal imports, reducing the environmental and climatic impact of animal production, and improving animal health and welfare [2]. Key steps towards the fulfillment of these aspirations may involve raising the portion of local animal production at the expense of importing live animals. However, within the global animal production system, as in the case of beef cattle, countries that do not produce sufficient fresh meat rely, to a great extent, on an imported supply of live animals to accomplish their needs.

In Israel, ~60% of the sources for fresh beef come from the import of live animals, mainly from Portugal, Australia and Eastern Europe, and the rest stems from free-range beef herds and local dairy farms [3]. During the past several years, trends in the Israeli preference for fresh beef consumption have changed; driven by health awareness, environmental consciousness and ethical considerations, locally produced fresh beef has been favored at the expense of imported meat [3]. Still, more than 200,000 live beef animals are imported to Israel annually. To encourage more sustainable beef production in Israel, sound arguments should be presented for stakeholders and policy makers to control the portion of imported animals. Since free-range beef animals constitute only a small fraction of the fresh meat production chain, and due to the limitation of space, which can hardly exceed its current capacity, the vast majority of fresh meat supplies could originate from fattened male calves and culled cows from local dairy farms. However, as their potential to produce competitive, high-quality beef products is still undefined, many of these animals are transferred out of the country.

Holstein is the premier dairy breed in Israel. Although these animals have primarily been selected for milk production, in many parts of the world they constitute a significant portion of the beef production chain [4]. In the United States alone, dairy breeds (most notably Holstein) make up a substantial quantity of the local feedlot cattle, with as many as 3 to 4 million calves annually grown to contribute approximately 15–20% of the nation's beef supply [5]. In Ireland, of the 1.4 million calves born every year in the local dairy herd, approximately 350,000 are Holstein males, which invariably find their way into the beef sector for rearing and finishing [6]. In comparison with traditional beef breeds, Holstein animals are often criticized for their inferior dressing percentage, as a result of their lower muscle-to-bone ratio [7,8], larger fat deposits (e.g., omental and mesenteric fat) and internal organs (e.g., liver), in order to support their greater lactation requirements [9,10]. In spite of the above, the meat sensory qualities (juiciness, tenderness, flavor, shear force and overall acceptability) of Holstein cows and male calves may be evaluated as indistinguishable from, or even superior to, those of traditional beef breeds [10–16].

Moreover, the genetic architecture of the Holstein breed might highlight its potential to contribute to meat quality phenotypes. Reports from the Animal Quantitative Trait Locus (QTL) database [17] point out the presence of Holstein QTLs associated with milk production traits in the vicinity of the QTLs for meat quality and carcass traits. For example, a QTL associated with milk protein yield on BTA7 in US Holstein cows [18] neighbors two QTLs for meat fat content [19] and meat tenderness [20]. Other QTLs for somatic cell count on BTA24 of Danish Holstein cow [21] overlaps with several QTLs associated with health, production, reproductive traits, and meat and carcass phenotypes [17].

However, the non-supported, yet widespread belief, in scientific studies, that beef of dairy origin is inferior to beef produced from traditional breeds, alongside the rising need, in Israel, to encourage sustainable beef production systems, serves to justify the current study. Herein, we compare and report the meat quality traits of local Holstein vs. imported Australian (*Bos indicus* X *Bos taurus* crosses) male calves.

Our findings indicate the superior meat quality characteristics of local Holstein beef over that of imported Australian calves and, thus, may lay the foundations for the facilitation of sustainable beef production in Israel.

2. Materials & Methods

2.1. Collaboration

The part of the study that involved the selection of animals and meat samples was carried out in collaboration with Bakar Tnuva Ltd. (Beit Shean, Israel). Bakar Tnuva is a major beef-producing stakeholder in Israel. It possesses feedlots, abattoirs and meat factories throughout the country, and employs nutritionists, veterinarians, economists, and food technologists, thus controlling the entire fresh beef supply chain of local and imported animals, for the local consumption.

2.2. Selection of Animals and Meat Samples

Three groups of animals were the source for the meat samples in the present study: *(i)* Israeli Holstein male calves (N = 205), reared and fattened, from weaning to slaughter, in two farms located in the northern part of Israel: Farm 1 (F1; N = 62) and Farm 2 (F2; N = 143), at the age of ~12 months. *(ii)* Farm 3 (F3; N = 169), Australian male calves, imported to Israel at the age of 8–12 weeks. Their genetic background included a mix of *Bos indicus* (mostly Brahman) and *Bos taurus*. The three farms were located in each other's vicinity. The three farms used similar rearing protocols and dietary design, provided by Bakar Tnuva. The diet was composed of ground corn (38.97%), gluten feed (8.13%), cotton seeds (pima; 3.75%), DDG (7.71%), Ca—salt (1.22%), Vit mix (0.17%), Whey (30.87%), wheat hay (4.38%), and wheat straw (4.82%). At the age of 12 months, one day prior to slaughter, the animals were transferred to Bakar Tnuva abattoir, and slaughtered on the following morning under similar conditions, as a single group. The carcasses were then trimmed and gradually chilled, initially at 18 °C, for several hours (±8 h), to avoid cold shortening, then hanged overnight, at 1 °C, in the chilling room. The individual records of live body weight (BW) at slaughter and carcass weight were provided in real time.

On the following morning, the carcasses were cut between the 12 and 13th ribs, and boned out from the rib to the lumbar sacral junction. The *longissimus dorsi et lumborum (LL)* muscle (the posterior side of the *longissimus dorsi* muscle) was taken off the left half-carcass of each animal, and the subcutaneous fat and epimysium were removed from the muscle. The muscles were delivered to the laboratory in isothermal containers at refrigerated temperature, within 1.5 h, for phenotyping of the meat quality characteristics, as described below.

2.3. Muscle Preparation

In the laboratory, the *LL* muscles were immediately cut into 280–300 g steaks. Two of the steaks were sealed in plastic bags and kept in a dry aging refrigerator (at 0–2 °C) for 48 h, then kept at −20 °C for cooking loss (CKL) and Warner-Bratzler Shear Force (WBSF) analysis. A third steak was used for the determination of the pH and color, followed by analyses of chemical composition, sarcomere length (SL), water holding capacity (WHC), total collagen content and fatty acid (FA) profile.

2.4. pH and Color

Meat pH was evaluated at 24 (pH ultimate) and 48 h post-slaughtering (p.s.), using a calibrated pH-meter equipped with a spear-head electrode (Meat pH-meter; Hannah instruments, Model #HI99163; Serial #B0083102). While the pH at 24 h was measured directly on the carcasses, between the 12/13th ribs, in the mid region, the pH at 48 was measured in the aged steaks.

The meat color was measured on the steaks 24 h p.s., following the exposure of their surface to room temperature. A Konica Minolta Chroma Meter CR-410 (KONICA MINOLTA) was used to measure the attributes of lightness (L^*), redness (a^*) and yellowness (b^*). The Chroma meter was operated using illuminant C mode. Prior to the measurements, the device was calibrated, using a white tile standard. Ten replicates were taken from every steak, with special care taken to avoid areas of connective tissue or intramuscular fat.

2.5. Chemical Composition

The chemical analyses were performed on 70 g pieces of meat. These included the determination of intra-muscular fat content (IMF%) using the Soxhlet method [22], crude protein (CP) content via the Kjeldahl method [23], moisture, and ash [24].

2.6. Water Holding Capacity

The water holding capacity was determined according to the Grau-Hamm method [25], with modifications [26]. Briefly, ~0.3 g of ground meat were weighed and placed on laminated plastic white paper, covered with a Whatman filter (No.1). This "cassette" was set between two Plexiglass plates and subjected to a constant pressure of 1 kg for 10 min. The content of the WHC was measured according to the following equation:

$$\% \, WHC = [((X * moisture \, \%) - (X - Y))/X] * 100 \tag{1}$$

where X is initial weight of the meat before pressing (g), and Y is the final weight of the meat after pressing (g).

2.7. Warner Bratzler Shear Force

2.7.1. Sample Preparation

The determination of shear force (SF) was performed according to AMSA 1995 [27] and Wheeler et al. [28]. Briefly, extra fat was removed from the surrounding muscle, and the steaks were frozen at −20 °C, in PA/EVOH/PE plastic bags, following 48 h of ageing. Prior to the analysis, the samples were thawed in the plastic bags, under circulating water, then moved to a 72 °C pre-warmed bath for cooking, until their core temperature reached 70 °C [27]. A temperature probe, HI 9061 (Hanna Food care Digital Thermometer, Bedfordshire, England), placed in the geometric center of a steak, was used to monitor the temperature. Following the cooking process, juices were poured out of the bag (for CKL measurement; see Section 2.8). The meat samples were cooled down and stored overnight at 4 °C.

2.7.2. Coring and SF Measurement

Six cores with a diameter of 1.27 cm (0.5 inch) diameter were cut, on the following morning, from the chilled steaks, in parallel to the longitudinal orientation of the muscle fibers, enabling the shearing action to be perpendicular to the longitudinal orientation of the fibers. The cores were sheared using a V-shaped shear blade with a triangular aperture of 60°, attached to an INSTRON Universal Testing Machine (Model 3343 Instron, UK Ltd. High Wycombe, UK), equipped with a 500 N loading cell, at a crosshead speed of 200 mm/min [27,28]. The Warner Bratzler SF values were calculated based on the average of the 6 cores, using Bluehill software. The peak force required to cut through the fibers was expressed in Newtons (N).

2.8. Cooking Loss and Thawing Loss

Cooking loss was expressed as the percentage of weight difference before and after cooking, according to the following equation [29]:

$$\% \, Cooking \, Loss = \left[\frac{X - Y}{X}\right] * 100 \tag{2}$$

where X = weight of raw steak and Y = weight of cooked steak.

The thawing loss (TL), the loss of meat fluids due to thawing, was determined as described by Honikel 1998 [30].

2.9. Sarcomere Length (SL)

The sarcomere length was determined on thawed meat samples, according to the method used by Cross et al. [31]. The solutions for fiber fixation were prepared according to Koolmees et al. [32]. Briefly, samples without tendons were selected in triplicates from each tissue, and excised in small pieces (2.0 cm × 1.0 cm × 1.0 cm), in a longitudinal orientation of the fibers. An incision was made with a scalpel in the middle of each sample. The pieces were placed in 50 mL tubes fixed with 30 mL of 5% glutaraldehyde solution, for 4 h at 40 °C, followed by overnight fixation at 40 °C, with a 30 mL 0.2 M sucrose solution. Thereafter, flat tweezers were used to gently separate long and thin fibers from the samples. The separated fibers (about 15–20) were placed in a mortar that contained 3–4 mL 0.2 M sucrose solution, and ground with the pestle to a consistent "soup". The sarcomere length was determined by laser diffraction, using a neon-helium laser (HeNe Laser; λ = 632.8 nm), which was mounted on an optics bench with a specimen-holding device and a screen, as previously described by Cross et al. [31]. The length of at least 10 projected sarcomeres was measured with a ruler for each biological sample. The SL was calculated by the following equation, as provided by Cross et al. [31]:

$$\mu = \frac{0.6328 \times D \times \sqrt{\left(\frac{T}{D}\right)^2 + 1}}{T} \quad (3)$$

in which D was the distance, in mm, from the specimen to the diffraction pattern screen and T referred to half of the separation distance (in mm) between the diffraction bands.

2.10. Total Collagen Content

The determination of the total collagen content was based on AOAC 990.26 [33], with adaptations from Starkey et al. [34]. Briefly, 20 g of meat were removed and trimmed of external fat and connective tissue. The meat was minced into a paste, and frozen in petri dishes for 3 h, at −20 °C, prior to lyophilization. The lyophilized samples were ground into powder, using a mortar and pestle. Triplicates of freeze dried muscle powder weighing 0.10 g were mixed with three ml of 3.5 M H2SO4 for subsequent hydrolyzation at 105 °C, for 16 h. Hydrolysis was terminated by the addition of 1.5 M NaOH to the hydrolyzed filtrate, prior to the determination of the hydroxyproline content, from a standard curve, as in Starkey et al. [34].

The content of Hydroxyproline (H) in the sample was calculated: H = h × 0. 25/m.

In which the h-hydroxyproline content as read from the calibration curve; 0.25—coefficient, based on the dilution factor and transition between the units; and m—weight of sample portion.

To convert hydroxyproline to total collagen, the following equation was used: total collagen, mg/g = H × 8 (with collagenous connective tissue containing 12.5% hydroxyproline, if nitrogen-to-protein factor is 6.25).

2.11. Fatty Acid Profile

The analysis of the FA profile was performed on the lyophilized muscles, as previously described [35]. Lipids were extracted from 1 g sample powder in a hexane: isopropanol solvent mixture, in accordance with Hara and Radin [36]. An aliquot of 40 mg of the lipid fraction was trans-methylated in accordance with Christie (1982) [37], with modifications [38]. Gas chromatography of the fatty acid methyl esters (FAME) was performed with a Hewlett Packard 6890 system, equipped with HP Chemstation software for peak integration. We used a Supelco SP-2560, 100-m fused silica capillary column of 0.25 mm i.d., with ultra-high purity helium carrier gas, at a flow rate of 20 mL/min. The injector and flame-ionization detector (FID) temperatures were 250 °C and 260 °C, respectively. The splitting ratio to the detector was 1:50. The oven temperature schedule was as follows: 140 °C for 5 min, T increase to 175 °C at 4 °C/min, constant 175 °C for 25 min, T increase to 220 °C at 4 °C/min, and constant 220 °C for 20 min. The total run time was 70 min. Standard FAME preparations (Sigma-Aldrich) were injected separately to relate the peaks to the FA. The FAME preparations used were methyl esters of: C10:0, C12:0, C14:0, C14:1, C16:0, C16:1, C18:0, C18:1t9, C18:1t10, C18:1t12, C18:1c9, C18:1c11, C18:1c12, C18:2c9c12, C18:3c6c9c12 (γ-linolenic), C18:3c9c12c15 (α-linolenic), C18:2t10c12 and C18:2c9t11 (conjugated linoleic acid; CLA), and C20:4c5c8c11c14 (arachidonic).

2.12. Statistical Analysis

All the variables met our assumptions of normality and were compared among the three farms or two breeds, using a one-way ANOVA, followed by a Bonferroni Multiple Comparison Test ($p < 0.05$). Pearson correlations between the meat quality phenotypes were calculated using the CORR procedure. Letters are used in the figures/tables to indicate pairwise differences identified through this analysis. All the statistical comparisons were conducted using SPSS version 21.0.

3. Results & Discussion

Sustainable food systems are designed to provide healthy and nutritious food that is available, accessible, and affordable to everyone for generations to come [39]. At the same time, sustainable systems, as engines of growth, nourish a continuous dialog between social, economic, and environmental components by: *(i)* encouraging local production and distribution infrastructures; *(ii)* protecting farmers and other workers (e.g., paying their salaries), consumers and entrepreneurs (e.g., profits or returns on assets); *(iii)* minimizing their negative effect on the natural environment. [40].

Many of these aspects should indeed be taken into consideration while aiming to promote sustainable beef production in Israel. However, in order to encourage stakeholders and decision-makers to set policies that will encourage positive transformations towards sustainable beef production, identifying the "intrinsic" properties of the food system that will ensure that its essential outcomes are continuously maintained [41] is a prerequisite. A cardinal obstacle ahead of this enterprise is the massive import of live beef animals to Israel [3]. Thus, an initial step to facilitate the above initiative would be through uncovering the advantages of local over imported beef production.

In the current study, we compared between key meat quality phenotypes, in the *LL* muscle, of Israeli Holstein and imported Australian male calves.

3.1. Carcass Production

Live bodyweight (BW), carcass weight, and dressing percentage are presented in Table 1. Although live BW did not differ among farms, nor between breeds, carcass weight and dressing percentage were significantly higher in the Australian calves (F3) in comparison with Holstein (F1 and F2; $p \leq 0.0001$; Table 1). Although a slight significant difference in dressing percentage was revealed between F1 and F2 animals, statistical adjustment to the breed effect highlighted the superior carcass yield of the Australian calves ($p \leq 0.0001$; Table 1). Indeed, crosses of beef X beef or beef X dairy animals are expected to produce heterogeneous progeny with higher growth rates and dressing percentage compared to dairy-bred cattle [42–44]. On the other hand, dairy-selected animals (e.g., Holstein Frisian) are known for their higher proportions of non-carcass parts, as external (head/feet/tail) and internal organs, offal fats and gastrointestinal tract [45], resulting from their engagement in the process of milk production [9,46–48].

Table 1. Carcass production of local Holstein (HOL) and imported Australian (AUS) male calves, adjusted to farm and breed effects. Farm 1 (F1; HOL; N = 62); Farm 2 (F2; HOL; N = 143); Farm 3 (F3; AUS; N = 169).

	FARM			*BREED*		*p-Value*	
Trait	**F1**	**F2**	**F3**	**HOL**	**AUS**	**FARM**	**BREED**
Live BW weight (kg)	519.6 ± 38.0^{a}	534.8 ± 58.8^{a}	536.9 ± 91.8^{a}	530.2 ± 55.5^{a}	536.9 ± 91.8^{a}	0.269	0.380
Carcass weight (kg)	279.1 ± 22.6^{a}	293.0 ± 32.2^{b}	309.5 ± 58.3^{c}	288.8 ± 30.3^{a}	309.5 ± 58.3^{b}	<0.0001	<0.0001
Dressing percentage (%)	53.7 ± 1.2^{a}	54.8 ± 2.5^{b}	57.5 ± 2.1^{c}	54.5 ± 2.2^{a}	57.5 ± 2.1^{b}	<0.0001	<0.0001

Different letters indicate significant differences between farms or breeds ($p < 0.0001$). BW = body weight; Dressing percentage was calculated as the ratio between hot carcass weight and live BW.

3.2. Technological Parameters of Raw and Cooked Meat

3.2.1. pH and Color

Among others, ultimate pH (pH_u; measured 24 h post-slaughter) is a major technical attribute that drives consumers' purchasing decisions about meat. It is influenced by different factors, such as individual cows' genetic background, their on-farm nutritional regime, and a variety of biochemical events occurring pre-and post-slaughter (e.g., the level of stress prior to slaughter and post-slaughter processing) [49].

Differences in initial pH and the rate of its decline mostly affect sarcomere shrinkage, protein denaturation and myofibrillar lattice spacing [50].

In the current study, the pH_u values ranged from 5.74 ± 0.12 (F3) to 5.88 ± 0.28 (F1), and were affected by both farm ($p < 0.0001$) and breed ($p = 0.0002$; Table 2). While the pH_u of the Holstein calves from F1 and F2 did not differ, statistical adjustment to breed revealed higher values in the Holstein meat ($p = 0.0002$; Table 2). Similar breed and farm effects were also determined 48 h post-slaughter ($p < 0.0001$; Table 2). These results, typical for meat without DFD or PSE syndromes, were in agreement with the pH values obtained in the *LL* muscle in other studies [51–53].

Table 2. Effects of farm and breed on pH, color attributes, thawing loss (TL), and water holding capacity (WHC) of raw meat, and cook loss (CKL) of thermally treated meat, from Holstein (HOL) and Australian (AUS) male calves. Farm 1 (F1; HOL; N = 62); Farm 2 (F2; HOL; N = 143); Farm 3 (F3; AUS; N = 169).

	FARM			*BREED*		*p-Value*	
	F1	F2	F3	HOL	AUS	FARM	BREED
Raw Beef							
pH_u †	5.88 ± 0.28 ^a^	5.82 ± 0.21 ^a^	5.74 ± 0.12 ^b^	5.85 ± 0.24 ^a^	5.74 ± 0.12 ^b^	<0.0001	0.0002
pH_{48h} ††	5.70 ± 0.40 ^a^	5.69 ± 0.20 ^a^	5.53 ± 0.16 ^b^	5.69 ± 0.28 ^a^	5.53 ± 0.16 ^b^	<0.0001	<0.0001
Color							
L^*	38.87 ± 3.34 ^a^	38.68 ± 4.14 ^a^	40.97 ± 3.50 ^b^	38.70 ± 3.91 ^a^	40.97 ± 3.50 ^b^	<0.0001	<0.0001
a^*	15.85 ± 1.73 ^a^	15.83 ± 1.97 ^a^	14.63 ± 1.93 ^b^	15.83 ± 1.90 ^a^	14.63 ± 1.93 ^b^	<0.0001	<0.0001
b^*	3.25 ± 0.98 ^a^	3.02 ± 1.18 ^a^	2.59 ± 1.28 ^b^	3.08 ± 1.10 ^a^	2.59 ± 1.28 ^b^	0.0002	<0.0001
TL (%)	3.30 ± 1.05 ^a^	4.14 ± 2.00 ^b^	5.82 ± 2.84 ^c^	3.88 ± 1.05 ^a^	5.82 ± 2.84 ^b^	<0.0001	<0.0001
WHC (%)	45.01 ± 4.65 ^a^	43.57 ± 3.13 ^b^	43.11 ± 3.40 ^b^	43.68 ± 3.75 ^a^	43.11 ± 3.40 ^a^	0.002	
Thermally treated beef							
CKL (%)	22.14 ± 4.40 ^a^	22.12 ± 3.30 ^a^	23.26 ± 3.18 ^b^	22.12 ± 3.66 ^a^	23.26 ± 3.18 ^b^	0.007	0.0017

Different letters indicate significant differences between farms or breeds ($p < 0.001$). † pH ultimate measured in the carcasses 24 h post-slaughter; †† pH measured in the steaks 48 h post-slaughter; * color attributes measured in the steaks 24 h post-slaughter.

The variation in pH_u mostly affected the meat color, an important technological and visual property of meat quality [4]. The light reflected from the surface of the meat is of primary importance, as it affects, to a great extent, consumers' perceptions and, hence, their purchasing decisions [50].

In the present study, the meat color was determined 24 h post-slaughter (Table 2), and included attributes of brightness (L^*), redness (a^*) and yellowness (b^*). These attributes did not differ between Holstein calves from F1 and F2, but varied significantly when compared to Australian calves from F3 (Table 2). Statistical adjustment to breed revealed differences in meat color characteristics (L^*, a^* and b^*) between the two breeds ($p \leq 0.0001$; Table 2). More specifically, while the Holstein meat had higher redness and yellowness scores, the Australian meat was brighter ($p < 0.0001$; Table 2). The color attributes reported herein were only in relative agreement with those presented by others [4,53–55], presumably due to environmental variations, such as on-farm rearing management and dietary regime, especially towards the end of the growing period. However, the most plausible effect seems technological; while in many studies color attributes are determined 14 days p.s., the data presented in the current study refer to 24 h p.s. Nevertheless, the attributes most appreciable to consumers favored the local Holstein meat, predicting its possible preference over imported Australian meat.

3.2.2. Thawing Loss and Cooking Loss

The flow of exudates from the raw meat of the Australian calves was significantly stronger than the Holstein, as evidenced by the measurement of TL (5.82 ± 2.84% and 3.88 ± 1.05%, respectively; $p < 0.0001$; Table 2). Although the farm effect revealed a difference between the TL scores of the Holstein calves from F1 and F2, it could not obscure the significant distinction among breeds, highlighting the superiority of this characteristic in the meat of the Holstein calves. The loss of exudates following a thermal treatment is evaluated as the CKL. Here, the loss of exudates from the meat of Australian calves was higher compared to the meat of the Holstein animals, when both breed (p = 0.0017) and farm (p = 0.007) effects were studied (Table 2). The CKL values detected in Holstein *LL* muscle were in agreement with those reported by others [4,15,53,54]. Taken together, following major processes of thawing and cooking that indicated smaller proportions of exudate loss in Holstein samples, both the TL and the CKL parameters demonstrated advantages, from which the industry of local beef may benefit [56].

3.2.3. Water-Holding Capacity

Water-holding capacity is defined as the ability of fresh meat to retain its own water during cutting, heating, grinding and pressing and during transport, storage and cooking [57]. Poor WHC results in high drip and purge loss, which may represent a significant loss of weight from carcasses and cuts and may affect the yield and quality of processed meat [58,59].

While no differences in WHC were found when the breed effect was studied (Table 2), statistical adjustment to farm revealed higher WHC values in F1 compared to those in F2 and F3 calves (p = 0.002; Table 2). It is noteworthy that the WHC values of the Holstein calves were lower than those previously reported for that breed and muscle [4,53,54]. Based on the above, and as changes in WHC levels may be caused by differences in the volume of the myofibrils, resulting from variations in the muscle's inter-filament spacing [60], it is tempting to assume that management conditions on farms, rather than genetic factors, may affect this parameter.

As indicated above, WHC is commonly found in association with pH values, post-mortem. Specifically, the power of muscle proteins to bind water becomes weaker when pH declines, due to their 'movement' towards their isoelectric point [61]. On the other hand, at higher pH, WHC increases due to an increase in the overall negative charge of proteins, resulting in repulsion of the filaments and more space for the water molecules [61,62].

The above trend was also exemplified in our study, where the pH_u rate was positively associated with WHC in the Holstein animals (R^2 = 0.15; $p \leq 0.01$, data not shown). Within this association, F1 were characterized by higher WHC (45.01 $\pm$ 4.65%) compared to F3 calves (43.11 $\pm$ 3.40%), which did not differ from the F2 animals (Table 2).

3.2.4. Chemical Composition

The chemical composition of muscles is relatively constant and includes about 75% water, 19–25% proteins, and 1–2% minerals and glycogen. The lipid fraction of muscle, however, may vary greatly between species, individuals and muscles, as well as cuts from the same animal [63,64].

The lipid fraction of muscle tends to vary highly between species, individuals, muscles, and cuts from the same animal [63,64]. Generally, *Bos taurus* types present higher marbling than *Bos indicus* breeds [65]. Examples include Brahman feedlot-fed steers, with an IMF content of 3.1%, in comparison to *Bos taurus* breeds, such as Angus (6.2%) [66,67] and Hereford (8.3%; fed forage with or without supplementation of high energy diet) [68], at similar ages and in the same muscle. Regardless of the divergence between *taurus* and *indicus*, Asian cattle breeds are known for their high IMF content [64]; Wagyu (Japanese Black cattle) and Hanwoo (Korean) feedlot steers had an IMF content of 34.3% and 13.3%, respectively, in their *longissimus thorasis* muscle [69,70].

These inter-breed differences in IMF were also demonstrated in the current study, in which the *LL* muscles of Holstein calves exhibited higher IMF content than those of their Australian counterparts (p = 0.002; Table 3). These findings are not surprising in light of the reported comparative marbling physiology, referring to the greater proportion of the marbling fleck area and number of marbling flecks of Holstein beef, at a similar slaughter age—12 months—to that reported here, even in comparison to typical beef breeds [67].

With respect to the positive effects of IMF on meat organoleptic characteristics, such as flavor, juiciness and tenderness, firmness, and overall acceptability by consumers [71], the findings presented herein rank the local breed in an advantageous position to satisfy consumers' sensory choices [16].

Table 3. Effects of farm and breed on proximate composition of meat from Holstein (HOL) and Australian (AUS) male calves. Farm 1 (F1; HOL; N = 62); Farm 2 (F2; HOL; N = 143); Farm 3 (F3; AUS; N = 169).

	FARM			*BREED*		*p-Value*	
Trait (%)	**F1**	**F2**	**F3**	**HOL**	**AUS**	**FARM**	**BREED**
Moisture	73.90 ± 0.76 [a]	73.59 ± 1.00 [a]	73.20 ± 1.10 [b]	73.7± 0.94 [a]	73.20 ± 1.10 [b]	<0.0001	<0.0001
Protein	22.92 ± 1.05 [a]	22.37 ± 0.85 [b]	22.27 ± 0.85 [b]	22.47 ± 0.96	22.27 ± 0.85	<0.0001	0.305
IMF	2.78 ± 0.96 [a,b]	2.82 ± 1.02 [a]	2.51 ± 0.80 [b]	2.80 ± 1.00 [a]	2.51 ± 0.80 [b]	0.009	0.002
Ash	1.20 ± 0.07 [a]	1.30 ± 0.18 [b]	1.27 ± 0.11 [b]	1.27 ± 0.16	1.27 ± 0.11	<0.0001	0.955

Different lower case letters indicate significant differences between the two breeds for each measurement ($p < 0.01$).

3.3. Characteristics of Meat Tenderness: Shear Force, Sarcomere Length and Total Collagen

Tenderness is the most important sensory attribute by which consumers judge the quality of their meat [16]. It is a variable phenotype, mostly influenced by genetic factors, muscle characteristics at slaughter [72], or post-mortem changes induced by ageing [16]. Indeed, inconsistency in meat tenderness is considered a major obstacle facing the beef production industry [73]. The phenotyping of tenderness is performed via sensorial panel testing or instrumental measurements [74], but most often the two are conducted in parallel, and correlations between them are determined [75].

In our study, tenderness, evaluated instrumentally, was significantly higher in the meat of Holstein calves, as judged by the lower SF values, statistically adjusted to the farm and breed effects (41.5 ± 9.66 N vs. 46.5 ± 9.27 N, respectively; $p < 0.0001$; Table 4). Differences in meat tenderness among *Bos taurus* and *Bos indicus* breeds, with the advantage held by the former, are well documented [76]. This phenomenon mostly stems from genetic variations in the gene encoding the calpastatin proteolytic enzyme, which are found in association with the rate and extent of muscle proteolysis postmortem [76–79]. Both the calpastatin's rate of activity and meat tenderness are moderately-to-highly-inheritable, and genetically correlated [78,80].

Table 4. Effects of farm and breed on tenderness characteristics of the *longissimus lumborum* (*LL*) muscle of Holstein (HOL) and Australian (AUS) male calves. Farm 1 (F1; HOL; N = 62); Farm 2 (F2; HOL; N = 143); Farm 3 (F3; AUS; N = 169).

	FARM			*BREED*		*p-Value*	
Trait	**F1**	**F2**	**F3**	**HOL**	**AUS**	**FARM**	**BREED**
SF (N)	41.3 ± 10.76 [a]	41.6 ± 9.17 [a]	46.5 ± 9.27 [b]	41.5 ± 9.66 [a]	46.5 ± 9.27 [b]	<0.0001	<0.0001
SL (μM)	2.22 ± 0.30 [a]	2.10 ± 0.27 [b]	1.98 ± 0.32 [c]	2.14 ± 0.29 [a]	1.98 ± 0.32 [b]	<0.0001	<0.0001
Total collagen (mg/g)	2.88 ± 0.64 [a]	2.81 ± 0.84 [a]	2.70 ± 0.83 [a]	2.82 ± 0.82 [a]	2.70 ± 0.83 [a]	0.410	0.220

Different lower case letters indicate significant differences between the two breeds for each measurement ($p < 0.0001$).

A comparative analysis of the meat quality characteristics among 15 muscles categorized *LL* as one of the most tender (WBSF < 35 N) muscles in Holstein male calves [4], making it a legitimate target for inter-breed comparisons. Accordingly, and based on tenderness classification, established upon the relationship between instrumental measurements and consumer perception [73], the *LL* muscle of the Holstein calves in the current study, may be ranked as tender, while that of Australian may be ranked as intermediate. Similarly, other studies of Holstein animals observed relatively high tenderness in the same muscle [4,16].

Moreover, a comparison between the meat quality characteristics of Holstein and Salers (dual breed, often used for beef) cull cows, following 14 days of ageing, did not reveal differences in sensory qualities, such as tenderness and juiciness [16]. These findings, which are in agreement with Monso'n, et al. [81], indicate that by applying the ageing process to local Holstein meat, its organoleptic characteristics may be improved beyond their genetic potential.

Another attribute of meat tenderness is SL. It is often used as a post-rigor indicator [82], representing a positive association with meat tenderness and WHC [83]. In our study, SL differed significantly ($p < 0.0001$) between the two breeds (Table 4). When the farm effect was studied, longer sarcomeres were measured in the muscles of Holstein calves from both farms (2.22 ± 0.30 μM and 2.10 ± 0.27 μM, respectively), compared to the SL of Australian calves from F3 (1.98 ± 0.32 μM; $p < 0.0001$).

Total collagen content is a key parameter for the evaluation of meat tenderness [84]. The increase in the stability of cross-linking between collagen molecules, which is determined by growth rate, nutrition and genetics [84], affects the toughness of meat [85]. Studies in cattle have shown a large variation in the content of collagen, which differs between muscles, breeds, and animals of different ages. No differences in the content of total collagen were found in the current study, between farms or breeds. The values ranged between 2.70 ± 0.83–2.88 ± 0.64 mg/g (Table 4), and were comparable to or lower than those in other breeds [86,87], including Holstein, that were exclusively reared on grass pasture [88]. The lower amounts of total collagen reported herein might have resulted from the relatively younger age of the calves (~12 months). Indeed, the proportion of maturity to reducible crosslinks increases with age, resulting in less tender meat in older animals [89]. Nevertheless, Archile-Contreras et al. [90] suggested that variations in collagen turnover might be affected by the position of the muscle in the animal's body and could, therefore, influence meat tenderness. Accordingly, less positional muscles, such as the *longissimus dorsi*, are characterized by reduced collagen concentrations and cross-links, and therefore, produce tenderer meat compared to locomotive muscles, such as *semimembranosus* (SM) [91].

3.4. Fatty Acid Composition

The nutritional, health and sensory qualities of meat are, to a great extent, determined by its FA composition, which is influenced by various factors, such as diet, breed, age and the level of fat content in the muscle [92]. With respect to their nutritional significance, the fact that meat is a major source of dietary SFA, which is implicated in diseases associated with modern life [93], has triggered increased interest in ways of manipulating the FA composition of meat. In particular, efforts have been made to increase the dietary ratio of polyunsaturated fatty acids (PUFA) to SFA to above 0.4, and to decrease the ratio of n-6:n-3 (i.e., alpha-linolenic to linoleic acid) PUFA to less than 4 [93]. In spite of some clear effects of diet on the FA composition of tissues [94], a combination of bovine genetic background and dietary manipulation could favor particular FAs [95].

In the current study, we estimated the effects of farm and breed on the profile of FA in the *LL* muscle of Holstein and Australian calves. The proportions of short chain saturated FAs, including capric (C:10; $p = 0.0012$), lauric (C12:0; $p < 0.0001$), myristic (C14:0; $p = 0.0002$), pentadecanoic (C15:0; $p < 0.0001$), palmitic (C16:0; $p = 0.0052$) and heptadecanoic (C17:0; $p < 0.0001$) acids, were significantly higher in the fat of Australian calves, when the breed effect was studied (Table 5). Surprisingly, in spite of the improved meat tenderness of Holstein calves (Table 4), the proportion of their stearic acid (C18:0) was significantly higher ($p = 0.0008$) (Table 5). Stearic acid is indeed known for its high and positive correlation with the melting point of fat, which in turn reflects the firmness of meat. However, the best prediction of firmness was provided by the ratio of stearic acid to linoleic acid (18:0:18:2) [93], which in the present study was lower for the Holstein calves (3.15 ± 0.86 vs. 3.97 ± 1.07; $p < 0.0001$). The total proportion of mono-unsaturated FAs (MUFAs) did not differ between breeds (Table 5). However, while oleic (C18:1n9c; $p = 0.0261$) and C18:1n10c ($p = 0.016$) FA were higher in the meat of Australian calves, vaccenic (C18:1n11c; <0.0001) and C18:1n12c ($p < 0.0001$) FAs were higher in the meat of Holstein calves (Table 5). Vaccenic acid is formed in the rumen as a result of partial bio-hydrogenation, and is a precursor for tissue-conjugated linoleic acid (CLA), a well-recognized, health-promoting FA [96], whose proportions as observed in this study were not in favor of the Holstein calves, presumably due to breed-specific differences in the activity of tissue stearoyl CoA desaturase (SCD). Unlike CLA, the total proportion of PUFA was significantly higher in the *LL* muscle of the Holstein calves ($p < 0.0001$). Specifically, higher proportions were revealed for linolelaidic (C18:2n6t; $p < 0.001$), linoleic (C18:2n6c; $p < 0.0001$), α-linolenic (C18:3n3; $p = 0.0015$), Eicosatrienoic (C20:3n6; $p < 0.0015$) and arachidonic (C20:4n6; $p = 0.0047$) acids (Table 5).

Table 5. Proportion of fatty acids in the *longissimus dorsi et lumborum (LL)* muscle of Holstein (HOL N = 110) and Australian (AUS; N = 100) calves.

Proportion of Fatty Acids	HOL	S.D.	*AUS*	S.D.	*p-Value*
C10:0	0.040	0.0002	0.035	0.0003	1.6×10^{-1}
C12:0	0.042	0.0003	0.058	0.0004	1.5×10^{-3}
C14:0	2.716	0.0047	3.242	0.0074	9.2×10^{-9}
C14:1	0.488	0.0013	0.632	0.0023	1.4×10^{-7}
C15:0	0.304	0.0006	0.398	0.0012	2.7×10^{-11}
C16:0	25.56	0.0180	26.27	0.0222	1.2×10^{-2}
C16:1	3.266	0.0047	3.539	0.0063	4.8×10^{-4}
C17:0	0.802	0.0028	1.175	0.0024	2.2×10^{-20}
C17:1	0.428	0.0016	0.723	0.0019	4.9×10^{-26}
C18:0	17.13	0.0189	16.32	0.0244	7.4×10^{-3}
C18:1n9t	3.376	0.0193	2.372	0.0093	2.6×10^{-6}
C18:1n9c	35.02	0.0302	36.12	0.0343	1.5×10^{-2}
C18:1n10c	1.715	0.0032	1.952	0.0031	1.2×10^{-7}
C18:1n11c	0.553	0.0019	0.323	0.0016	3.0×10^{-18}
C18:1n12c	0.317	0.0013	0.387	0.0018	1.6×10^{-3}
C18:2n6t	0.130	0.0015	0.109	0.0010	2.2×10^{-1}
C18:2n6c	5.862	0.0174	4.467	0.0172	1.9×10^{-8}
C20:0	0.041	0.0006	0.000	0.0000	4.5×10^{-12}
C18:3n3	0.277	0.0008	0.358	0.0015	3.5×10^{-6}
CLA c9,t11	0.239	0.0015	0.273	0.0011	5.7×10^{-2}
C22:0	0.237	0.0017	0.000	0.0000	1.7×10^{-28}
C20:3n6	0.067	0.0016	0.198	0.0015	9.8×10^{-9}
C20:4n6	1.017	0.0073	0.837	0.0047	3.4×10^{-2}
C22:2	0.061	0.0013	0.000	0.0000	3.1×10^{-6}
C24:0	0.000	0.0000	0.086	0.0010	1.0×10^{-12}
C22:6n3	0.202	0.0026	0.117	0.0022	1.1×10^{-2}
Short FA	0.082	0.0004	0.093	0.0006	1.2×10^{-1}
SFA	46.88	0.0217	47.58	0.0415	1.3×10^{-1}
MUFA	45.16	0.0256	45.75	0.0440	2.5×10^{-1}
PUFA	7.854	0.0262	6.359	0.0229	1.6×10^{-5}
PUFA/SFA	0.170	0.0610	0.135	0.0530	2.7×10^{-5}
C18:0 (stearic): C18:2n6c (linoleic)	3.150	0.8561	3.973	1.0693	7.33×10^{-9}

FA: Fatty Acid; SFA: saturated fatty acids; MUFA: mono unsaturated fatty acids; PUFA: polyunsaturated fatty acids; S.D: standard deviation.

PUFAs are known to possess anticarcinogenic and hypolipidemic properties [97], and to act as modulators of different transcription factors, providing, at least partially, the metabolic link between dietary PUFA intake, health, and the progression of chronic diseases [98]. However, taking into account the health-promoting capacity of PUFA and the deleterious dietary potential of SFA, nutritionists have suggested that the desired ratio of PUFA to SFA should exceed 0.4, while in some meats, naturally, it is around 0.1 [93]. While in accordance with this notion, the PUFA-to-SFA ratio reported herein again highlights the superiority of local Holstein over imported Australian meat ($p < 0.0001$; Table 5).

4. Conclusions

In summary, from a scientific perspective, the present study provides an understanding of the 'beefy' qualities of the dairy Israeli Holstein, via the characterization of the organoleptic, physical and technological properties of its meat. Aiming to lay the foundations for sustainable beef production system in Israel, the current study evaluated the comparative potential of local Israeli Holstein and genetically mixed *Bos indicus* X *Bos taurus* calves, imported from Australia, to produce qualitatively fresh meat. It was found that the meat produced by the local breed was superior according to quality- and health-related characteristics. Specifically, while the Australian calves demonstrated a superior dressing percentage, the Holstein meat was characterized by higher tenderness, greater IMF content, longer sarcomeres, improved PUFA-to-SFA ratio (Figure 1) and superior technological parameters of raw and thermally treated meat.

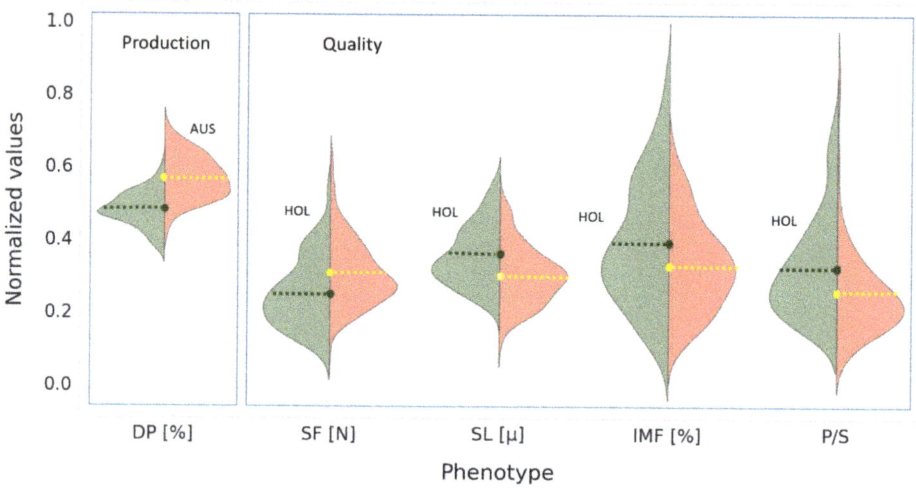

Figure 1. An illustrative violin plot of production (dressing percentage; DP) and meat quality characteristics (shear force, SF; sarcomere length, SL; intra-muscular fat content, IMF; polyunsaturated-to-saturated fatty acid ratio, P/S), comparatively evaluated in Holstein (green; HOL) and Australian (pink; AUS) male calves. The values of each trait are normalized (0,1). Green and yellow dots with dashed lines indicate the mean value of each variable, for Holstein and Australian plots, respectively. The two box plots outlined in blue differentiate between the advantages in production of AUS (left) and meat quality of HOL (quality; right) characteristics.

The results presented herein may provide sound arguments for stakeholders and policy makers to facilitate sustainable local beef production in Israel. Once implemented, such production may comply with the 'farm-to-fork' approach by emphasizing potential motives, such as improved animal welfare, traceable, transparent and shorter supply chains, positive environmental impact, the production of nutritious, healthy, safe and sufficient foods, and fairer economic returns for primary producers.

In a broader sense, the current study may serve as an example of how the 'farm-to-fork' approach may also be implemented in other parts of the world. In developing countries, for instance, fulfilling the productive potential of imported animals depends upon favorable but unsustainable conditions. Hence, adjusting to an integrative model for sustainable indigenous breed production, which is based on their economic and biological efficiencies, might orient local food systems towards a positive environmental impact, improved product quality and animal health and welfare. Moreover, it may enable smallholder farmers to maintain their animals in the long run, and provide an income for poor farmers while maintaining their cultural identity.

Author Contributions: Conceptualization and methodology, A.S. and M.C.-Z.; formal analyses, A.O., R.A., E.S.-S. and O.T. (Olena Trofimyuk and Ofir Tal); validation, A.S. and M.C.-Z.; resources, A.S. and M.C.-Z.; writing, review and editing, A.S. and M.C.-Z. All authors have read and agreed to the published version of the manuscript.

Funding: This study was supported by the Israeli Chief Scientist, Ministry of Agriculture, Project # 362-0486-16, and by the Israeli Dairy Board Foundation, Project #.362-0457-16.

Institutional Review Board Statement: Ethical review and approval were waived for this study, since it did not involve living animals; meat cuts (steaks) and relevant data on carcass and live weight performances were provided by Bakar Tnuva Ltd, Beit Shean, Israel.

Informed Consent Statement: Research does not involve human studies.

Data Availability Statement: The current study does not report any supporting data.

Acknowledgments: We thank Wassim Geraisy, the Chief Veterinarian in "Bakar Tnuva Ltd." abattoir for his collaboration, which enabled to obtain meat samples, and Ido Izhaki, for his assistance in the statistical analyses.

Conflicts of Interest: The authors declare no conflict of interest.

References

1. Food and Agriculture Organization of the United Nations. *Meat Market Review*; FAO: Rome, Italy, 2020.
2. European Commission. *Food Safety: A Farm to Fork Strategy for a Fair, Healthy and Environmentally Friendly Food System*; European Comission: Brussels, Belgium, 2020.
3. Hershkowitch, R.; Osman, E. *The Beef Market of Israel—A Report, Summary of 2018*; State of Israel, Ministry of Agriculture and Rural Development, Division of Research, Economy and Strategy: Jerusalem, Israel, 2019.
4. Moon, S.S.; Seong, P.N.; Jeong, J.Y. Evaluation of meat color and physiochemical characteristics in forequarter muscles of holstein steers. *Korean J. Food Sci. Anim. Res.* **2015**, *35*, 646–652. [CrossRef]
5. Drouillard, J.S. Current situation and future trends for beef production in the United States of America—A review. *Asian Aust. J. Anim. Sci.* **2018**, *31*, 1007–1016. [CrossRef]
6. Teagasc Ireland. *Sustainable Grass-Based Production—Advancing Knowledge for an Evolving Industry*; Teagasc Ireland: Carlow, Ireland, 2019.
7. Buege, D.R. Pricing and value of Holstein beef. *Meat Facts Anal.* **1988**, *88*, 7.
8. Nogalski, Z. Effect of slaughter weight on the carcass value of young crossbred ('Polish Holstein Friesian' x 'Limousin') steers and bulls. *Chil. J. Agric. Res.* **2014**, *74*, 59–66. [CrossRef]
9. Nour, A.Y.M.; Thonney, M.L.; Stouffer, J.R.; White, W.R.C. Changes in primal cut yield with increasing weight of large and small cattle. *J. Anim. Sci.* **1983**, *57*, 1166–1172. [CrossRef]
10. Reynolds, C.K.; Durst, B.; Lupoli, B.; Humphries, D.J.; Beever, D.E. Visceral Tissue Mass and Rumen Volume in Dairy Cows During the Transition from Late Gestation to Early Lactation. *J. Dairy Sci.* **2004**, *87*, 961–971. [CrossRef]
11. Armbruster, G.; Nour, A.Y.M.; Thonney, M.L.; Stouffer, J.R. Changes in cooking losses and sensory attributes of Angus and Holstein beef with increasing carcass weight, marbling score or longissimus ether extract. *J. Food Sc.* **1983**, *18*, 835–840. [CrossRef]
12. Schaefer, D.M. Yield and Quality of Holsten Beef. 1986. Available online: https://fyi.extension.wisc.edu/wbic/files/2010/11/Yield-and-Quality-of-Holstein-Beef.pdf (accessed on 5 August 2021).
13. Thonney, M.L.; Perry, T.C.; Armbruster, G.; Beermann, D.H.; Fox, D.G. Comparison of steaks from Holstein and Simmental x Angus steers. *J. Anim. Sci.* **1991**, *69*, 4866–4870. [CrossRef]
14. McKenna, D.R.; Roebert, D.L.; Bates, P.K.; Schmidt, T.B.; Hale, D.S.; Griffin, D.B.; Savell, J.W.; Brooks, J.C.; Morgan, J.B.; Montgomery, T.H.; et al. National Beef Quality Audit-2000: Survey of targeted cattle and carcass characteristics related to quality, quantity, and value of fed steers and heifers. *J. Anim. Sci.* **2002**, *80*, 1212–1222. [CrossRef]
15. Santos, P.V.; Wagner, P.; Glasenapp de Menezes, L.F.; Vonz, D.; da Silveira, M.F.; Tubin, J. Carcass physical composition and meat quality of Holstein calves, terminated in different finishing systems and slaughter weights. *Ciênc. Agrotec. Lavras* **2013**, *37*, 443–450. [CrossRef]
16. Jurie, C.; Picard, B.; Hocquette, J.-F.; Dransfield, E.; Micol, D.; Listrat, A. Muscle and meat quality characteristics of Holstein and Salers cull cows. *Meat Sci.* **2007**, *77*, 459–466. [CrossRef] [PubMed]
17. Animal QTL Database. Available online: https://www.animalgenome.org/cgi-bin/QTLdb/BT/index (accessed on 5 August 2021).

18. Cole, J.B.; Wiggans, G.R.; Ma, L.; Sonstegard, T.S.; Lawlor, T.J.; Crooker, B.A.; Van Tassell, C.P.; Yang, J.; Wang, S.; Matukumalli, L.K.; et al. Genome-wide association analysis of thirty one production, health, reproduction and body conformation traits in contemporary US Holstein cows. *BMC Genom.* **2011**, *12*, 1–17. [CrossRef] [PubMed]
19. Cesar, A.S.; Regitano, L.C.; Mourão, G.B.; Tullio, R.R.; Lanna, D.P.; Nassu, R.T.; A Mudado, M.; Oliveira, P.S.; Nascimento, M.L.D.; Chaves, A.S.; et al. Genome-wide association study for intramuscular fat deposition and composition in Nellore cattle. *BMC Genet.* **2014**, *15*, 39. [CrossRef] [PubMed]
20. McClure, M.C.; Ramey, H.R.; Rolf, M.M.; McKay, S.D.; Decker, J.E.; Chapple, R.H.; Kim, J.W.; Taxis, T.M.; Weaber, R.L.; Schnabel, R.D.; et al. Genome-wide association analysis for quantitative trait loci influencing Warner-Bratzler shear force in five taurine cattle breeds. *Anim. Genet.* **2012**, *43*, 662–673. [CrossRef] [PubMed]
21. Lund, M.; Guldbrandtsen, B.; Buitenhuis, A.; Thomsen, B.; Bendixen, C. Detection of Quantitative Trait Loci in Danish Holstein Cattle Affecting Clinical Mastitis, Somatic Cell Score, Udder Conformation Traits, and Assessment of Associated Effects on Milk Yield. *J. Dairy Sci.* **2008**, *91*, 4028–4036. [CrossRef]
22. AOAC. 991.36. Fat (Crude) in Meat and Meat Products. In *Official Methods of Analysis*, 18th ed.; Association of Analytical Communities: Gaithersburg, MD, USA, 2006.
23. AOAC. 992.15. Proximate Analysis and Calculations Crude Protein Meat and Meat Products. In *Official Methods of Analysis*, 18th ed.; Association of Analytical Communities: Gaithersburg, MD, USA, 2006.
24. AOAC. *Official Methods of Analysis*, 17th ed.; The Association of Official Analytical Chemists: Gaithersburg, MD, USA, 2000.
25. Grau, R.; Hamm, E. Eine einfache Methode zur Bestimmung der Wasserbindung im Fleisch. *Fleischwirt* **1952**, *4*, 295–297.
26. Kovalenok, V.A.; Gilman, Z.D.; Orlov, A. Metodicheskie Recomendazii po ozenke myasnoi produktivnosti, kachestvu myasa I podkozhnogo zhira swinei. *Moscow Vashniil.* **1987**, *7*, 17.
27. AMSA. *Research Guidelines Cookery, Sensory Evaluation and Instrumental Tenderness Measurements of Fresh Meat*; AMSA: Savoy, IL, USA, 1995.
28. Wheeler, T.L.; Shackelford, S.D.; Koohmaraie, M. Sampling, cooking, and coring effects on Warner-Bratzler shear force values in beef. *J. Anim. Sci.* **1996**, *74*, 1553–1562–1562. [CrossRef]
29. Shackelford, S.; Koohmaraie, M.; Whipple, G.; Wheeler, T.; Miller, M.; Crouse, J.; Reagan, J. Predictors of Beef Tenderness: Development and Verification. *J. Food Sci.* **1991**, *56*, 1130–1135. [CrossRef]
30. Honikel, K.O. Reference methods for the assessment of physical characteristics of meat. *Meat Sci.* **1998**, *49*, 447–457. [CrossRef]
31. Cross, H.; West, R.; Dutson, T. Comparison of methods for measuring sarcomere length in beef semitendinosus muscle. *Meat Sci.* **1981**, *5*, 261–266. [CrossRef]
32. Koolmees, P.A.; Korteknie, F.; Smulders, F.J.M. Accuracy and utility of sarcomere-length assessment by laser diffraction. *Food Struct.* **1986**, *5*, 9.
33. AOAC. Hydroxyproline in Meat and Meat Products, 990.26. In *Official Methods of Analysis*, 14th ed.; The Association of Official Analytical Chemists: Gaithersburg, MD, USA, 1993.
34. Starkey, C.P.; Geesink, G.H.; Oddy, V.H.; Hopkins, D. Explaining the variation in lamb longissimus shear force across and within ageing periods using protein degradation, sarcomere length and collagen characteristics. *Meat Sci.* **2015**, *105*, 32–37. [CrossRef]
35. Cohen-Zinder, M.; Orlov, A.; Trofimyuk, O.; Agmon, R.; Kabiya, R.; Shor-Shimoni, E.; Wagner, E.K.; Hussey, K.; Leibovich, H.; Miron, J.; et al. Dietary supplementation of Moringa oleifera silage increases meat tenderness of Assaf lambs. *Small Rumin. Res.* **2017**, *151*, 110–116. [CrossRef]
36. Hara, A.; Radin, N.S. Lipid extraction of tissues with a low-toxicity solvent. *Anal. Biochem.* **1978**, *90*, 420–426. [CrossRef]
37. Christie, W.W. A simple procedure for rapid transmethylation of glycerolipids and cholesteryl esters. *J. Lipid Res.* **1982**, *23*, 1072–1075. [CrossRef]
38. Chouinard, P.Y.; Corneau, L.; Barbano, D.M.; Metzger, L.E.; Bauman, D.E. Conjugated Linoleic Acids Alter Milk Fatty Acid Composition and Inhibit Milk Fat Secretion in Dairy Cows. *J. Nutr.* **1999**, *129*, 1579–1584. [CrossRef] [PubMed]
39. Story, M.; Hamm, M.W.; Wallinga, D. Food Systems and Public Health: Linkages to Achieve Healthier Diets and Healthier Communities. *J. Hunger. Environ. Nutr.* **2009**, *4*, 219–224. [CrossRef] [PubMed]
40. United States, FAO. Sustainable Food Systems—Concept and Framework. 2018. Available online: http://www.fao.org/3/ca2079en/CA2079EN.pdf (accessed on 5 August 2021).
41. Allen, T.; Prosperi, P. Modeling Sustainable Food Systems. *Environ. Manag.* **2016**, *57*, 956–975. [CrossRef]
42. Morris, S.T.; Parker, W.J.; Purchas, R.W.; McCutcheon, S.N. Dairy crossbreeding alternatives to improve New Zealand beef production. *Proc. N. Z. Grassl. Assoc.* **1992**, *54*, 19–22.
43. Zotto, R.D.; Penasa, M.; De Marchi, M.; Cassandro, M.; López-Villalobos, N.; Bittante, G. Use of crossbreeding with beef bulls in dairy herds: Effect on age, body weight, price, and market value of calves sold at livestock auctions1,2. *J. Anim. Sci.* **2009**, *87*, 3053–3059. [CrossRef]
44. Martín, N.; Schreurs, N.; Morris, S.; López-Villalobos, N.; McDade, J.; Hickson, R. Sire Effects on Carcass of Beef-Cross-Dairy Cattle: A Case Study in New Zealand. *Animals* **2021**, *11*, 636. [CrossRef]
45. Keane, M.G. *Beef Cross Breeding of Dairy and Beef Cows*; Occasional Series No. 8; Grange Beef Research Centre, Teagasc, Dunsany, Co.: Meath, Ireland, 2011; Available online: https://core.ac.uk/download/pdf/84886318.pdf (accessed on 5 August 2021).
46. Taylor, S.C.S.; Murray, J.I. Effect of feeding level, breed and milking potential on body tissues and organs of mature, non-lactating cows. *Anim. Sci.* **1991**, *53*, 27–38. [CrossRef]

47. Baldwin, R.; McLeod, K.; Capuco, A. Visceral Tissue Growth and Proliferation During the Bovine Lactation Cycle. *J. Dairy Sci.* **2004**, *87*, 2977–2986. [CrossRef]
48. Pfuhl, R.; Bellmann, O.; Kühn, C.; Teuscher, F.; Ender, K.; Wegner, J. Beef versus dairy cattle: A comparison of feed conversion, carcass composition, and meat quality. *Arch. Anim. Breed.* **2007**, *50*, 59–70. [CrossRef]
49. Warner, R.D.; Ferguson, D.M.; McDonagh, M.B.; Channon, H.A.; Cottrell, J.; Dunshea, F. Acute exercise stress and electrical stimulation influence the consumer perception of sheep meat eating quality and objective quality traits. *Aust. J. Exp. Agric.* **2005**, *45*, 553–560. [CrossRef]
50. Hughes, J.; Oiseth, S.; Purslow, P.; Warner, R. A structural approach to understanding the interactions between colour, water-holding capacity and tenderness. *Meat Sci.* **2014**, *98*, 520–532. [CrossRef]
51. Klont, R.E.; Barnier, V.M.H.; Van Dijk, A.; Smulders, F.J.M.; Hoving-Bolink, A.H.; Hulsegge, B.; Eikelenboom, G. Effects of rate of pH fall, time of deboning, aging period, and their interaction on veal quality characteristics. *J. Anim. Sci.* **2000**, *78*, 1845–1851. [CrossRef]
52. Nian, Y.; Kerry, J.P.; Prendiville, R.; Allen, P. The eating quality of beef from young dairy bulls derived from two breed types at three ages from two different production systems. *Ir. J. Agric. Food Res.* **2017**, *56*, 31–44. [CrossRef]
53. Modzelewska-Kapituła, M.; Tkacz, K.; Nogalski, Z.; Karpińska-Tymoszczyk, M.; Draszanowska, A.; Pietrzak-Fiecko, R.; Purwin, C.; Lipiński, K. Addition of herbal extracts to the Holstein-Friesian bulls' diet changes the quality of beef. *Meat Sci.* **2018**, *145*, 163–170. [CrossRef]
54. Cho, S.; Kang, S.M.; Seong, P.; Kang, G.; Choi, S.; Kwon, E.; Moon, S.; Kim, D.; Park, B. Physico-chemical Meat Qualities of Loin and Top Round Beef from Holstein Calves with Different Slaughtering Ages. *Food Sci. Anim. Resour.* **2014**, *34*, 674–682. [CrossRef]
55. Zhang, Y.M.; Hopkins, D.L.; Zhao, X.X.; van de Ven, R.; Mao, Y.W.; Zhu, L.X.; Han, G.X.; Luo, X. Characterization of pH decline and meat color development of beef carcasses during the early postmortem period in a Chinese beef cattle abattoir. *J. Integr. Agric.* **2018**, *17*, 1691–1695. [CrossRef]
56. Jama, N.; Muchenje, V.; Chimonyo, M.; Strydom, P.E.; Dzama, K.; Raats, J.G. Cooking loss components of beef from Nguni, Bonsmara and Angus steers. *Afr. J. Agric. Res.* **2008**, *3*, 416–420.
57. Pearce, K.L.; Rosenvold, K.; Andersen, H.J.; Hopkins, D.L. Water distribution and mobility in meat during the conversion of muscle to meat and ageing and the impacts on fresh meat quality attributes—A review. *Meat Sci.* **2011**, *89*, 111–124. [CrossRef] [PubMed]
58. Aaslyng, M.D. Quality indicators for raw meat. In *Meat Processing*; Kerry, J.P., Kerry, J.F., Ledward, D., Eds.; Woodhead Publishing Ltd.: Cambridge, UK, 2002; pp. 157–174.
59. Woelfel, R.L.; Owens, C.M.; Hirschler, E.M.; Martinez-Dawson, R.; Sams, A.R. The characterization and incidence of pale, soft, and exudative broiler meat in a commercial processing plant. *Poult. Sci.* **2002**, *81*, 579–584. [CrossRef] [PubMed]
60. Offer, G.; Trinick, J. On the mechanism of water holding in meat: The swelling and shrinking of myofibrils. *Meat Sci.* **1983**, *8*, 245–281. [CrossRef]
61. Huff-Lonergan, E.; Lonergan, S.M. Mechanisms of water-holding capacity of meat: The role of postmortem biochemical and structural changes. *Meat Sci.* **2005**, *71*, 194–204. [CrossRef]
62. Fischer, K. Drip loss in pork: Influencing factors and relation to further meat quality traits. *J. Anim. Breed. Genet.* **2007**, *124*, 12–18. [CrossRef]
63. Hocquette, J.F.; Gondret, F.; Baéza, E.; Médale, F.; Jurie, C.; Pethick, D.W. Intramuscular fat content in meat-producing animals: Development, genetic and nutritional control, and identification of putative markers. *Animal* **2010**, *4*, 303–319. [CrossRef]
64. Park, S.J.; Beak, S.H.; Da Jin Sol Jung, S.Y.; Kim, I.H.J.; Piao, M.Y.; Kang, H.J.; Fassah, D.M.; Na, S.W.; Yoo, S.P.; Maik, M. Genetic, management, and nutritional factors affecting intramuscular fat deposition in beef cattle—A review. *Asian-Australas. J. Anim. Sci.* **2018**, *31*, 1043–1061. [CrossRef]
65. Rodrigues, R.T.D.S.; Chizzotti, M.L.; Vital, C.; Baracat-Pereira, M.C.; Barros, E.; Busato, K.; Gomes, R.A.; Ladeira, M.; Martins, T. Differences in Beef Quality between Angus (*Bos taurus taurus*) and Nellore (*Bos taurus indicus*) Cattle through a Proteomic and Phosphoproteomic Approach. *PLoS ONE* **2017**, *12*, e0170294. [CrossRef] [PubMed]
66. Krone, K.G.; Ward, A.K.; Madder, K.M.; Hendrick, S.; McKinnon, J.J.; Buchanan, F.C. Interaction of vitamin A supplementation level with ADH1C genotype on intramuscular fat in beef steers. *Animal* **2016**, *10*, 403–409. [CrossRef] [PubMed]
67. Dinh, T.; Blanton, J.R.; Riley, D.G.; Chase, C.C.; Coleman, S.W.; Phillips, W.A.; Brooks, J.C.; Miller, M.F.; Thompson, L.D. Intramuscular fat and fatty acid composition of longissimus muscle from divergent pure breeds of cattle. *J. Anim. Sci.* **2010**, *88*, 756–766. [CrossRef] [PubMed]
68. Greenwood, P.L.; Siddell, J.P.; Walmsley, B.J.; Geesink, G.H.; Pethick, D.W.; McPhee, M.J. Postweaning substitution of grazed forage with a high-energy concentrate has variable long-term effects on subcutaneous fat and marbling in Bos taurus genotypes1. *J. Anim. Sci.* **2015**, *93*, 4132–4143. [CrossRef] [PubMed]
69. Lbrecht, E.; Gotoh, T.; Ebara, F.; Xu, J.; Viergutz, T.; Nürnberg, G.; Maak, S.; Wegner, J. Cellular conditions for intramuscular fat deposition in Japanese Black and Holstein steers. *Meat Sci.* **2011**, *89*, 13–20. [CrossRef] [PubMed]
70. Irie, M.; Kouda, M.; Matono, H. Effect of ursodeoxycholic acid supplementation on growth, carcass characteristics, and meat quality of Wagyu heifers (Japanese Black cattle). *J. Anim. Sci.* **2011**, *89*, 4221–4226. [CrossRef]
71. Robelin, J. Growth of adipose tissues in cattle; partitioning between depots, chemical composition and cellularity. A review. *Livest. Prod. Sci.* **1986**, *14*, 349–364. [CrossRef]

72. Maltin, C.; Balcerzak, D.; Tilley, R.; Delday, M. Determinants of meat quality: Tenderness. *Proc. Nutr. Soc.* **2003**, *62*, 337–347. [CrossRef]
73. Destefanis, G.; Brugiapaglia, A.; Barge, M.; Molin, E.D. Relationship between beef consumer tenderness perception and Warner–Bratzler shear force. *Meat Sci.* **2008**, *78*, 153–156. [CrossRef]
74. Gadisa, B.; Yusuf, Y.; Yousuf, M. Evaluation of eating quality in sensory panelist and instrumental tenderness of beef from Harar, Arsi and Bale cattle breeds in Oromia, Ethiopia. *Int. J. Agric. Sc. Food Technol.* **2019**, *5*, 35–42.
75. Cierach, M.; Majewska, K. Comparison of instrumental and sensory evaluation of texture of cured and cooked beef meat. *Nahr. Food* **1997**, *41*, 366–369. [CrossRef]
76. O'Connor, S.F.; Tatum, J.D.; Wulf, D.M.; Green, R.D.; Smith, G.C. Genetic effects on beef tenderness in *Bos indicus* composite and *Bos taurus* cattle. *J. Anim. Sci.* **1997**, *75*, 2826. [CrossRef]
77. Whipple, G.; Koohmaraie, M.; Dikeman, M.E.; Crouse, J.D.; Hunt, M.C.; Klemm, R.D. Evaluation of attributes that affect longissimus muscle tenderness in Bos taurus and Bos indicus cattle. *J. Anim. Sci.* **1990**, *68*, 2716–2728. [CrossRef]
78. Shackelford, S.D.; Koohmaraie, M.; Miller, M.F.; Crouse, J.D.; Reagan, J.O. An evaluation of tenderness of the longissimus muscle of Angus by Hereford versus Brahman crossbred heifers. *J. Anim. Sci.* **1991**, *69*, 171–177. [CrossRef]
79. Curi, R.A.; Chardulo, L.A.L.; Giusti, J.; Silveira, A.C.; Martins, C.L.; de Oliveira, H.N. Assessment of GH1, CAPN1 and CAST polymorphisms as markers of carcass and meat traits in Bos indicus and Bos taurus–Bos indicus cross beef cattle. *Meat Sci.* **2010**, *86*, 915–920. [CrossRef]
80. Wulf, D.M.; Tatum, J.D.; Green, R.D.; Morgan, J.B.; Golden, B.L.; Smith, G.C. Genetic influences on beef longissimus palatability in charolais- and limousin-sired steers and heifers. *J. Anim. Sci.* **1996**, *74*, 2394–2405. [CrossRef] [PubMed]
81. Monsón, F.; Sañudo, C.; Sierra, I. Influence of cattle breed and ageing time on textural meat quality. *Meat Sci.* **2004**, *68*, 595–602. [CrossRef] [PubMed]
82. Weaver, A.D.; Bowker, B.C.; Gerrard, D.E. Sarcomere length influences postmortem proteolysis of excised bovine semitendinosus muscle. *J. Anim. Sci.* **2008**, *86*, 1925–1932. [CrossRef]
83. Ertbjerg, P.; Puolanne, E. Muscle structure, sarcomere length and influences on meat quality: A review. *Meat Sci.* **2017**, *132*, 139–152. [CrossRef] [PubMed]
84. Weston, A.; Rogers, R.; Althen, T. Review: The Role of Collagen in Meat Tenderness. *Prof. Anim. Sci.* **2002**, *18*, 107–111. [CrossRef]
85. Subramaniyan, S.A.; Hwang, I. Biological Differences between Hanwoo longissimus dorsi and semimembranosus Muscles in Collagen Synthesis of Fibroblasts. *Food Sci. Anim. Resour.* **2017**, *37*, 392–401. [CrossRef]
86. Moon, S.S. The Effect of Quality Grade and Muscle on Collagen Contents and Tenderness of Intramuscular Connective Tissue and Myofibrillar Protein for Hanwoo Beef. *Asian-Australasian J. Anim. Sci.* **2006**, *19*, 1059–1064. [CrossRef]
87. Christensen, M.; Ertbjerg, P.; Failla, S.; Sañudo, C.; Richardson, R.I.; Nute, G.R.; Olleta, J.L.; Panea, B.; Albertí, P.; Juárez, M.; et al. Relationship between collagen characteristics, lipid content and raw and cooked texture of meat from young bulls of fifteen European breeds. *Meat Sci.* **2011**, *87*, 61–65. [CrossRef] [PubMed]
88. Silva, C.; Rego, O.; Simões, E.; Rosa, H.J.D. Consumption of high energy maize diets is associated with increased soluble collagen in muscle of Holstein bulls. *Meat Sci.* **2010**, *86*, 753–757. [CrossRef] [PubMed]
89. Jurie, C.; Martin, J.-F.; Listrat, A.; Jailler, R.; Culioli, J.; Picard, B. Effects of age and breed of beef bulls on growth parameters, carcass and muscle characteristics. *Anim. Sci.* **2005**, *80*, 257–263. [CrossRef]
90. Archile-Contreras, A.; Mandell, I.; Purslow, P. Disparity of dietary effects on collagen characteristics and toughness between two beef muscles. *Meat Sci.* **2010**, *86*, 491–497. [CrossRef]
91. Dubost, A.; Micol, D.; Meunier, B.; Lethias, C.; Listrat, A. Relationships between structural characteristics of bovine intramuscular connective tissue assessed by image analysis and collagen and proteoglycan content. *Meat Sci.* **2013**, *93*, 378–386. [CrossRef]
92. De Smet, S.; Raes, K.; Demeyer, D. Meat fatty acid composition as affected by fatness and genetic factors: A review. *Anim. Res.* **2004**, *53*, 81–98. [CrossRef]
93. Wood, J.; Richardson, I.; Nute, G.; Fisher, A.; Campo, M.M.; Kasapidou, E.; Sheard, P.; Enser, M. Effects of fatty acids on meat quality: A review. *Meat Sci.* **2004**, *66*, 21–32. [CrossRef]
94. Nieto, G.; Ros, G. Modification of fatty acid composition in meat through diet: Effect on lipid peroxidation and relationship to nutritional quality—A review. *Lipid Peroxidation* **2012**, *12*, 239–258.
95. Costa, A.S.; Silva, M.P.; Alfaia, C.P.; Pires, V.M.; Fontes, C.M.; Bessa, R.J.; Prates, J.A. Genetic background and diet impact beef fatty acid composition and stearoyl-CoA desaturase mRNA expression. *Lipids* **2013**, *48*, 369–381. [CrossRef] [PubMed]
96. Jéssica, L.O.; Suellen, M.G.; Josevan, S.; Ana, S.M.; Figueirêdo, R.D.; Rita, C.R.; Marta, S.M. Conjugated linoleic acid (Cla) concentration and fatty acid composition of brazilian fermented dairy products. *J. Nutr. Food Sci.* **2015**, *5*, 1–3. [CrossRef]
97. Vahmani, P.; Mapiye, C.; Prieto, N.; Rolland, D.C.; McAllister, T.A.; Aalhus, J.L.; Dugan, M.E. The scope for manipulating the polyunsaturated fatty acid content of beef: A review. *J. Anim. Sci. Biotechnol.* **2015**, *6*, 1–13. [CrossRef] [PubMed]
98. Minihane, A.M.; Lovegrove, J.A. Health benefits of polyunsaturated fatty acids (PUFAs). In *Improving the Fat Content of Foods*; Woodhead Publishing: Sawston, UK, 2006; pp. 107–140.

Article

Influence of Maternal Carbohydrate Source (Concentrate-Based vs. Forage-Based) on Growth Performance, Carcass Characteristics, and Meat Quality of Progeny

Erin R. Gubbels [1,*], Janna J. Block [2], Robin R. Salverson [1], Adele A. Harty [1], Warren C. Rusche [1], Cody L. Wright [1], Kristi M. Cammack [1], Zachary K. Smith [1], J. Kyle Grubbs [1], Keith R. Underwood [1], Jerrad F. Legako [3], Kenneth C. Olson [1] and Amanda D. Blair [1]

1 Department of Animal Science, South Dakota State University, Brookings, SD 57007, USA; robin.salverson@sdstate.edu (R.R.S.); adele.harty@sdstate.edu (A.A.H.); warren.rusche@sdstate.edu (W.C.R.); cody.wright@sdstate.edu (C.L.W.); kristi.cammack@sdstate.edu (K.M.C.); zachary.smith@sdstate.edu (Z.K.S.); judson.grubbs@sdstate.edu (J.K.G.); keith.underwood@sdstate.edu (K.R.U.); Kenneth.olson@sdstate.edu (K.C.O.); amanda.blair@sdstate.edu (A.D.B.)

2 Hettinger Research Extension Center, North Dakota State University, Hettinger, ND 58639, USA; janna.block@ndsu.edu

3 Department of Animal and Food Sciences, Texas Tech University, Lubbock, TX 79409, USA; jerrad.legako@ttu.edu

* Correspondence: erin.gubbels@sdstate.edu

Citation: Gubbels, E.R.; Block, J.J.; Salverson, R.R.; Harty, A.A.; Rusche, W.C.; Wright, C.L.; Cammack, K.M.; Smith, Z.K.; Grubbs, J.K.; Underwood, K.R.; et al. Influence of Maternal Carbohydrate Source (Concentrate-Based vs. Forage-Based) on Growth Performance, Carcass Characteristics, and Meat Quality of Progeny. *Foods* **2021**, *10*, 2056. https://doi.org/10.3390/foods10092056

Academic Editors: Benjamin W. B. Holman and Eric Nanthan Ponnampalam

Received: 29 June 2021
Accepted: 27 August 2021
Published: 31 August 2021

Publisher's Note: MDPI stays neutral with regard to jurisdictional clai-ms in published maps and institutio-nal affiliations.

Abstract: The objective of this research was to investigate the influence of maternal prepartum dietary carbohydrate source on growth performance, carcass characteristics, and meat quality of offspring. Angus-based cows were assigned to either a concentrate-based diet or forage-based diet during mid- and late-gestation. A subset of calves was selected for evaluation of progeny performance. Dry matter intake (DMI), body weight (BW), average daily gain (ADG), gain to feed (G:F), and ultrasound measurements (muscle depth, back fat thickness, and intramuscular fat) were assessed during the feeding period. Carcass measurements were recorded, and striploins were collected for Warner-Bratzler shear force (WBSF), trained sensory panel, crude fat determination and fatty acid profile. Maternal dietary treatment did not influence ($p > 0.05$) offspring BW, DMI, ultrasound measurements, percent moisture, crude fat, WBSF, or consumer sensory responses. The forage treatment tended to have decreased ($p = 0.06$) 12th rib backfat compared to the concentrate treatment and tended to have lower ($p = 0.08$) yield grades. The concentrate treatment had increased ($p < 0.05$) a^* and b^* values compared to the forage treatment. These data suggest variation in maternal diets applied in this study during mid- and late-gestation has limited influence on progeny performance.

Keywords: beef; carcass characteristics; carbohydrate source; fetal programming; maternal nutrition; meat quality

1. Introduction

Recent advances in fetal programming research indicate that altering maternal nutrition during the fetal stage can result in altered offspring productivity measures, including growth, feed intake, feed efficiency, muscle development, and meat quality [1]. Within the first two months of conception in the ruminant, development of adipocytes (fat tissue) and fibroblasts (connective tissue) occur along with development of skeletal muscle cells, all of which are primarily derived from mesenchymal stem cells [2].

Development of intramuscular fat, or marbling, is of great economic importance to the U.S. beef industry. Adipogenesis is initiated around the fourth month of gestation, partially overlapping with the second wave of myogenesis [2]. This stage of development represents an opportunity for maternal nutrition to positively or negatively affect stem cell differentiation [2]. Since the number of mesenchymal stem cells decrease as cattle mature, strategies to increase marbling during early life could be advantageous to improving meat quality. After 250 days of age, marbling is primarily enhanced only through the growth of preexisting adipocytes and nutritional influences have little impact on adipocyte development [3]. Further, different regulatory processes control fatty acid synthesis in intramuscular and subcutaneous adipose tissue, indicating that it may be possible to increase marbling without proportional increases in backfat that could negatively impact yield grades [4]. Thus, the fetal stage may be of key importance to programming carcass quality.

Volatile fatty acids (VFA) are the main products of the digestion of feed by bacteria in the rumen, provide a majority of the energy required by ruminants, and serve as substrates for synthesis of glucose and fat [5,6]. Major VFA produced by rumen microorganisms include acetate, propionate, and butyrate [6]. Various dietary carbohydrates ferment in the rumen to yield differing proportions of specific short- and long-chain fatty acids. Forage-based diets result in VFA composition of approximately 65 to 70% acetate, 15 to 25% propionate, and 5 to 10% butyrate in cattle [7]. Grain-based diets high in readily fermentable carbohydrate (starch) reduce acetate by 10 to 15% and increase propionate by 20 to 25% [7]. Propionate is the only VFA that contributes directly to the net synthesis of glucose, which is a major energy substrate utilized by uterine and placental tissues for fetal growth [5]. Although ruminal VFA production in gestating cows was not determined in the present study, it is plausible that diets based on nonstructural carbohydrates (starch), found in concentrate-based diets, rather than structural carbohydrates (fiber), found in forage-based diets, could influence fetal development and subsequent composition of the developing calf by way of altered VFA production profiles.

From a production perspective, management decisions made in response to drought, availability of feedstuffs, or cost of feedstuffs can alter the gestational environment, potentially leading to changes in fetal development. Previous literature has shown that providing first-calf heifers and mature cows with a high-energy diet 100 d prepartum increased body weight before parturition and calf birth weight [8]. In the study by Corah et al. [8] subsequent weaning weight was heavier for calves from cows consuming the high-energy diet. However, it has been reported feeding corn to dams in late pregnancy resulted in offspring with reduced marbling scores, a tendency towards reduced intramuscular fat percentage, and more carcasses grading United States Department of Agriculture (USDA) Select compared to offspring from hay-fed cows [9]. Because fetal adipocyte differentiation and growth is initiated during mid-gestation, it is possible that different responses would be observed if maternal dietary treatments had been implemented earlier. Based on these results, there may be differences in nutrient utilization and performance of offspring from cows fed forage or concentrate-based diets. We hypothesized that variations in the proportion of volatile fatty acids produced in the rumen of the gestating cow caused by differing dietary carbohydrate sources during mid- and late- gestation would differentially influence fetal development and offspring composition, leading to alterations in performance and meat quality of offspring. The objective of this study was to investigate the effects of maternal prepartum dietary carbohydrate source (forage- vs. concentrate-based) during mid- and late-gestation on growth performance, carcass characteristics, and meat quality of offspring.

2. Materials and Methods

2.1. Cow Management

All animal care and experimental protocols were approved by the South Dakota State University (SDSU) Animal Care and Use Committee (approval number 18-081E). Mature, Angus-based, spring-calving cows (n = 131) from the SDSU Antelope Range and Livestock Research Station were evaluated for pregnancy in the fall of 2017 and assigned to dietary treatments based on cow age and body condition score (BCS). Groups were randomly assigned to a forage-based or concentrate-based dietary treatment and allotted to two pens based on treatment (Forage (n = 64) or Concentrate (n = 65)). The uterine environment created by differing VFA profiles within each cow was considered the experimental unit. Dietary composition of the treatment diets is provided in Table 1.

Table 1. Dietary components (dry matter basis) consumed by cows receiving a forage-based (For) or concentrate-based (Conc) diet during mid- and late-gestation.

Ingredient	Conc [1]	For [1]
Wheat Straw, %	24.1	71.9
Grass/Alfalfa Hay, %	0.0	21.8
Corn Silage, %	0.0	3.7
Suspension Supplement [2], %	4.6	2.6
Corn Grain, %	56.6	0.0
Modified Distiller's Grain w/Solubles, %	13.3	0.0
Limestone, %	1.4	0.0
	Diet Composition	
Dry Matter Intake, kg	6.4	10.73
Dry Matter Intake, % BW	0.98	1.65
Roughage Intake, % BW	0.30	1.58
Crude Protein, % of DM	12.02	7.55
TDN, % of DM	73.18	50.88
NEm (Mcal/kg)	1.67	0.99
NEg (Mcal/kg)	1.05	0.46

[1] Diets formulated based on NRC (2000) requirements. [2] Suspension supplement: 20% Crude Protein (≤20% Non-protein nitrogen), 3.55–4.55% Ca, 0.20% P, 0.30% Mg, 1% K, 528.63 ppm Mn, 12.65 ppm Co, 480 ppm Cu, 5.50 ppm Se, 1440 ppm Zn, 88184 IU/kg Vit. A, 24912 IU/kg Vit. D3, 165 IU/kg Vit. E, 400 g/ton monensin.

Feed intake was controlled so that cows in both treatments consumed equal amounts of protein and energy. Cows were provided the treatment diets beginning at approximately day 94 of gestation and continuing until approximately 30 days prior to calving. Both diets were formulated to maintain cow body condition. Body weight (BW) and BCS from the beginning (day 0) and end (day 98) of the treatment period were used to monitor the influence of dietary carbohydrate source on cow performance. Initial BW was recorded after a two-week diet adaptation period to account for differences in gut fill (cows were provided treatment diets that varied in digestibility and intake compared to the pre-treatment diet). Average initial BW of the cows was 598 ± 49.4 kg and 666 ± 52.4 kg for concentrate and forage treatments, respectively (likely due to differences in rumen fill), and average BCS was 5.2 ± 0.39 and 5.3 ± 0.31 for concentrate and forage treatments, respectively. At the completion of the treatment period the average BW of the cows was 639 ± 60.7 kg and 635 ± 57.4 kg, and average BCS was 5.4 ± 0.57 and 5.1 ± 0.38 for concentrate and forage treatments, respectively. At the end of the treatment period, cows were returned to native range pastures and managed as a common group through weaning.

2.2. Offspring Management

At approximately 60 d of age, all calves were vaccinated with a killed vaccine for clostridial diseases (Vision 7 Somnus with SPUR, Merck Animal Health, Madison, NJ, USA). At approximately 110 days of age, all calves were administered a modified-live vaccine for prevention of bovine rhinotracheitis (IBR), bovine viral diarrhea (BVD), bovine respiratory syncytial virus (BRSV) Types 1 and 2, and parainfluenza-3 (PI_3), Haemophilus somnus, and *Mannheimia haemolytica* (Pyramid 5+ Presponse SQ, Boehringer Ingelheim Vetmedica, Inc., St. Joseph, MO, USA). At weaning, all calves were administered an anthelmintic (Dectomax Pour-On Solution, Zoetis, Parsippany, NJ, USA) and were provided boosters of the clostridial disease and respiratory disease vaccines. At this time, a subset of 96 calves (n = 24 heifers/treatment, n = 24 steers/treatment) closest to the mean weaning weight were shipped to the SDSU Cottonwood Field Station. Calves were fed a common receiving diet consisting of grass hay and dried distiller's grains with solubles during an 83-days backgrounding period. On day 36 postweaning, calves were weighed and ultrasounded to determine backfat thickness (BF), muscle depth of the *longissimus dorsi*, and intramuscular fat (IMF) measured at the 12th and 13th rib.

At the conclusion of the backgrounding phase, all calves were transported approximately 526 km to Brookings, SD for the finishing phase of the study. Upon arrival, calves were vaccinated against clostridia perfringens type A (Clostridium Perfringens Type A Toxoid; Elanco, Greenfield, IN, USA). The calves were finished in an Insentec monitoring system (Insentec, Marknesse, The Netherlands) to monitor individual feed intake (steers and heifers were fed separately in two pens) at the SDSU Cow-Calf Education and Research facility. Calves were stepped up to their finishing diets over 14-days; final diets are shown in Table 2. Diet ingredients were sampled weekly and monthly composites were used to determine the dry matter [10], crude protein [11], neutral detergent fiber [12], acid detergent fiber [13], ash [14], crude fat [15]. Tabular values for diet ingredients were used to calculate energy content of diets.

Table 2. Dietary components and nutrient composition of finishing diet [1] consumed by offspring of cows receiving a forage-based or concentrate-based diet during mid- and late-gestation.

Ingredient	% DM Basis
Grass Hay	10.77
Earlage	11.20
Dry Rolled Corn	53.85
Dried Distiller's Grains w/Solubles [2]	17.66
Suspension Supplement [3]	6.51
	Nutrient Composition of Diet
DM %	72.00
CP %	14.61
ADF %	10.32
NDF %	20.74
Crude Fat %	3.74
Ash %	3.41
NEm (Mcal/kg)	2.05
NEg (Mcal/kg)	1.36

[1] Diet formulated based on NRC (2000) requirements. [2] Dried distiller's grains w/solubles fed to heifers included melengestrol acetate (MGA, Zoetis, Parsippany, NJ, USA) at a rate sufficient to provide 0.50 mg·hd^{-1}·day^{-1}; steers received dried distiller's grains w/solubles without MGA. [3] Suspension supplement: 30.8% protein (26.6% non-protein nitrogen), 8% Ca, 0.2% P, 0.4% Mg, 7.1% K, 15.6 ppm Co, 337.6 ppm Cu, 33.8 ppm I, 723.8 ppm Mn, 3.2 ppm Se, 1107.8 ppm Zn, 9502 IU/kg Vit A, 2381 IU/kg Vit D3, 848 IU/kg Vit E, 512.3 g/ton monensin.

Cattle were weighed at 28-days intervals during the finishing period to monitor performance (hereafter referred to as Period 1, Period 2, etc.). Calves were administered an initial growth promoting implant on day 23 of the finishing period containing 100 mg trenbolone acetate (TBA) and 14 mg estradiol benzoate (EB) (Synovex-Choice, Zoetis Inc., Parsippany, NJ, USA). Cattle were re-implanted with 100 mg TBA and 14 mg EB (Synovex-Choice, Zoetis Inc., Parsippany, NJ, USA) and a second ultrasound was conducted on day 80 of the finishing period. Ultrasound measures collected during the backgrounding period and finishing period were compared to determine changes in composition. The second ultrasound was also used to predict harvest date. The harvest target was determined when the predicted BF was approximately 1.27 cm, resulting in three harvest dates at day 131, day 145, and day 180 of the finishing period. Cattle were weighed the morning of slaughter to determine final live BW and shipped 235 km to a commercial harvest facility.

2.3. Carcass Evaluation and Sample Collection

All cattle were tracked individually through the harvest process. Following carcass chilling (approximately 24 h), hot carcass weight (HCW), ribeye area (REA), 12th rib BF, USDA Yield Grade, marbling score, carcass maturity, and USDA Quality Grade were evaluated according to the United States Standards for Grades of Carcass Beef [16]. Objective color measurements (L^*, a^*, and b^*) were also recorded at the exposed REA of each carcass using a handheld Minolta colorimeter (Model CR-310, Minolta Corp., Ramsey, NJ, USA; 50 mm diameter measuring space, D65 illuminant). A strip loin (IMPS #180) was collected from each carcass and transported to the SDSU Meat Science Laboratory, portioned into 2.54-cm steaks, and vacuum packaged. Four steaks were aged for either 3, 7, 14, or 21 days at 4 °C and then frozen at −10 °C for evaluation of Warner-Bratzler shear force (WBSF). Additional steaks were utilized to determine fatty acid profile using Fatty Acid Methyl Ether (FAME) synthesis, crude fat percentage using ether extraction, and consumer palatability of 14-d aged samples using a trained sensory panel.

2.4. Warner-Bratzler Shear Force

Steaks designated for WBSF determination were thawed for 24 h at 4 °C then cooked on an electric clamshell grill (George Foreman, Model GRP1060B, Middleton, WI, USA) to an internal temperature of 71 °C. A thermometer (Model 35140, Cooper-Atkins Corporation, Middlefield, CT, USA) was used to record the peak internal temperature. Cooked steaks were cooled at 4 °C for 24 h before removing 6 cores (1.27 cm diameter) parallel to the muscle fiber orientation [17]. A single, peak shear force measurement was obtained for each core using a texture analyzer (Shimadzu Scientific Instruments Inc., Lenexa, KS, USA, Model EZ-SX) with a Warner-Bratzler attachment. Measurements of the peak shear force value were averaged to obtain a single WBSF value per steak.

2.5. Ether Extract

At 3 days postmortem, the anterior face of each striploin was removed during fabrication and frozen at −20 °C and later used to determine percent crude fat using the ether extract method described by Mohrhauser et al. [18]. Steaks were thawed slightly and all exterior fat, epimysial connective tissue, and additional muscles were removed leaving the longissimus muscle for evaluation. Samples were minced, immersed in liquid nitrogen, and powdered for 15 s using a Waring commercial blender (Waring Products Division, Model 51BL32, Lancaster, PA, USA). Homogenized samples were weighed in duplicate 5-g samples into dried aluminum tins, covered with dried filter papers, and dried in an oven at 100 °C for 24 h. Dried samples were then placed into a desiccator and were reweighed after cooling. Samples were extracted using petroleum ether in a side-arm Soxhlet extractor (Thermo Fischer Scientific, Rockville, MD, USA) for 60 h followed by drying at room temperature and subsequent drying in an oven at 100 °C for 4 h. Dried extracted samples were placed into a desiccator for 1 h and were cooled and then reweighed. Crude fat was calculated by subtracting the pre-extraction weight from the post-extraction sample weight and expressed as a percentage of the pre-extraction sample weight.

2.6. Fatty Acid Composition

A sub-sample of 60 steaks (n = 30 steaks closest to the mean marbling score of each treatment) were selected to evaluate fatty acid profile using direct FAME synthesis. Steaks were thawed slightly and external fat, epimysial connective tissue, and additional muscles were trimmed from the longissimus muscle. Samples were minced, immersed in liquid nitrogen, and powdered for 15 s using a Waring commercial blender (Waring Products Division, Model 51BL32, Landcaster, PA, USA). Duplicate 1 g samples were weighed and processed to generate FAMEs according to procedures of O'Fallon et al. [19]. Fatty acids were identified through comparison with retention times of an authentic fatty acid standard mixture (GLC-463, Nu-Check Prep Inc., Elysian, MN, USA). Quantities were computed as mg/g of raw wet tissue through an internal standard calibration method where C13:0 served as the internal standard. Final contents were then summed and %, g/100 g total fatty acids was produced after summing all fatty acids.

2.7. Trained Sensory Panel

The human sensory panel utilized in this study was approved by the Institutional Review Board of South Dakota State University (IRB-1911019-EXM). Eight sensory panelists were trained to evaluate meat quality attributes of strip loin steaks according to the American Meat Science Association training guidelines appropriate for the study [17]. Panelists were 18 years or older, had no food allergies or sensitivities, and had consumed any type of meat products at least once a year. Strip loin samples were evaluated for juiciness (1 = extremely dry; 18 = extremely juicy), tenderness (1 = extremely tough; 18 = extremely tender), and beef flavor (1 = extremely bland; 18 = extremely intense) on an anchored unmarked line scale. Steaks were cooked on an electric clamshell grill (George Foreman, Model GRP1060B, Middleton, WI) to an internal temperature of 71 °C. After cooking, steaks were rested for five minutes and then cut into 2.5 × 1 × 1-cm samples. Two cubes were placed into a prelabeled plastic cup, covered with a plastic lid in order to retain heat and moisture, and held in a warming oven (Metro HM2000, Wilkes-Barre, PA, USA) at 60 °C until served. Evaluations were performed according to American Meat Science Association guidelines [17]. Ten samples were evaluated in each session, one session per d, for a total of 10 sessions. Samples evaluations were alternated by treatment to reduce first and last order bias. Samples were served to panelists in a randomized fashion, in private booths, under red lights to limit observation of visual differences.

2.8. Statistical Analyses

Response variables were analyzed using generalized linear mixed model procedures (SAS GLIMMIX, SAS Inst. Inc., Cary, NC, USA) in a completely randomized design. The intrauterine environment was considered the experimental unit for ultrasound measurements, carcass characteristics, and meat quality data and was designated as a random effect. Treatment, sex, and their interaction were included in the model as fixed effects. For carcass characteristics and meat quality data, harvest date was included in the model as a fixed effect to absorb variation due to this effect (data not shown). For WBSF, aging period was added to the model as a repeated measure and peak cooking temperature was included as a covariate. Separation of least squares means was conducted using protected LSD. Treatment by sex interactions were evaluated and discussed if significant.

3. Results

3.1. Growth Performance

Animal performance and growth data are reported in Table 3. Maternal dietary treatment did not influence ($p > 0.05$) offspring BW, or DMI. In Period 1 (day 0–23) of the finishing phase, offspring from dams fed a forage-based diet tended ($p = 0.079$) to have an improved ADG compared to the offspring from dams fed a concentrate-based diet.

Table 3. Growth performance for progeny of dams fed a prepartum dietary carbohydrate source consisting of concentrate-based (Conc) or forage-based (For) diet during mid- and late-gestation.

	Treatment [1]			Sex			*p*-Value [2]		
	Conc	For	SEM [3]	Heifers	Steers	SEM [3]	Trmt	Sex	T × S
Backgrounding Phase									
Weaning BW, kg	281	277	3.7	272	286	3.7	0.475	0.009	0.951
Day 1–36									
Ending BW, kg	280	280	3.2	274	286	3.2	0.830	0.012	0.748
ADG [4], kg	−0.04	0.09	0.067	0.06	−0.01	0.067	0.166	0.495	0.735
Day 37–83									
BW, kg	321	321	3.4	309	333	3.4	0.994	<0.001	0.909
ADG [4], kg	0.86	0.84	0.042	0.74	0.96	0.042	0.738	<0.001	0.743
Finishing Phase									
Period 1 (day 0–23)									
Ending BW, kg	354	357	3.7	346	365	3.7	0.544	<0.001	0.618
ADG [4], kg	1.46	1.60	0.055	1.60	1.45	0.055	0.079	0.051	0.246
DMI [5], kg	6.47	6.02	0.271	6.94	5.56	0.271	0.243	<0.001	0.743
G:F [6]	0.25	0.26	0.002	0.22	0.29	0.002	0.825	0.105	0.148
Period 2 (day 23–51)									
Ending BW, kg	402	403	4.5	385	421	4.5	0.915	<0.001	0.255
ADG [4], kg	1.72	1.65	0.055	1.37	2.00	0.055	0.312	<0.001	0.054
DMI [5], kg	7.40	7.22	0.328	7.35	7.26	0.328	0.706	0.843	0.960
G:F [6]	0.21	0.21	0.004	0.17	0.27	0.004	0.566	<0.001	0.065
Period 3 (day 51–78)									
Ending BW, kg	448	451	5.0	428	471	5.0	0.651	<0.001	0.629
ADG [4], kg	1.68	1.77	0.054	1.60	1.84	0.054	0.224	0.002	0.071
DMI [5], kg	8.47	8.36	0.378	8.39	8.48	0.374	0.881	0.852	0.973
G:F [6]	0.19	0.20	0.004	0.18	0.21	0.004	0.435	0.033	0.319
Period 4 (day 78–106)									
Ending BW, kg	502	507	5.27	486	524	5.27	0.499	<0.001	0.612
ADG [4], kg	1.96	2.02	0.057	2.07	1.91	0.057	0.416	0.047	0.874
DMI [5], kg	11.13	11.07	0.275	10.81	11.39	0.275	0.866	0.143	0.880
G:F [6]	0.18	0.17	0.004	0.19	0.16	0.004	0.902	0.007	0.727
Period 5 [7]									
Ending BW, kg	579	590	6.95	555	614	6.86	0.241	<0.001	0.660
ADG [4], kg	1.43	1.49	0.046	1.43	1.47	0.046	0.416	0.764	0.067
DMI [5], kg	14.02	14.00	0.190	14.04	13.99	0.190	0.964	0.862	0.253
G:F [6]	0.10	0.10	0.003	0.10	0.10	0.003	0.263	0.505	0.307

[1] Diets formulated based on NRC (2000) requirements for dams fed either a concentrate or forage diet during mid- and late-gestation. [2] Probability of difference among least square means. [3] Standard error of the mean. [4] ADG calculated from end of previous period to end of current period. [5] DMI: Dry matter intake. [6] G:F. gain to feed ratio. [7] Final BW, ADG, DMI, and G:F calculated based on when each animal was harvested at either d 131, d 145, or d 180. however, no differences ($p > 0.05$) in ADG were detected between treatment groups in subsequent periods.

A tendency ($p = 0.054$) for a treatment × sex interaction was detected for ADG in Period 2 (Figure 1a). Steers from the concentrate treatment had greater ($p < 0.04$) ADG compared with steers from the forage treatment, while ADG of heifers did not differ ($p > 0.05$) between treatments. A tendency ($p = 0.071$) for a treatment × sex interaction was also detected for ADG in Period 3 (Figure 1b).

Steers from the forage treatment had greater ($p < 0.04$) ADG than steers from the concentrate treatment as well as the heifers from either treatment, which were similar ($p > 0.05$). A tendency ($p = 0.067$) for a treatment × sex interaction was observed for ADG in Period 5 (Figure 1c). Steers from both treatments had similar ($p > 0.05$) ADG, and had similar ($p > 0.05$) ADG compared to both forage and concentrate heifers; however, forage heifers tended to have improved ($p = 0.06$) ADG compared to concentrate heifers.

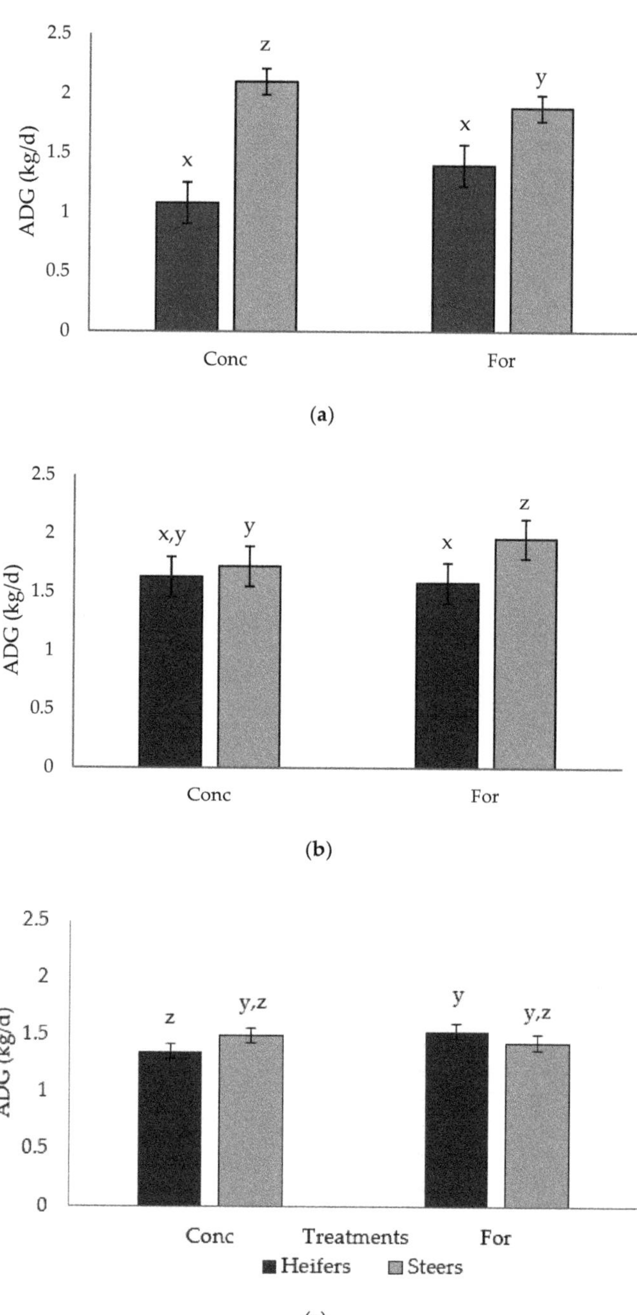

Figure 1. Treatment by sex interaction for ADG (kg/d) of progeny in: (**a**) Period 2 (p = 0.054), (**b**) Period 3 (p = 0.071), and (**c**) Period 5 (p = 0.067) from dams fed a concentrate-based (Conc) or forage-based (For) diet during mid- and late-gestation. Diets formulated based on NRC (2000) requirements for dams fed either a concentrate or forage diet during mid- and late-gestation. x, y, z LSmeans lacking a common superscript differ ($p \leq 0.05$).

No differences ($p > 0.05$) in G:F were observed between treatment groups; however, a tendency ($p = 0.065$) for a treatment × sex interaction was detected for G:F in Period 2 (Figure 2). Steers from both treatments had similar ($p > 0.05$) G:F, and had improved ($p < 0.05$) G:F compared to heifers from both treatments, however the forage heifers tended to have improved ($p = 0.09$) G:F compared to the concentrate heifers.

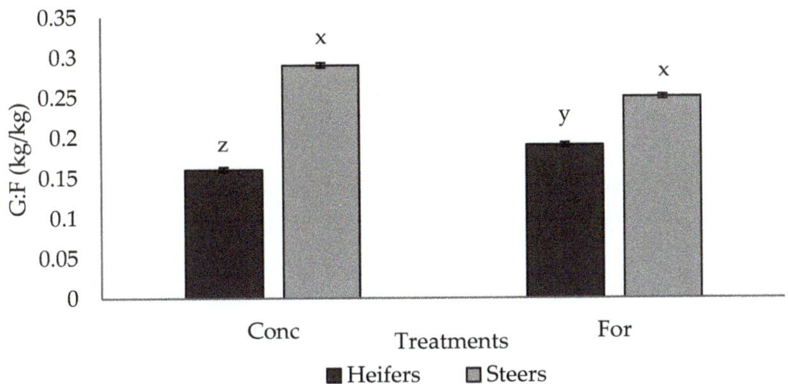

Figure 2. Treatment by sex interaction ($p = 0.065$) for G:F (kg/kg) of progeny in Period 2 from dams fed a concentrate-based (Conc) or forage-based (For) diet during mid- and/or late-gestation [1]. [1] Diets formulated based on NRC (2000) requirements for dams fed either a concentrate or forage diet during mid- and late-gestation. [x, y, z] LSmeans lacking a common superscript differ ($p \leq 0.05$).

As expected, steers had greater ($p < 0.05$) BW compared to heifers at all time periods and had an increased ($p < 0.05$) ADG from day 37–83. Steers also had increased ($p < 0.05$) ADG in Period 2 (day 23–51) and Period 3 (day 51–78) compared to heifers. However, heifers had an increased ($p < 0.05$) ADG in Period 4 (day 78–106) and tended to have an increased ($p = 0.051$) ADG compared to steers in Period 1 (day 0–23). Heifers had greater ($p < 0.05$) DMI during Period 1, however, DMI did not differ ($p > 0.05$) between steers and heifers for the remainder of the finishing period. Steers had improved ($p < 0.05$) G:F during Period 2 (day 23–51) and 3 (day 51–78), while heifers had improved ($p < 0.05$) G:F during Period 4 (day 78–106). It is likely that differences in G:F were driven by differences in ADG rather than DMI.

3.2. Ultrasound Measurements

Ultrasound measurements are reported in Table 4. Maternal treatment did not influence ($p > 0.05$) offspring BF, IMF percentage or muscle depth during the finishing phase. A treatment × sex interaction ($p = 0.028$) was detected for muscle depth during the backgrounding phase (Figure 3). Heifers from the concentrate treatment tended to have increased ($p = 0.07$) muscle depth compared with heifers from the forage treatment, while muscle depth of steers did not differ ($p > 0.05$) between treatments. Heifers had increased ($p < 0.05$) BF compared to steers at the initial ultrasound during the backgrounding phase.

Table 4. Least square means for ultrasound measurements of progeny from dams fed a prepartum dietary carbohydrate source consisting of concentrate-based (Conc) or forage-based (For) diet during mid- and late-gestation.

	Treatment [1]			Sex			p-Value [3]		
	Conc	For	SEM [2]	Heifers	Steers	SEM [2]	Trmt	Sex	T × S
Initial ultrasound during backgrounding phase									
Backfat, mm	3.94	3.82	0.124	4.06	3.70	0.124	0.503	0.046	0.502
Muscle Depth, mm	40.18	39.66	0.926	39.67	40.17	0.926	0.692	0.700	0.028
Intramuscular fat,%	5.07	4.98	0.1104	5.07	4.97	0.110	0.557	0.539	0.486

Table 4. *Cont.*

	Treatment [1]			Sex			p-Value [3]		
	Conc	For	SEM [2]	Heifers	Steers	SEM [2]	Trmt	Sex	T × S
Ultrasound during finishing phase									
Backfat, mm	6.69	6.52	0.249	6.69	6.52	0.249	0.663	0.622	0.265
Muscle Depth, mm	50.76	50.93	0.877	50.64	51.05	0.877	0.890	0.743	0.926
Intramuscular fat,%	4.25	4.28	0.065	4.31	4.22	0.064	0.711	0.339	0.172
Change between ultrasound periods									
Backfat, mm	2.75	2.69	0.226	2.63	2.81	0.226	0.802	0.576	0.405
Muscle Depth, mm	10.58	11.31	1.276	10.97	10.91	1.276	0.684	0.974	0.127
Intramuscular fat,%	−0.82	−0.72	0.123	−0.76	−0.78	0.123	0.546	0.945	0.975

[1] Diets formulated based on NRC (2000) requirements for dams fed either a concentrate or forage diet during mid- and late-gestation. [2] Standard error of the mean [3] Probability of difference among least square means.

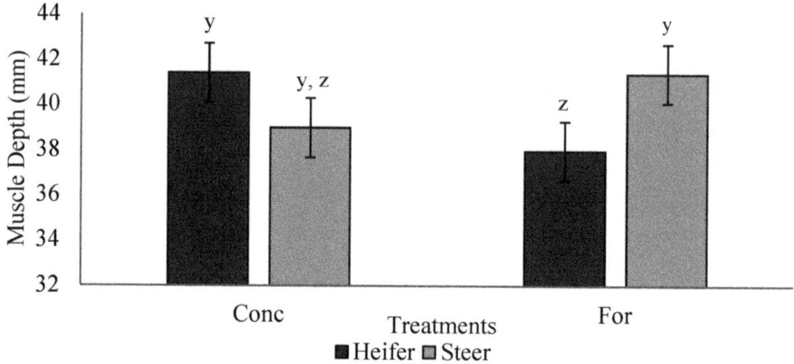

Figure 3. Treatment by sex interaction ($p = 0.028$) for muscle depth measured via ultrasound during backgrounding (d 36) of progeny from dams fed a concentrate-based (Conc) or forage-based (For) diet during mid- and/or late-gestation. Diets formulated based on NRC (2000) requirements for dams fed either a concentrate or forage diet during mid- and late-gestation. [y,z] LSmeans lacking a common superscript differ ($p \leq 0.05$).

3.3. Carcass Characteristics

Carcass measurements are reported in Table 5. Maternal treatment did not influence ($p > 0.05$) offspring HCW, REA, marbling score, L^* values, or the proportion of carcasses in each USDA Quality and Yield Grade category. Offspring from the forage treatment tended to have decreased ($p = 0.060$) 12th rib fat thickness and tended to have lower ($p = 0.084$) USDA Yield Grades compared to offspring from the concentrate treatment. Offspring from the concentrate treatment had increased ($p < 0.05$) a^* and b^* values compared to the forage treatment. As expected, steers had heavier ($p < 0.05$) HCW and larger ($p < 0.05$) REA than heifers. Heifers had increased ($p < 0.05$) BF and marbling scores, as well as increased ($p < 0.05$) a^* and b^* values and tended ($p = 0.070$) to have higher USDA Yield Grades.

Table 5. Least square means for carcass characteristics and meat quality of progeny from dams fed a prepartum dietary carbohydrate source consisting of concentrate-based (Conc) or forage-based (For) diet during mid- and late-gestation.

	Treatment [1]			Sex			p-Value [2]		
	Conc	For	SEM [3]	Heifers	Steers	SEM [3]	Trmt	Sex	T × S
Hot carcass weight, kg	349	351	4.4	335	366	4.4	0.710	<0.001	0.299
Ribeye area, cm ^{2}	85.8	87.7	1.23	83.2	89.7	1.35	0.271	0.006	0.889
12th rib fat thickness, cm	1.22	1.14	0.041	1.27	1.09	0.046	0.060	0.002	0.304

Table 5. *Cont.*

	Treatment [1]			Sex			p-Value [2]		
	Conc	For	SEM [3]	Heifers	Steers	SEM [3]	Trmt	Sex	T × S
USDA Yield grade	3.0	2.8	0.08	3.0	2.8	0.09	0.084	0.070	0.811
Marbling score [4]	537	539	13.9	563	513	15.7	0.909	0.013	0.699
L^* [5]	42.05	41.83	0.277	41.99	41.90	0.314	0.534	0.838	0.826
a^* [5]	25.27	24.59	0.138	25.25	24.60	0.156	<0.001	0.002	0.921
b^* [5]	10.45	10.03	0.093	10.46	10.02	0.105	<0.001	0.001	0.660
USDA Quality Grade [6]									
Prime, %	5.22	9.14	4.821	9.21	5.17	4.454	0.588	0.615	0.963
Upper 2/3 Choice, %	53.00	50.66	8.423	65.66	37.72	9.184	0.865	0.272	0.864
Low Choice, %	36.19	30.95	8.136	20.16	50.18	10.510	0.715	0.267	0.635
USDA Yield Grade [6]									
Yield Grade 2, %	57.55	61.62	8.006	50.95	67.69	8.328	0.761	0.384	0.556
Yield Grade 3, %	40.50	36.50	7.849	46.59	30.96	8.184	0.761	0.399	0.794

[1] Diets formulated based on NRC (2000) requirements for dams fed either a concentrate or forage diet during mid- and late-gestation. [2] Probability of difference among least square means [3] Standard error of the mean. [4] Marbling score: 200 = Traces ^{0}, 300 = Slight ^{0}, 400 = Small ^{0}, 500 = Modest ^{0}. [5] Recorded 3 d postmortem; L^*: 0 = Black, 100 = White; a^*: Negative values = green; Positive values = red; b^*: Negative values = blue; Positive values = yellow. [6] Calculated proportions of USDA Quality and Yield Grade (data did not converge for a quality grade of USDA Select, or USDA Yield Grade less than a 2 or greater than a 3).

3.4. Meat Quality Characteristics

Meat quality characteristics are reported in Table 6. Maternal treatment did not influence ($p > 0.05$) crude fat percentage, moisture content, WBSF, or sensory characteristics of steaks from offspring. Heifers had decreased ($p < 0.05$) moisture and increased crude fat content compared to steers and tended ($p = 0.068$) to have improved WBSF values compared to steers. No differences ($p > 0.05$) were detected between steers and heifers for sensory characteristics of steaks. As expected, WBSF improved ($p < 0.05$) with each aging period (4.75 ± 0.152 kg, 3.79 ± 0.112 kg, 2.98 ± 0.088 kg, and 2.65 ± 0.064 kg for steaks aged 3, 7, 14, and 21 days, respectively).

Table 6. Least square means for meat characteristics of progeny from dams fed a prepartum dietary carbohydrate source consisting of concentrate-based (Conc) or forage-based (For) diet during mid- and late-gestation.

	Treatment [1]			Sex			p-Value [2]		
	Conc	For	SEM [3]	Heifers	Steers	SEM [3]	Trmt	Sex	T × S
Crude Fat, %	6.31	6.24	0.339	7.17	5.39	0.384	0.865	<0.001	0.621
Moisture, %	71.48	71.50	0.264	70.69	72.29	0.299	0.945	<0.001	0.728
WBSF [4], kg	3.48	3.60	0.128	3.38	3.71	0.137	0.480	0.068	0.637
Tenderness [5]	12.43	12.85	0.285	12.87	12.41	0.318	0.263	0.284	0.833
Juiciness [5]	10.98	11.49	0.295	11.33	11.14	0.330	0.192	0.665	0.328
Flavor [5]	9.83	9.64	0.228	9.84	9.64	0.255	0.531	0.555	0.232

[1] Diets formulated based on NRC (2000) requirements for dams fed either a concentrate or forage diet during mid- and late-gestation. [2] Probability of difference among least square means. [3] Standard error of the mean. [4] Warner-Bratzler Shear Force. [5] Strip loin samples were evaluated for juiciness (1 = extremely dry; 18 = extremely juicy), tenderness (1 = extremely tough; 18 = extremely tender), and beef flavor (1= extremely bland; 18 = extremely intense).

3.5. Fatty Acid Composition

Fatty acid composition data is reported in Tables 7 and 8. The concentration (mg/g wet raw tissue; Table 7) of arachidonic (C20:4n6), nervonic (C20:1n9), and docosapentaenoic (C22:5n3) acids were increased ($p < 0.05$) in samples from the concentrate treatment; however, treatment did not influence ($p > 0.05$) concentration of other fatty acids. The concentration of capric (C10:0), myristic (C14:0), myristoleic (C14:1n5), palmitoleic (C16:1n7), and heptadecenoic (C17:1) acids were increased ($p < 0.05$) in samples from heifers compared with steers. Sex did not influence ($p > 0.05$) concentration of other fatty acids.

Table 7. Least squares means for the fatty acid composition (mg/g raw wet tissue) of progeny from dams fed a prepartum dietary carbohydrate source consisting of concentrate-based (Conc) or forage-based (For) diet during mid- and late-gestation.

	Treatment [1]			Sex			p-Value [2]		
Fatty Acid	Conc	For	SEM [3]	Heifer	Steer	SEM [3]	Trmt	Sex	T × S
C10:0	0.03	0.03	0.003	0.03	0.02	0.003	0.710	0.013	0.290
C12:0	0.04	0.04	0.003	0.05	0.04	0.003	0.540	0.100	0.466
C14:0	2.15	2.06	0.154	2.34	1.87	0.172	0.663	0.042	0.348
C15:0	0.29	0.30	0.024	0.32	0.27	0.027	0.846	0.105	0.629
C16:0	19.37	19.43	1.410	20.58	18.23	1.572	0.974	0.264	0.477
C17:0	0.86	0.89	0.079	0.94	0.81	0.088	0.742	0.250	0.853
C18:0	10.33	10.73	0.788	10.45	10.61	0.879	0.697	0.896	0.495
C20:0	0.05	0.04	0.006	0.05	0.04	0.007	0.452	0.103	0.660
C14:1n5	0.57	0.50	0.042	0.62	0.46	0.047	0.204	0.017	0.402
C16:1n7	2.15	1.95	0.134	2.35	1.76	0.150	0.264	0.005	0.295
C16:1trans	0.24	0.25	0.014	0.25	0.24	0.016	0.723	0.698	0.566
C18:1n9	27.24	27.33	1.909	29.34	25.23	2.128	0.970	0.152	0.593
C18:1trans	2.58	2.41	0.203	2.47	2.52	0.226	0.517	0.853	0.467
C18:1n7	0.94	1.10	0.104	1.16	0.89	0.116	0.230	0.088	0.603
C18:2trans	0.004	0.003	0.0001	0.004	0.003	0.0006	0.628	0.596	0.245
C18:2n6	2.96	2.63	0.170	2.80	2.79	0.190	0.147	0.978	0.657
C18:3n6	0.02	0.02	0.001	0.01	0.02	0.001	0.766	0.201	0.806
C18:3n3	0.27	0.24	0.012	0.25	0.25	0.014	0.051	0.916	0.948
C20:2	0.06	0.05	0.004	0.06	0.05	0.005	0.638	0.240	0.921
C20:3n6	0.01	0.01	0.001	0.01	0.01	0.001	0.210	0.901	0.749
C20:4n6	0.55	0.46	0.025	0.493	0.524	0.028	0.009	0.405	0.547
C22:3	0.01	0.01	0.001	0.01	0.01	0.001	0.056	0.721	0.855
C24:1n9	0.02	0.01	0.002	0.01	0.01	0.002	0.011	0.530	0.224
C22:5n3	0.02	0.01	0.003	0.02	0.02	0.003	0.007	0.329	0.544
C22:6n3	0.03	0.03	0.003	0.03	0.03	0.003	0.514	0.811	0.888
SFA	33.12	33.52	2.410	34.77	31.87	2.688	0.897	0.419	0.477
MUFA	34.45	34.21	2.248	36.97	31.69	2.506	0.937	0.119	0.651
PUFA	3.93	3.47	0.192	3.69	3.71	0.214	0.068	0.958	0.767

[1] Diets formulated based on NRC (2000) requirements for dams fed either a concentrate or forage diet during mid- and late-gestation. [2] Probability of difference among least square means. [3] Standard error of the mean.

Table 8. Least squares means for the fatty acid composition (%, g/100 g total fatty acids) of progeny from dams fed a prepartum dietary carbohydrate source consisting of concentrate-based (Conc) or forage-based (For) diet during mid- and late-gestation.

	Treatment [1]			Sex			p-Value [2]		
Fatty Acid	Conc	For	SEM [3]	Heifer	Steer	SEM [3]	Trmt	Sex	T × S
C10:0	0.04	0.04	0.004	0.05	0.04	0.004	0.863	0.130	0.303
C12:0	0.06	0.06	0.004	0.07	0.06	0.004	0.689	0.348	0.349
C14:0	2.97	2.90	0.082	3.08	2.79	0.092	0.508	0.021	0.202
C15:0	0.40	0.42	0.017	0.43	0.39	0.019	0.464	0.096	0.988
C16:0	26.84	27.18	0.380	27.18	26.83	0.424	0.491	0.540	0.403
C17:0	1.18	1.24	0.058	1.25	1.17	0.065	0.410	0.324	0.564
C18:0	14.38	14.80	0.360	13.76	15.41	0.401	0.373	0.003	0.886
C20:0	0.07	0.06	0.007	0.07	0.06	0.008	0.569	0.330	0.269
C14:1n5	0.81	0.73	0.041	0.82	0.71	0.045	0.158	0.082	0.389
C16:1n7	3.05	2.85	0.121	3.15	2.75	0.135	0.194	0.032	0.313
C16:1trans	0.34	0.34	0.010	0.33	0.35	0.011	0.670	0.083	0.867
C17:1	0.99	0.96	0.038	1.05	0.89	0.042	0.497	0.008	0.593
C18:1n9	38.03	38.32	0.616	38.76	37.59	0.0687	0.715	0.203	0.925
C18:1trans	3.62	3.41	0.170	3.27	3.76	0.190	0.343	0.057	0.094
C18:1n7	1.43	1.53	0.152	1.61	1.35	0.170	0.581	0.254	0.609

Table 8. *Cont.*

	Treatment [1]			Sex			p-Value [2]		
Fatty Acid	Conc	For	SEM [3]	Heifer	Steer	SEM [3]	Trmt	Sex	T × S
C18:2trans	0.005	0.005	0.0006	0.005	0.005	0.0007	0.814	0.847	0.213
C18:2n6	4.29	3.88	0.204	3.83	4.35	0.228	0.128	0.095	0.461
C18:3n6	0.02	0.02	0.002	0.02	0.03	0.002	0.346	0.078	0.348
C18:3n3	0.42	0.36	0.031	0.37	0.41	0.034	0.134	0.304	0.769
C20:2	0.09	0.08	0.007	0.08	0.10	0.008	0.428	0.936	0.720
C20:3n6	0.02	0.02	0.001	0.02	0.02	0.002	0.456	0.371	0.808
C20:4n6	0.83	0.70	0.053	0.70	0.83	0.059	0.057	0.120	0.912
C22:3	0.02	0.02	0.002	0.02	0.02	0.002	0.046	0.481	0.797
C24:1n9	0.02	0.01	0.003	0.02	0.02	0.003	0.003	0.639	0.323
C22:5n3	0.04	0.02	0.004	0.03	0.03	0.004	0.007	0.497	0.906
C22:6n3	0.05	0.05	0.006	0.04	0.05	0.007	0.384	0.229	0.936
SFA	45.94	46.70	0.681	45.89	46.75	0.681	0.390	0.397	0.516
MUFA	48.28	48.15	0.627	49.00	47.43	0.699	0.876	0.096	0.649
PUFA	5.78	5.15	0.275	5.11	5.82	0.307	0.080	0.086	0.568
PUFA:SFA	0.13	0.11	0.007	0.11	0.13	0.007	0.088	0.163	0.560
n6:n3	11.21	11.55	0.598	11.17	11.60	0.666	0.655	0.628	0.605
All Lipid	71.50	71.21	4.669	75.43	67.27	5.206	0.962	0.242	0.567

[1] Diets formulated based on NRC (2000) requirements for dams fed either a concentrate or forage diet during mid- and late-gestation. [2] Probability of difference among least square means. [3] Standard error of the mean.

When analyzed as a percentage of total fatty acids (%, g/100 g total fatty acids; Table 8), docosatrienoic (C22:3), nervonic (C24:1n9), and docosapentaenoic (C22:5n3) acids were increased ($p < 0.05$) in samples from the concentrate treatment compared with the forage treatment. Treatment did not influence ($p < 0.05$) the percentage of other fatty acids. The percentage of myristic (C14:0), palmitoleic (C16:1n7), and heptadecenoic (C17:1) acids were increased ($p < 0.05$) in samples from heifers compared with steers, but the percentage of stearic (C18:0) acid was increased ($p < 0.05$) in samples from steers. Sex did not influence ($p > 0.05$) the percentage of other fatty acids.

4. Discussion

The majority of fetal muscle and adipose tissue growth and development occurs during mid- and late-gestation [2]. Alterations to fetal development imposed by maternal stressors, such as maternal nutrient restriction have been shown to have long term impacts on offspring growth and performance [18,20,21]. Dietary carbohydrate sources (i.e., fiber vs. starch) alter molar proportions of ruminal VFA and overall production of VFA's [4]. While this is well documented in the literature, ruminal VFA production was not determined in gestating cows used in the present study, presenting a limitation to the results presented herein. In the present study, drought conditions in 2017 resulted in limited forage availability at the SDSU Antelope Range and Livestock Research Station. Therefore, a management decision was made to transport a portion of the cow herd to a drylot from November 2017 through February 2018 to take advantage of lower cost feedstuffs and preserve range conditions. Based on feed prices in 2017, dams in the concentrate-based treatment were fed a diet that cost approximately $0.90/day and the forage-based treatment were fed a diet that cost approximately $1.07/day. Others have evaluated dietary energy source during late gestation [9], but to date literature concerning the effects of maternal dietary carbohydrate source (forage vs. concentrate) during mid- and late-gestation on offspring performance and meat quality traits is limited.

In agreement with the present study, Radunz et al. [9] reported that maternal energy source did not influence feedlot receiving BW, DMI, ADG, G:F, or final BW of offspring. Taylor et al. [22] also reported that maternal energy status (positive or negative energy status) during mid-gestation did not influence offspring BW, ADG, DMI, or G:F during the finishing phase. However, studies investigating maternal protein supplementation in late gestation have reported differences in offspring performance. Larson et al. [23] investigated the effects of winter grazing system and crude protein supplementation to dams during late gestation, and offspring weaning BW, BW at feedlot entry, reimplant BW, ADG, and DMI were all increased when dams were supplemented with protein during late gestation [23]. Summers et al. [24] compared dams provided a supplement with a high level of rumen undegradable protein (RUP) or a low level of RUP during late gestation to a non-supplemented control. Offspring from dams supplemented with a high level of RUP had increased BW at feedlot entry compared to progeny from non- supplemented dams. However, progeny from non-supplemented dams tended to have greater ADG and had greater DMI during the reimplant period as well as greater overall DMI [24]. Differences in growth performance between studies is likely due to differences in nutrients evaluated (energy vs. protein), timing of maternal dietary treatments during gestation, and varying degrees of restriction or supplementation. However, these studies indicate that growth performance of offspring is sensitive to changes in the maternal diet.

There was a tendency for muscle depth of heifers from the concentrate treatment to be greater (9% increase) compared to heifers from the forage treatment at the initial ultrasound during the backgrounding phase. As ultrasound measures were recorded shortly after the weaning event, this result may indicate that heifers from the forage treatment required longer to adjust to the backgrounding environment, hindering their muscle growth. However, no differences were detected at the finishing period ultrasound, which may be attributed to recovery of muscle growth via compensatory growth. Radunz et al. [9] provided dams either hay-based, corn-based, or dried corn distiller's grains-based diets during late gestation and evaluated carcass measures of progeny via ultrasound at 24 to 72 h after birth and 84 d into the finishing phase. However, unlike the present study, no differences were reported in ultrasound measures of progeny carcass traits. Differences in diet composition, timing of dietary treatments during gestation, and timing of ultrasound evaluation may explain the differences between the findings of Radunz et al. [9] and the present study.

Backfat thickness of offspring from forage fed dams tended to be decreased by 7% and USDA Yield Grades also tended to be 7% lower than offspring from concentrate fed dams. While no direct comparisons with the present study are available in the literature, other research has demonstrated that offspring fat depots may be especially sensitive to alterations in the maternal diet. When fed to a common BF endpoint, Radunz et al. [9] reported that offspring from dams fed a fiber-based diet (hay) in late gestation had increased marbling scores and no carcasses that graded USDA Select compared to offspring from dams fed a starch-based diet (corn). Underwood et al. [21] reported that BF and adjusted 12th rib BF were increased in offspring from dams grazing improved pasture that provided more crude protein than offspring form dams grazed on native range during mid gestation. Wilson et al. [25] observed a tendency for progeny from dams provided a distiller's grain supplement during late gestation to have decreased backfat thickness compared to progeny from dams that were not supplemented. Steers from dams fed supplemental protein during late gestation were reported to have increased marbling scores, as well as a greater proportion of carcasses grading USDA Choice or higher compared to steers from dams not supplemented protein [23]. Mohrhauser et al. [18] reported a tendency for decreased BF and lower USDA Yield Grades, with no influence on marbling score, in offspring from dams in a negative maternal energy status during mid-gestation compared to offspring from dams in a positive maternal energy status. Summers et al. [24] also observed decreased 12th rib fat thickness with no differences in marbling score in progeny from dams that were supplemented a diet with low RUP in late gestation compared to progeny from dams not supplemented with RUP.

Heifers had increased BF (14%) and USDA Yield Grade (7%) compared to steers but decreased HCW (9%) and REA (8%). Mohrhauser et al. [18] also reported steers to have heavier HCW, reduced marbling scores, and larger ribeye areas. However, in contrast to the present study, steers were reported to have higher a^* values and tended to have higher L^* values compared to heifers [18]. In addition, the marbling score of heifers was greater (9%) compared to steers. This is consistent with other studies suggesting heifers have greater amounts of marbling when compared to steers and bulls [26].

Because there were no differences in marbling scores between treatment groups, the lack of difference in crude fat and moisture content is not unexpected. Other studies investigating alterations in maternal energy have evaluated WBSF and also reported no differences in this objective measure of tenderness [9,18]. However, studies investigating alterations in maternal protein levels reported steaks from offspring of dams with restricted protein intake during mid-gestation had increased WBSF values (less tender meat) compared to offspring of dams with adequate protein intake [20,21]. Other studies investigating the effects of maternal nutrition during gestation on sensory characteristics of steaks are lacking. Heifers had increased crude fat (25%) and decreased moisture content (2%) compared to steers, which is likely attributed to heifers having greater amounts of marbling.

There is limited information on the effects of maternal diet on the fatty acid composition of meat from offspring. Webb et al. [20] reported that arachidonic acid was sensitive to changes in maternal diet. Offspring of dams provided adequate protein during mid-gestation produced offspring with increased concentrations of arachidonic acid compared with protein restricted dams. A study by Chail et al. [27] evaluated the effects of finishing diet on fatty acid composition in the *gluteus medius* and *triceps brachii* and observed increased concentration of arachidonic acid when cattle were fed a grain-based diet as compared to a forage-based diet. In a recent review, Ponnampalam et al. [28] outlined that concentrate-based diets are common sources of omega-6 (n-6) polyunsaturated fatty acids compared to forage-based diets, which are common sources of omega-3 (n-3) polyunsaturated fatty acids. This is an important concern as current human dietary recommendations suggest a n6:n3 of 1–4:1. In the present study no differences were observed between treatment groups when n6:n3 fatty acid levels of progeny were evaluated. However, results from the present study suggest that maternal diet can influence fatty acid composition of steaks from progeny and warrants further investigation.

5. Conclusions

Results from this study suggest variation in maternal carbohydrate source during mid- and late-gestation has limited influence on progeny performance. Collectively, these data suggest a forage-based diet provided to cows during mid- and late-gestation differentially influences deposition of subcutaneous fat without compromising marbling score or tenderness. As dams in the present study were fed to meet nutrient requirements during mid- and late-gestation, mechanisms by which carbohydrate source in mid- to late-gestation can affect growth rate of progeny might be minimized when energy needs of the cow are met. Provided that nutrient requirements are met, it appears that utilizing alternative diets for the beef cow herd does not significantly influence progeny performance and beef product quality. Based on this study, cattle producers have flexibility to feed their gestating cows available carbohydrate sources during drought and/or variable growing conditions without concern for offspring performance or carcass traits.

Author Contributions: Conceptualization, A.D.B., K.C.O., J.J.B., J.K.G., W.C.R., R.R.S., A.A.H. and K.R.U.; methodology, E.R.G., R.R.S., W.C.R., J.J.B., C.L.W., A.A.H., Z.K.S., J.F.L., J.K.G., K.R.U., K.C.O. and A.D.B.; validation, E.R.G., K.M.C., Z.K.S., J.F.L., J.K.G., K.R.U., K.C.O. and A.D.B.; formal analysis, E.R.G., K.M.C., K.C.O., J.F.L. and A.D.B.; investigation, E.R.G., K.C.O. and A.D.B.; resources, K.C.O., K.M.C., Z.K.S., C.L.W., J.F.L., J.K.G., K.R.U. and A.D.B.; data curation, E.R.G., R.R.S., J.J.B., K.C.O., J.K.G., K.R.U. and A.D.B.; writing—original draft preparation, E.R.G.; writing—review and editing, J.J.B., R.R.S., A.A.H., W.C.R., C.L.W., K.M.C., J.F.L., Z.K.S., J.K.G., K.R.U., K.C.O. and A.D.B.; visualization, E.R.G. and A.D.B.; supervision, A.D.B.; project administration, A.D.B.; funding acquisition, A.D.B., K.C.O., J.J.B., W.C.R., R.R.S., A.A.H., J.K.G. and K.R.U. All authors have read and agreed to the published version of the manuscript.

Funding: This research was supported by state and federal funds appropriated to South Dakota State University including support by the USDA National Institute of Food and Agriculture, Hatch project (accession no. 1020088), and from the Beef Checkoff through the South Dakota Beef Industry Council (Grant no. 3X9408).

Institutional Review Board Statement: All animal care and experimental protocols were approved by the South Dakota State University Animal Care and Use Committee (approval number 18-081E). The human sensory panel utilized in this study was approved by the Institutional Review Board of South Dakota State University (IRB-1911019-EXM, 14 November 2019).

Informed Consent Statement: Informed consent was obtained from all subjects involved in the study.

Data Availability Statement: The data presented in this study are available on request from the corresponding author.

Acknowledgments: The authors express their gratitude to the South Dakota Beef Industry Council for support of this project.

Conflicts of Interest: The authors declare no conflict of interest.

References

1. Funston, R.N.; Summers, A.F.; Roberts, A.J. Alpharma beef cattle nutrition symposium: Implications of nutritional management for beef cow-calf systems. *J. Anim. Sci.* **2012**, *90*, 2301–2307. [CrossRef] [PubMed]
2. Du, M.; Tong, J.; Zhao, J.; Underwood, K.R.; Zhu, M.; Ford, S.P.; Nathanielsz, P. Fetal programming of skeletal muscle development in ruminant animals. *J. Anim. Sci.* **2010**, *88*, E51–E60. [CrossRef]
3. Du, M.; Huang, Y.; Das, A.K.; Yang, Q.; Duarte, M.S.; Dodson, M.V.; Zhu, M.-J. Meat science and muscle biology symposium: Manipulating mesenchymal progenitor cell differentiation to optimize performance and carcass value of beef cattle. *J. Anim. Sci.* **2013**, *91*, 1419–1427. [CrossRef] [PubMed]
4. Smith, S.B.; Crouse, J.D. Relative contributions of acetate, lactate and glucose to lipogenesis in bovine intramuscular and subcutaneous adipose-tissue. *J. Nutr.* **1984**, *114*, 792–800. [CrossRef]
5. Ferrell, C.L.; Ford, S.P.; Prior, R.L.; Christenson, R.K. Blood Flow, Steroid Secretion and Nutrient Uptake of the Gravid Bovine Uterus and Fetus. *J. Anim. Sci.* **1983**, *56*, 656–667. [CrossRef] [PubMed]
6. Bell, A.W.; Bauman, D.E. Adaptations of Glucose Metabolism during Pregnancy and Lactation. *J. Mammary Gland. Biol. Neoplasia* **1997**, *2*, 265–278. [CrossRef] [PubMed]
7. Penner, G.B.; Taniguchi, M.; Guan, L.; Beauchemin, K.; Oba, M. Effect of dietary forage to concentrate ratio on volatile fatty acid absorption and the expression of genes related to volatile fatty acid absorption and metabolism in ruminal tissue. *J. Dairy Sci.* **2009**, *92*, 2767–2781. [CrossRef]
8. Corah, L.R.; Dunn, T.G.; Kaltenbach, C.C. Influence of Prepartum Nutrition on the Reproductive Performance of Beef Females and the Performance of Their Progeny. *J. Anim. Sci.* **1975**, *41*, 819–824. [CrossRef] [PubMed]
9. Radunz, A.E.; Fluharty, F.L.; Relling, A.E.; Felix, T.L.; Shoup, L.M.; Zerby, H.N.; Loerch, S.C. Prepartum dietary energy source fed to beef cows: II. Effects on progeny postnatal growth, glucose tolerance, and carcass composition. *J. Anim. Sci.* **2012**, *90*, 4962–4974. [CrossRef] [PubMed]
10. AOAC. *Official Methods of Analysis*, 19th ed.; Association of Official Analytical Chemist: Gaithersburg, MD, USA, 2012.
11. AOAC International. *AOAC Official Method 990.03 Protein (Crude) in Animal Feed. Combustion Method. Official Methods of Analysis of AOAC International*, 19th ed.; Association of Official Analytical Chemist: Gaithersburg, MD, USA, 2012.
12. Ankom. *Ankom Technology Method 6. Neutral Detergent Fiber in Feeds. Filter Bag Technique. Ankom Technology. Macedon, New York. This Method Is a Modification of AOAC Official Method 2002.04. Amylase-Treated Neutal Detergent Fiber in Feeds. Using Refluxing in Beakers or Crucibles. Official Methods of Analysis of AOAC International*, 19th ed.; Association of Official Analytical Chemist: Gaithersburg, MD, USA, 2006.
13. Ankom. *Ankom Technology Method 5. Acid Detergent Fiber in Feeds. Filter Bag Technique. Ankom Technology. Macedon, New York. This Method Is a Modification of AOAC Official Method 973.18. Fiber (Acid Detergent) and Lignin (H2SO4) in Animal Feed. Official Methods of Analysis of AOAC International*, 19th ed.; Association of Official Analytical Chemist: Gaithersburg, MD, USA, 2006.
14. AOAC International. *AOAC Official Method 942.05. Ash of Animal Feed. Official Methods of Analysis of AOAC International*, 19th ed.; Association of Official Analytical Chemist: Gaithersburg, MD, USA, 2012.
15. AOAC International. *AOAC Official Method 2003.06. Crude Fat in Feeds, Cereal Grains, and Forages. Official Methods of Analysis of AOAC International*, 19th ed.; Modified for Extraction with Petroleum Ether; Association of Official Analytical Chemist: Gaithersburg, MD, USA, 2012.
16. USDA. *United States Standards for Grades of Carcass Beef*; United States Department of Agriculture, Agriculture Marketing Service: Washington, DC, USA, 2017.
17. American Meat Science Association (AMSA). *Research Guidelines for Cookery, Sensory Evaluation, and Instrumental Measurements of Meat*; AMSA: Savoy, IL, USA, 2015.

18. Mohrhauser, D.A.; Taylor, A.R.; Underwood, K.R.; Pritchard, R.H.; Wertz-Lutz, A.E.; Blair, A.D. The influence of maternal energy status during midgestation on beef offspring carcass characteristics and meat quality. *J. Anim. Sci.* **2015**, *93*, 786–793. [CrossRef] [PubMed]
19. O'Fallon, J.V.; Busboom, J.R.; Nelson, M.L.; Gaskins, C.T. A direct method for fatty acid methyl ester synthesis: Application to wet meat tissues, oils, and feedstuffs. *J. Anim. Sci.* **2007**, *85*, 1511–1521. [CrossRef] [PubMed]
20. Webb, M.; Block, J.; Funston, R.; Underwood, K.; Legako, J.; Harty, A.; Salverson, R.; Olson, K.; Blair, A. Influence of maternal protein restriction in primiparous heifers during mid- and/or late-gestation on meat quality and fatty acid profile of progeny. *Meat Sci.* **2019**, *152*, 31–37. [CrossRef]
21. Underwood, K.; Tong, J.; Price, P.; Roberts, A.; Grings, E.; Hess, B.; Means, W.; Du, M. Nutrition during mid to late gestation affects growth, adipose tissue deposition, and tenderness in cross-bred beef steers. *Meat Sci.* **2010**, *86*, 588–593. [CrossRef] [PubMed]
22. Taylor, A.R.; Mohrhauser, D.A.; Pritchard, R.H.; Underwood, K.R.; Wertz-Lutz, A.E.; Blair PAS, A.D. The influence of maternal energy status during mid-gestation on growth, cattle performance, and the immune response in the resultant beef progeny. *Prof. Anim. Sci.* **2016**, *32*, 389–399. [CrossRef]
23. Larson, D.M.; Martin, J.L.; Adams, D.C.; Funston, R.N. Winter grazing system and supplementation during late gestation influence performance of beef cows and steer progeny. *J. Anim. Sci.* **2009**, *87*, 1147–1155. [CrossRef] [PubMed]
24. Summers, A.F.; Blair, A.D.; Funston, R.N. Impact of supplemental protein source offered to primiparous heifers during gestation on II. Progeny performance and carcass characteristics. *J. Anim. Sci.* **2015**, *93*, 1871–1880. [CrossRef]
25. Wilson, T.B.; Schroeder, A.R.; Ireland, F.A.; Faulkner, D.B.; Shike, D.W. Effects of late gestation distillers grains supplementation on fall-calving beef cow performance and steer calf growth and carcass characteristics. *J. Anim. Sci.* **2015**, *93*, 4843–4851. [CrossRef] [PubMed]
26. Park, S.J.; Beak, S.-H.; Jung, D.J.S.; Kim, S.Y.; Jeong, I.H.; Piao, M.Y.; Kang, H.J.; Fassah, D.M.; Na, S.W.; Yoo, S.P.; et al. Genetic, management, and nutritional factors affecting intramuscular fat deposition in beef cattle—A review. *Asian-Australas. J. Anim. Sci.* **2018**, *31*, 1043–1061. [CrossRef] [PubMed]
27. Chail, A.; Legako, J.F.; Pitcher, L.R.; Ward, R.E.; Martini, S.; MacAdam, J.W. Consumer sensory evaluation and chemical composition of beef gluteus medius and triceps brachii steaks from cattle finished on forage or concentrate diets. *J. Anim. Sci.* **2017**, *95*, 1553–1564. [PubMed]
28. Ponnampalam, E.N.; Sinclair, A.J.; Holman, B.W.B. The Sources, Synthesis and Biological Actions of Omega-3 and Omega-6 Fatty Acids in Red Meat: An Overview. *Foods* **2021**, *10*, 1358. [CrossRef] [PubMed]

Article

Effect of Aging and Retail Display Conditions on the Color and Oxidant/Antioxidant Status of Beef from Steers Finished with DG-Supplemented Diets

Manuela Merayo [1,2], Sergio Aníbal Rizzo [1,3], Luciana Rossetti [1], Dario Pighin [1,3,*] and Gabriela Grigioni [1,3]

1 Instituto Tecnología de Alimentos—Instituto de Ciencia y Tecnología de Sistemas Alimentarios Sustentables, UEDD Instituto Nacional de Tecnología Agropecuaria—Consejo Nacional de Investigaciones Científicas y Técnicas, CC 25, Castelar 1712, Argentina; merayo.manuela@inta.gob.ar (M.M.); rizzo.sergio@inta.gob.ar (S.A.R.); rossetti.luciana@inta.gob.ar (L.R.); grigioni.gabriela@inta.gob.ar (G.G.)

2 Facultad de Ciencias Médicas, Pontificia Universidad Católica Argentina, Alicia M. de Justo 1500, Ciudad Autónoma de Buenos Aires 1425, Argentina

3 Escuela Superior de Ingeniería, Informática y Ciencias Agroalimentarias, Universidad de Morón, Cabildo 134, Morón 1708, Argentina

* Correspondence: pighin.dario@inta.gob.ar; Tel.: +54-911-6885-7665

Citation: Merayo, M.; Rizzo, S.A.; Rossetti, L.; Pighin, D.; Grigioni, G. Effect of Aging and Retail Display Conditions on the Color and Oxidant/Antioxidant Status of Beef from Steers Finished with DG-Supplemented Diets. *Foods* **2022**, *11*, 884. https://doi.org/10.3390/foods11060884

Academic Editors: Benjamin W.B. Holman and Eric Nanthan Ponnampalam

Received: 10 January 2022
Accepted: 10 February 2022
Published: 20 March 2022

Abstract: The aim of this work was to study the effect of finishing diets including distiller grains (DG) on color and oxidative stability of beef after being exposed to aerobic retail display conditions, with or without previous aging. For this purpose, beef samples from animals fed with finishing diets including 0%, 15%, 30%, and 45% DG (on a dry matter basis), which had been exposed to aerobic retail display conditions, with or without previous aging under vacuum packaging, were evaluated. The content of γ-tocopherol, β-carotene, and lutein in diet samples increased with the level of DG. Beef evaluated at 72 h post-mortem showed greater content of γ-tocopherol and retinol as the DG level increased. Meat color was not affected by DG inclusion, but color parameters decreased with retail conditions. Meat from animals fed with DG showed the lowest values of thiobarbituric acid reactive substances (TBARS), independently of the retail display conditions. However, all samples were below the threshold associated with rancid aromas and above the a* value related to meat color acceptance. Thus, feeding diets including up to 45% of DG improved the antioxidant status of meat, preserving the color, and delaying lipid oxidation in meat samples under the display conditions evaluated.

Keywords: beef; distiller grains; antioxidants; oxidative stability; color

1. Introduction

The bioethanol industry has been an important contributor to country economies since it has supplied renewable energy sources and has demanded biobased feedstocks [1]. Some countries, like the USA, Brazil, and Canada, have a long history in bioethanol production and they shared, in 2021, 84% of world production [2], while other countries, like Argentina, have produced bioethanol more recently, using corn as a feedstock. Distiller grains (DG), which are the byproduct of the bioethanol industry, are considered a valuable supplement in animal feeding because they are rich in the fat, protein, and fiber portion of corn grain [3,4]. Several works have analyzed the effect of feeding with DG on beef pro-oxidative components, like unsaturated fatty acids and lipid oxidation [3]. However, few reports have focused on the effect of DG on beef anti and pro-oxidant balance and its relationship with beef color and lipid oxidation. During meat storage, lipid oxidation could be influenced by the packaging and ambient conditions, and intrinsic meat characteristics, like the balance of anti- and pro-oxidants content and the abundance of unsaturated fatty acids [5]. This process, together with microbial spoilage, is the main cause of meat quality deterioration, affecting color, flavor, and nutritional value [6,7]. Previous works have stated that there is a relationship between meat color deterioration and lipid oxidation since the biochemical reactions responsible for both myoglobin and lipid oxidation generate products that can accelerate oxidation processes reciprocally [5]. Moreover, it is known that meat visual appearance plays an important role in consumers' choices [8]. This visual appearance depends on physical and chemical factors, including the meat color, which is one of the most important attributes evaluated by consumers at the purchase point [8–10]. Meat color could be influenced by a variety of factors related to animal production and commercialization strategies, including aging and retail display conditions [9,11]. Meat visual appearance can be compromised by discoloration, which is related to oxymyoglobin oxidization to metmyoglobin [5,11].

Oxidative processes could be delayed through different strategies, including the addition of antioxidant compounds into the meat product, adding these compounds directly on the meat surface or through active packaging, and including antioxidant compounds in the feeding diets of the animals [6,12,13]. Among antioxidant compounds, α-tocopherol has been demonstrated to be effective in preserving color in beef, while it has an important effect on lipid oxidation delay [9]. In addition, dietary delivery of α-tocopherol has been shown to be more effective than its exogenous addition due to more efficient antioxidant incorporation through cell membranes [5,13,14]. Among feeding diets, there are plenty of strategies targeted either to improve meat quality, like supplementation with antioxidant compounds or the inclusion of feeding sources of n-3 polyunsaturated fatty acids (PUFAs) or to reduce the cost of the diet, like the inclusion of agro-industrial byproducts [7,15]. In this sense, several studies have demonstrated that DG supplementation increased PUFA content of beef and lipid oxidation during retail display conditions [16,17], while other studies have found no effect of DG inclusion on lipid oxidation [18,19]. These divergencies could be explained by the complex balance required between antioxidant and pro-oxidant compounds present in meat. Therefore, the aim of the present study was to evaluate the effect of finishing diets containing increasing levels (15, 30, and 45%) of wet DG with solubles on color and oxidative stability of *Longissimus thoracis et lumborum* (LTL) steaks after being exposed to aerobic retail display conditions, with or without previous aging under vacuum packaging.

Our findings indicate that beef from DG diets had greater antioxidant status than beef from diets without DG inclusion, which positively impacts the balance between anti- and pro-oxidant compounds and meat oxidative stability during simulated commercial conditions.

2. Materials and Methods

2.1. Experimental Design and Diets

Meat samples (LTL) were obtained from yearling Angus steers that had been fed with finishing diets including increasing levels of wet DG with solubles. For this purpose, a total of thirty-six weaned steers (initial live weight, 191 ± 12 kg) were randomly allocated to 12 pens (three steers per pen) and fed with a high-concentrate diet for three weeks to adapt to the feeding system. Then, the pens (3 pens per treatment) were randomly assigned to one of the four dietary treatments. Steers were fed for 70 days. The control diet consisted of cracked corn grain, soybean meal, and alfalfa hay. The dietary treatments consisted of four levels of inclusion of DG, replacing the corn and soybean meal: 0% (0DG), 15% (15DG), 30% (30DG), or 45% (45DG) (dry matter (DM) basis; Table 1). The diets were formulated to be isoenergetic (2.95 Mcal/kg DM) and isoproteic (14.9%), except for 45DG which supplied 17.9%. Three samples of each diet were collected during the feeding, vacuum packed, and stored in darkness until analysis. No effect of feeding diets on animal performance during the trial was observed and animals showed an average daily gain of 1.586 ± 0.045 kg/d. Details of the feeding diets and cattle performance have been previously reported [20,21].

Table 1. Composition of the finishing diets on a dry matter (DM) basis (g/100 g DM).

Item	Dietary Treatment [1]			
	0DG	15DG	30DG	45DG
Ingredient (% DM)				
Cracked corn grain	84	74	64	48
Distiller grains (DG)	0	15	30	45
Soybean meal	10	5	0	0
Alfalfa hay	6	6	6	6
Nucleus [2]	0.3	0.3	0.3	0.3
Chemical composition (% DM)				
Crude protein	11.72	12.82	14.02	16.80
Fat	4.16	5.37	6.44	7.56
Ash	2.87	3.14	3.27	3.73

[1] Dietary treatment: 0DG, control, 15DG 15% DG, 30DG 30% DG and 45DG 45% DG. [2] Nucleus composition (Vetifarma SA, La Plata, Argentina): vitamin A: 1,000,000 IU/kg; vitamin D3: 200,000 IU/kg; vitamin E: 6500 IU/kg; vitamin B1: 650 ppm; manganese: 12,000 ppm; zinc: 12,000 ppm; copper: 6000 ppm; cobalt: 40 ppm; selenium: 60 ppm; iodine: 200 ppm; with added calcium 0.05%.

When animals reached slaughter conditions (commercial endpoint based on visual appraisal and final body weight of 316 ± 14 kg on average), they were transported to a licensed commercial abattoir in a single group. Animals were slaughtered according to standard commercial procedures on the same day after mixing groups to avoid peri-slaughter effects. Carcasses were suspended through the Achilles tendon and were not electrically stimulated.

2.2. Sample Preparation

After slaughter, carcasses were individually graded and chilled at 4 °C. Forty-eight hours later, a section of the LTL from the 13th rib region was removed from the striploin of each carcass and transported under refrigeration to the Food Technology Institute of the National Institute of Agricultural Technology (INTA), Buenos Aires, Argentina. At 72 h post-mortem, two steaks (2.5 cm thick each) from each rib section were stored at one of the following retail display conditions that resembled commercial local practice:

- R1: the steaks were individually placed on Styrofoam trays, overwrapped with oxygen-permeable polyvinylchloride film, and placed under refrigeration at 4 ± 2 °C for four days in darkness and then for three days with 7 h of illumination (D65, 700 lux) per day.
- R2: the steaks were placed under vacuum-packed conditions at 1 ± 1 °C and darkness for 25 days, plus a retail display period under aerobic exposure. For this last stage, the steaks were placed on a Styrofoam tray overwrapped with oxygen-permeable polyvinylchloride film for three days under refrigeration at 4 ± 2 °C and with 7 h of illumination (D65, 700 lux) per day.

Other steaks from each rib section were stored at −80 °C until further analysis.

2.3. α- and γ-Tocopherol, β-Carotene, Retinol and Lutein Content

The contents of α-tocopherol, γ-tocopherol, β-carotene, α-carotene, retinol, and lutein from feedstuffs and meat samples were determined as described by Buttriss and Diplock [22] with modifications by Descalzo et al. [23]. Briefly, 5 g of lean tissue was placed in a plastic conical tube containing 10 mL of phosphate buffer (0.05 M; pH 7.7) and homogenized for 2 min at 3000 rpm with an Ultraturrax T25 homogenizer (IKA, Darmstadt, Germany). Aliquots of 1 g of homogenate were placed into a screwcap test tube with 3 mL of pyrogallol 1% in ethanol. Thereafter, 0.3 mL of KOH 12 N in water was added to each tube for saponification. The tube contents were mixed by vortexing for 30 s, and placed in a stirred water bath for 30 min at 70 °C. After cooling, 1 mL of water was added to each tube. Following the addition of 5 mL of n-hexane, the samples were mixed by vortexing for 2 min; the upper hexane layer was then transferred into a new screw cap tube and the aqueous phase was reextracted with 5 mL of n-hexane. The combined extracts were taken to dryness under a dry nitrogen gas stream, and the residue was dissolved in 500 μL of absolute ethanol (J.T. Baker, Mexico, HPLC grade) and filtered through a 0.45 μm-pore nylon membrane before injection of samples. All samples and standards (external standards for each vitamin) were analyzed by reverse-phase high-performance liquid chromatography (HPLC) using a Thermo Scientific Dionex UltiMate 3000 RS system consisting of a quaternary pump with a membrane vacuum degasser connected to an autosampler with an injection loop (10 to 100 μL) and a C18 column (250 × 4.6 mm i.d., Alltima, 5 μm particle size; Alltech, Argentina) fitted with a guard column (Security GuardAlltima C18, Alltech, Argentina). The mobile phase was ethanol:methanol (60:40, *v*/*v*) at a flow rate of 1 mL/min. The technique was optimized to determine tocopherols, carotenoids, and retinol within the same elution time of 25 min. Tocopherols were detected by fluorescence at λ_{ex} = 296 and λ_{em} = 330 nm. A diode array detector was set at λ = 445 nm and λ = 325 nm for the detection of carotenoids and retinol, respectively. Calibration curves were performed with DL-α-tocopherol (Merck, Darmstadt, Germany), γ-tocopherol, β-carotene, lutein, and retinol standards (Sigma-Aldrich, St. Louis, MO, USA) diluted in ethanol. Chromatograms were recorded using the Chromeleon 6 software.

For feedstuff samples, the methodology was identical, except that the samples were diluted 1 to 10 with ethanol before injection.

2.4. Muscle Ferric Reducing Antioxidant Power (FRAP)

The FRAP assay applied to meat samples measures endogenous ions that could react with tripyridyltriazine (TPTZ) and develop blue color. Following the procedure described by Ou et al. [24] and modified by Descalzo et al. [25], 5-g chopped meat samples were disrupted for 2 min at 3000 rpm with an Ultraturrax homogenizer (IKA, Staufen, Germany) in 10 mL of potassium phosphate buffer (0.05 M, pH 7.7). Homogenates were centrifuged at 10,000× *g* for 30 min at 4 °C and the supernatant was collected. Then, 83-μL aliquots of supernatant were added to 2.5 mL of FRAP buffer containing 10 mM TPTZ, 40 mM HCl and 20 mM Fe_2Cl_3 (Sigma-Aldrich, St. Louis, MO, USA) added to 300 mM acetate buffer. The reaction mixture was incubated for 5 min at 37 °C in a water bath and then cooled in an ice-water bath for 10 min. Immediately after, samples were measured at λ = 593 nm (Spectrometer UV–vis-BIO Lambda 20, Perkin Elmer). Another 83-μL aliquots of supernatant were added to 2.5 mL of TPTZ/HCl solution, without the addition of Fe_2Cl_3, to determine endogenous Fe^{+2} content. The FRAP activity of the samples was measured against a calibration curve made with ferrous sulfate ($Fe_2SO_4 \cdot 7H_2O$, Sigma-Aldrich de Argentina SA) within a range from 100 to 1000 μM, and results were expressed as Fe^{+2} equivalent in μM.

2.5. Thiobarbituric Acid Reactive Substances (TBARS)

TBARS were analyzed by the steam distillation method, as described by Pensel [26] with modifications of Descalzo et al. [23], and expressed as mg of malonaldehyde (MDA) per kg of lean muscle. Briefly, triplicate aliquots (5 g) of meat were chopped and processed in a stomacher-type homogenizer for 2 min in bags containing 12.5 mL of trichloroacetic acid (Merck, Darmstadt, Germany) solution (20% *w*/*v*) in 1.6% metaphosphoric acid. Then, 12.5 mL of water was added, and the mixture was processed for another 30 s. Slurries were filtered and aliquots of 5 mL were separated. An equal volume (5 mL) of 0.02 M 2-thiobarbituric acid (Sigma-Aldrich, St. Louis, MO, USA) was added. Samples were incubated at 80 °C for 1 h until a pink color was developed. Color intensity was determined at maximum absorption, λ = 530 nm, and TBARS concentrations were calculated from a calibration curve by using 1,1,3,3-tetraethoxypropane (Sigma-Aldrich, St. Louis, MO, USA) as standard within a range from 0 to 0.5 μM.

2.6. Meat Color

Meat color was assessed using the CIELab system, which provides the color parameters L* (lightness, from black to white), a* (redness, from green to red), and b* (yellowness, from blue to yellow). The chroma parameter was calculated according to the American Meat Science Association (AMSA) [27]. Measurements were carried out with a Minolta CR-400 colorimeter (Konica Minolta Sensing, Inc., Bergen, NJ, USA) as described by the AMSA [27]. The instrumental conditions used were artificial D65 illuminant, 8 mm port size, and a two-degree standard angle observer. The instrument was calibrated against a white plate (Y = 93.8, x = 0.3155, y = 0.3319). Each sample was allowed to bloom for 45 min at 4 °C prior to the first measurement, and six scans of each steak were averaged for statistical analysis.

2.7. Statistical Analysis

Statistical analysis was conducted using InfoStat Software version 2018e [28]. Data normality was checked with the Shapiro-Wilk test and homogeneity of variances with the Levene test. Data that did not show a normal distribution or homogeneity of variances (p-value < 0.05) were analyzed with the Kruskal-Wallis test. Data from diet composition were analyzed as a completely randomized design with the level of DG inclusion in the diets as the main effect. Data from meat quality at 72 h post-mortem were analyzed as a split-plot design where pens were the whole plot and the level of DG in the diets was the split-plot. Linear and quadratic relationships were detected by response curves. Data from the retail display were analyzed as a split-split plot design where pens were the whole plot, the level of DG in the diets was the split-plot and the conditions of the retail display were the split-split plot. The least significant differences were set at a 5% level and means were compared by Tukey's test. Principal component analysis (PCA) was applied to the data of beef parameters in each retail display condition, using SPSS 13.0, following [29].

3. Results and Discussion

3.1. Content and Animal Intake of Antioxidant Compounds in Feedstuffs

Corn is a good source of fat-soluble antioxidant compounds, such as vitamin E and carotenoids [30,31]. Carotenoids are composed of carotenes and xanthophylls, which provide yellow color to animal products [30,31]. The co-products obtained after corn processing, like DG, increase the concentration of corn nutrients, except for sugars [3]. Previous reports have indicated that samples of dried DG with solubles obtained from different plants of bioethanol production have, on average, two-fold higher content of α-tocopherol, γ-tocopherol, and lutein than corn grain [30].

In our study, the content of α-tocopherol was almost three times greater in the dietary treatments with DG inclusion (Table 2) than in the control diet. Other authors have reported that the content of α-tocopherol was 8.0 and 7.4 µg/g in corn-based diets supplemented with 20 and 40% DG respectively and no differences were seen when compared with dry-rolled corn control diet (9.3 µg/g) [32].

Table 2. Tocopherol and carotenoid contents (µg/g DM) in feeding diets [1].

Item	0DG [1]	15DG	30DG	45DG	SEM [2]	p-Value	L [3]	Q
γ-tocopherol	48.37 [c]	102.43 [b]	132.80 [a,b]	160.68 [a]	9.84	0.0002	<0.0001	NS
α-tocopherol	5.25 [b]	12.00 [a]	16.15 [a]	17.27 [a]	1.45	0.002	0.0003	NS
β-carotene	1.91 [c]	1.95 [b,c]	2.04 [a,b]	2.11 [a]	0.02	0.0002	<0.0001	NS
α-carotene	n.d. [4]	n.d.	n.d.	n.d.				
Lutein	9.62 [b]	9.53 [b]	13.18 [a]	15.01 [a]	0.75	0.002	0.0004	NS

[1] Dietary treatment: 0DG, control; 15DG, 15% DG; 30DG, 30% DG and 45DG, 45% DG (%DM basis). [2] SEM: standard error of the mean. [3] Linear (L) and quadratic (Q) response to DG level. Means in the same row having different letters are significant at the $p \leq 0.05$ level. NS: no significant. [4] n.d., not detectable.

Regarding γ-tocopherol, β-carotene, and lutein, their contents increased with the level of DG inclusion in the diets (Table 2). To our knowledge, previous works had analyzed only the content of these compounds in DG alone, as an ingredient, but not in feedstuffs that included DG. Therefore, the values obtained in this work were compared with values reported for complete diets based on corn grain. The values selected for the comparison were those reported by Blanco et al. [33] for total mixed ration (TMR) and corn diets and those by Pouzo et al. [34] for corn diets. The values of lutein observed were similar to those reported for TMR (10.1 μg/g DM), whereas those of γ-tocopherol observed in 15DG, 30DG, and 45DG were higher than the values reported for the corn diet (79.5 μg/g DM, [33] and 14.1 μg/g DM, [34]). Regarding β-carotene, the values observed in all dietary treatments in this study were greater than those reported for the corn diet (1.47 μg/g DM, [34]).

To estimate the vitamin intake of the animals, the composition of the feeding diets was weighted with the consumption of animals (Figure 1). The DM consumption of repetitions and treatments was measured every other week [20] and an average of consumption was calculated for each diet: 8.64 DM/day for 0DG, 8.77 DM/day for 15DG, 8.31 DM/day for 30DG, and 8.17 DM/day for 45DG. No statistical analysis was done on these data since DM intake was recorded on a group basis. Results showed that the intake of antioxidant compounds, except for that of β-carotene, numerically increased with the level of DG inclusion in the diets.

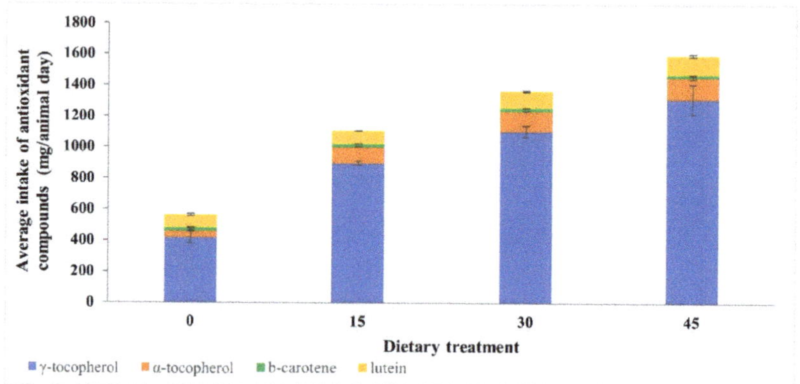

Figure 1. Average intake (mean ± SD) of antioxidant compounds weighted with the consumption of the animals, for each dietary treatment (0, 0% DG control; 15, 15% DG; 30, 30%DG; 45, 45% DG, DM basis).

3.2. Antioxidant Compounds in Fresh LTL Muscle

The diet composition influences animal metabolism, use, and storage of glycogen and accumulation of anti- and pro-oxidant compounds [13,35]. Thus, the feeding diets could affect the content of antioxidant compounds in meat and its susceptibility to oxidative reactions. In our study, the content of γ-tocopherol in meat samples increased with the level of DG (Table 3), which could be explained by the fact that the feeding diets supplied greater content of γ-tocopherol and that the intake of animals increased with the level of DG inclusion. Besides, the content of γ-tocopherol observed was greater than that reported by Pouzo et al. [34], when feeding diets were supplemented with flaxseed.

It has been reported that the supplementation of feeding diets with vitamin E increases the concentration of α-tocopherol in meat [13,32], and that feeding diets including less than 200 μg/g of α-tocopherol produce lamb meat with 1.97 μg/g of α-tocopherol, on average [36]. In our study, although feeding diets were not supplemented with vitamin E, a numerical increase of α-tocopherol was observed in meat samples as the level of DG increased. Salami et al. [37] have recently shown no effect of the inclusion of DG in finishing diets on the content of α-tocopherol in meat. These authors reported that the α-tocopherol content was 2.57 μg/g muscle, on average, i.e., greater than the values observed in our study. However, in their case, the supplementation of DG was in combination with *ad libitum* grass silage, which provides greater amounts of α-tocopherol than grain [13]. Other authors have reported that ground beef from animals fed with 20 and 40% of DG in corn-based diets had 1.70 and 1.79 μg/g of α-tocopherol, respectively [32]. In our work, the content of α-tocopherol was lower than the value reported by Da Silva Hampel et al. [36] in the meat of lambs fed with 0–200 μg/g of tocopherol level in the diet; it was only similar (1.41 μg/g) in beef samples from 45DG dietary treatment, which supplied 17.27 μg/g of α-tocopherol.

Table 3. Content of antioxidant vitamins and oxidation stability in fresh meat samples (LTL) muscle at 72 h post-mortem.

Item	0DG [1]	15DG	30DG	45DG	SEM [2]	*p*-Value	L [3]	Q
			Antioxidant compounds (μg/g meat)					
γ-tocopherol	0.83 [b]	1.02 [a,b]	1.05 [a,b]	1.44 [a]	0.14	0.04	0.007	0.39
α-tocopherol	0.78	0.98	0.98	1.41	0.22	NS	0.06	0.52
Retinol	0.017	0.025	0.030	0.024	0.003	0.099	0.12	0.93
			Oxidation stability					
FRAP (eq Fe^{+2}/μM)	204	229	226	216	17	NS	0.73	0.33
TBARS (mg MDA/kg meat)	0.39	0.29	0.25	0.22	0.06	NS	0.09	0.27

[1] Dietary treatment: 0DG, control; 15DG, 15% DG; 30DG, 30% DG and 45DG, 45% DG (%DM basis). [2] SEM: standard error of the mean. [3] Linear (L) and quadratic (Q) response to DG level. Means in the same row having different letters are significant at the $p \leq 0.05$ level. NS: no significant.

In addition to tocopherol, feeding diets supplied carotenoid compounds. These compounds are converted to retinol in the animal and exert antioxidant capacity and contribute to the yellowness of subcutaneous fat [6,12]. The content of carotenoid compounds in meat is highly variable since their incorporation depends on their content in the feeding diets, and the accumulation in adipose tissue, which is related to the type of muscle and the individual uptake capacity of the animal [13,35]. Previously, it has been reported that the content of carotenoids and their retinoid derivatives is at least one order below that of α-tocopherol [13]. In fact, Pouzo et al. [34] reported that meat samples from grazing systems supplemented with corn had 0.09 ug of retinol/g of fresh meat and 1.15 ug of α-tocopherol/g of fresh meat. In our study, the values of retinol observed in meat tended to be greater in samples from DG diets. Additionally, the values observed were one order below the values of α-tocopherol, independently of the dietary treatment (Table 3).

Neither the antioxidant capacity, measured as FRAP, which determines the total reducing capacity of antioxidant compounds nor the lipid oxidation, measured as TBARS, in meat samples at 72 h post-mortem were affected by the level of DG inclusion in the diets (Table 3). In general terms, no differences were seen in meat oxidation stability due to the dietary treatment. Additionally, other authors have observed a lack of effect on lipid oxidation in meat from steers fed 50% DG and aged for 2 days [38]. Interestingly, recently published data have shown that meat from cattle finished on a grass silage-based diet in combination with corn or wheat DG supplementation showed no differences in the FRAP values, after 14 days of aging, regardless of the source of DG [37].

3.3. Antioxidant Compounds, Oxidation and Color Stability of LTL under Storage Conditions

The content of antioxidant compounds observed in LTL samples stored at two retail display conditions is shown in Table 4. No significant interactions were observed between the main effects.

Table 4. Content of antioxidant vitamins and antioxidant stability in LTL muscle at the end of each storage and retail display condition.

Item	Dietary Treatment [1]					Retail Treatment			p-Value		
	0DG	15DG	30DG	45DG	SEM [2]	R1	R2	SEM	D*R [3]	D	R
				Antioxidant compounds (μg/g meat)							
γ-tocopherol	0.62 [b]	0.88 [a]	0.81 [a,b]	0.75 [a,b]	0.05	0.78	0.75	0.04	NS	0.04	NS
α-tocopherol	0.52	0.65	0.63	0.51	0.09	0.57	0.58	0.04	NS	NS	NS
Retinol	0.023 [b]	0.026 [a,b]	0.030 [a]	0.024 [a,b]	0.001	0.027	0.025	0.002	NS	0.03	NS
				Antioxidant capacity							
FRAP (eq Fe^{+2}/μM)	261	250	261	251	9.20	270 [a]	241 [b]	6.82	NS	NS	0.004

[1] D, Dietary treatment: 0DG, control; 15DG, 15% DG; 30DG, 30% DG and 45DG, 45% DG (%DM basis); R, retail treatment: R1, aerobic exposure for seven days; R2, vacuum-packed conditions for 25 days, plus aerobic exposure for three days. [2] SEM: standard error of the mean. [3] D*R, dietary treatment x retail treatment interaction. Means in the same row having different letters are significant at the $p \leq 0.05$ level. NS: no significant.

The content of α-tocopherol was not affected either by the feeding diet or the storage conditions, and the average concentration was 0.58 μg/g meat. Regarding γ-tocopherol, a higher content was observed in meat from steers fed with DG diets, while it was not affected by the storage conditions. However, in all cases, the values of α- and γ-tocopherol observed were lower than those observed at 72 h post-mortem. Previously, it has been reported that vitamin E affects lipid and color stability after refrigerated storage [13]. In the present study, during storage, α-tocopherol levels dropped around 34% in meat from all diets, except in those from 45DG, in which it decreased by 64%, whereas γ-tocopherol levels dropped around 20% in meat from all diets, except in those from 45DG, in which it decreased by 48%. Indeed, the FRAP assay showed that the total reducing activity in meat was greater for R1 than for R2, but also that it was not affected by the feeding diets. Other authors have reported that, after meat aging and aerobic exposure, both the FRAP values and the content of antioxidant vitamins decreased [34]. In our study, the effect of storage on FRAP was not observed on vitamin content.

Regarding retinol content, the meat from animals fed 30DG had greater content than that from animals fed 15DG and 45DG, while the meat from 0DG had the lowest content (Table 4). Retinol content was not affected by the retail display conditions.

Together with α-tocopherol, other compounds like the minor forms of vitamin E, carotenoids, and their retinoid derivatives, protect tissues against free radicals and prevent oxidation reactions [13]. All these compounds could be administered in the feeding diets, as stated before. However, pro-oxidant compounds, like unsaturated fatty acids, could also be administered in the diets. It has been reported that, during refrigerated storage, oxidative processes increase exponentially [13], since factors like temperature, the presence of light, and oxygen act as catalysts [6,9]. In our study, the values of TBARS measured at the end of the storage period in LTL samples showed a tendency ($p = 0.08$) to interact between dietary treatment and retail display conditions (Figure 2). Meat from DG diets showed similar TBARS values, independently of the storage condition. Interestingly, the meat from 0DG showed the highest TBARS values when exposed to R2 conditions. In this sense, greater TBARS values have been reported in meat samples from 0% DG diets, compared to meat samples from 50% DG diets, after 21 days of aging [38]. However, de Mello et al. [17] observed greater TBARS values in meat from animals fed 30% DG than in meat from animals fed 0 and 15% DG and aged for 42 days. These authors observed meat surface discoloration after aging followed by a retail display and related these events with a higher content of PUFAs in meat when feeding cattle with 30% DG.

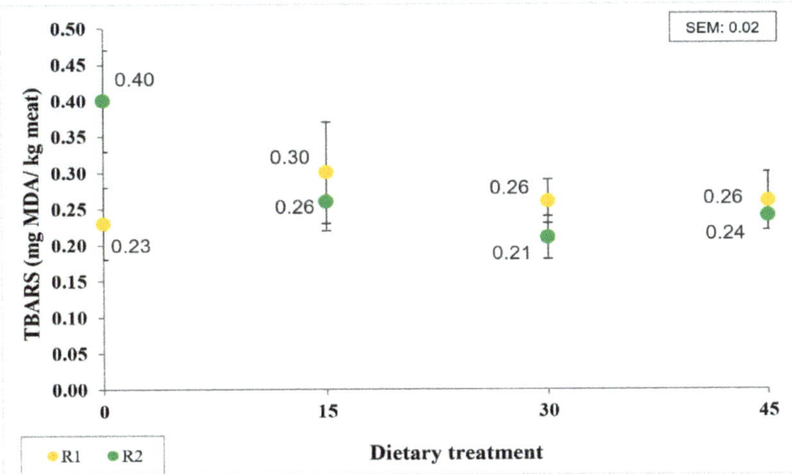

Figure 2. Lipid oxidation (mean ± SEM) measured as thiobarbituric acid reactive substances (TBARS) in LTL at the end of storage and retail display conditions for each dietary treatment (0, 0%DG control; 15, 15% DG; 30, 30%DG; 45, 45% DG, DM basis).

The TBARS assay is used as a marker of lipid oxidation and as a predictor of the perception of rancidity in meat. In different studies, values of 2–2.5 mg MDA/kg have been established as the accepted limit at which meat shows no rancidity [6]. In our study, all the TBARS values observed were below that threshold.

As stated above, anti- and pro-oxidants can be provided in the feeding diets, and their balance would be a determinant for the oxidative stability of meat. In our study, the feeding diets supplied liposoluble vitamins to meat, mostly γ-tocopherol. However, α-tocopherol is the isomer with the greatest antioxidant action and at least 3.0 to 3.5 μg/g of meat is needed to preserve tissues from oxidation and/or to maintain color stability [7,23,39]. Interestingly, it has been proposed that a dosage level of γ-tocopherol and a similar dosage level of α-tocopherol are approximately equally effective in preventing oxidation in red meat [39]. In our study, the sum of the content of antioxidant vitamins (1.39 μg/g meat on average) was below the optimal levels proposed to preserve tissues from oxidation, but the extent of oxidative damage was far below the threshold for rancidity. Therefore, it could be assumed that the antioxidant compounds present in meat samples were exerting a cooperative antioxidant activity.

It is known that the nutritional strategies that modify the fatty acid profile of meat could increase the susceptibility of meat to lipid oxidation by increasing n-3 PUFAs [7,13]. We showed that the content of PUFAs and n-6 PUFAs did not change with the DG level, while that of n-3 PUFAs decreased numerically [40]. This effect was statistically seen in the n-6/n-3 PUFA ratio, which increased as the DG level increased due to the decrease in n-3 PUFAs. In the present study, we observed that the inclusion of DG in the feeding diets supplied both more antioxidant compounds and fewer pro-oxidant compounds.

In this sense, it has been reported that the supplementation of DG diets with vitamin E does not improve the effect against lipid oxidation, since the meat from 30DG-diets, with and without vitamin E supplementation, had similar values of TBARS after 7 days of retail display [41].

The meat color parameters found at the end of the retail display conditions are presented in Table 5. In the case of the R2 storage condition, an extra measurement was done at the end of the vacuum-packed storage, prior to display conditions ($R2_a$). The color parameters analyzed showed no interaction between the main effects.

Table 5. Color parameters (CIELab) in LTL muscle at the end of each storage and retail display condition.

Item	Dietary Treatment [1]					Retail Treatment				*p*-Value		
	0DG	15DG	30DG	45DG	SEM [2]	R1	$R2_a$	$R2_b$	SEM	D*R [3]	D	R
L*	43.24	42.50	41.85	41.37	0.57	40.93 [b]	42.84 [a]	42.94 [a]	0.37	NS	NS	0.0002
a*	17.10	18.32	18.02	18.08	0.47	15.72 [c]	20.09 [a]	17.83 [b]	0.26	NS	NS	<0.0001
C*	19.96	21.45	20.98	21.01	0.45	19.03 [c]	22.71 [a]	20.82 [b]	0.31	NS	NS	<0.0001

[1] D, Dietary treatment: 0DG, control; 15DG, 15% DG; 30DG, 30% DG and 45DG, 45% DG (%DM basis); R, retail treatment: R1, aerobic exposure; $R2_a$, end of vacuum-packed storage (25 days); $R2_b$, end of aerobic exposure of R2. [2] SEM: standard error of the mean. [3] D*R, dietary treatment x retail treatment interaction. Means in the same row having different letters are significant at the $p \leq 0.05$ level. NS: no significant.

Meat color did not differ statistically with the inclusion of increasing levels of DG in the diets. These results are in agreement with those presented by de Mello et al. [17], who compared beef from DG diets with control diets (0% only corn, 15% and 30% DG). On the other hand, Depenbusch et al. [19] reported a linear decrease in redness (a* value), after 7 days of retail display conditions, in steaks from heifers fed increasing levels of DG (0% to 75%). Moreover, to analyze differences in beef color, delta color change (ΔE) was calculated, following AMSA procedures, between steaks from DG diets and steaks from the control diet. The values of ΔE obtained for 15DG, 30DG, and 45 DG were 1.58, 1.69, and 2.13, respectively. Some authors reported that values of ΔE from 2 to 10 indicate that differences in color are perceptible by the human eye at a quick look [42] while other authors reported that values between 1.5 to 3.0 in beef could be evident to consumers [43].

Meat color stability is the result of the balance between anti- and pro-oxidant compounds [44]. It could be assumed that meat with an increase in PUFA content, due to the inclusion of DG in the diet, would be more susceptible to lipid oxidation and discoloration. Thus, as previously reviewed, it is difficult to define a general effect on meat color due to the inclusion of DG in the diet [3].

During storage, the formation of metmyoglobin (MMb) modifies meat color with a decrease in both a* and C* parameters [45]. Besides, the rate at which MMb develops could be affected by the conditions of packaging and storage. In the present study, storage conditions influenced color parameters. Samples from R2 conditions were brighter and had higher values of a* and C* than steaks from R1 conditions. Indeed, samples from R1 and R2 showed a value of ΔE of 2.93, which indicated that differences could be evident to the human eye [42,43]. Additionally, a decrease in a* and C* values was observed in $R2_b$ with respect to $R2_a$.

Meat from R1 had lower values of color parameter than meat from $R2_b$ (Table 5). These results are in accordance with those of Jose and McGilchrist [46], who proposed that in low pH beef, aging increased bloom depth and made meat appear brighter and redder in color. However, other authors reported that aging time before retail display did not affect the redness (a* parameter) of meat samples from DG diets, while retail display conditions decreased the values of a* [41]. Similarly, in our study, a decrease in a* and C* values were observed when changing from vacuum to aerobic packaging in R2 ($R2_b$ with respect to $R2_a$). This decrease could be associated with an increase in MMb content when samples were exposed to aerobic conditions.

In this sense, it has been reported that during refrigeration storage, lipid oxidation occurs more slowly than discoloration [14]. McKenna et al. [47] reported that color-stable muscles had lower TBARS than less color-stable ones. These authors found values up to 0.35 mg/kg for *M. longissimus lumborum* and *M. longissimus thoracis* after 5 days of a retail display while preserving color traits. In agreement, in our study, the values of TBARS observed were lower than 0.35 mg/kg, except for meat from the 0DG treatment at R2 storage conditions. Furthermore, the values of a* measured for all samples in our study were above 14.5, the value considered as a threshold for beef color acceptability [48].

Moreover, a PCA was performed to depict the relationship between the oxidant/antioxidant status and the color parameters, at each storage condition (Figure 3). The new components identify where the maximum variance of the data occurs in a multidimensional space. For the R1 condition, the first component (PC1) accounted for 29.5% of the variance, the second component (PC2) accounted for 21.0% of the variance, and the third component (PC3) accounted for 19.3% of the variance. For R2 conditions, the principal components accounted for 36.3% (PC1), 28.2% (PC2), and 16.7% (PC3) of the variance. Interestingly, it was observed that the relationship between the variables differed with the type of storage. In R1 storage, retinol and FRAP showed a positive correlation with a* and C* and a negative correlation with TBARS. In R2 storage, γ-tocopherol and FRAP showed a positive correlation with a* and C* and a negative correlation with TBARS. For both storage conditions, a* and C* parameters showed a negative correlation with TBARS.

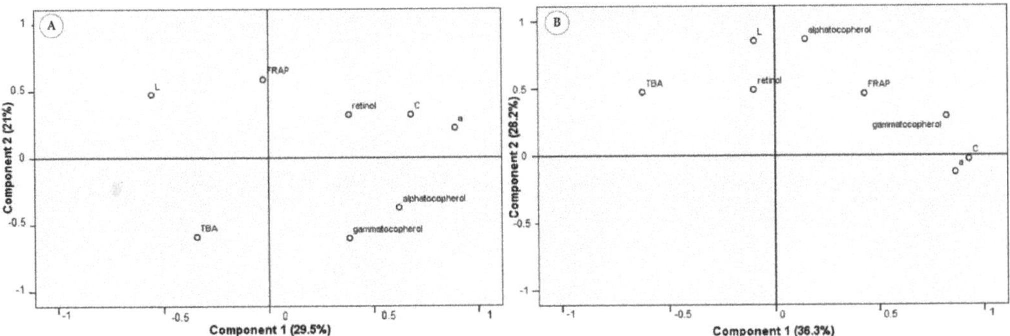

Figure 3. Principal component analysis (PCA) depicting the relationship between the oxidant/antioxidant status and the color parameters, at (**A**) R1 conditions (aerobic exposure for seven days) and at (**B**) R2 conditions (vacuum-packed storage for 25 days, plus aerobic exposure for three days).

4. Conclusions

The inclusion of DG in the feeding diets of beef steers increased the content of antioxidant compounds in the diets. This was reflected in a greater content of γ-tocopherol in beef from DG diets, and a numerical increase in retinol content. Although the values of α-tocopherol were below those indicated as optimal for oxidation delay, the values of TBARS observed were far lower than the ones associated with meat rancidity. Therefore, it could be assumed that the antioxidant compounds present in the meat from animals fed with DG exerted a cooperative antioxidant activity.

The meat exposed to refrigerated storage was affected by the conditions of retail display since FRAP and color parameters decreased. The inclusion of DG affected the extent of lipid oxidation. In this regard, the samples from DG-diets showed similar values of TBARS for both storage conditions, and these values were lower than those from 0DG and extended storage. Thus, DG inclusion in finishing diets positively impacts the balance between anti- and pro-oxidant compounds and meat oxidative stability during the display conditions evaluated.

Author Contributions: Conceptualization, M.M., D.P. and G.G.; methodology, M.M., S.A.R., L.R., D.P. and G.G.; software, M.M., S.A.R. and L.R.; validation, M.M., S.A.R., L.R., D.P. and G.G.; formal analysis, M.M., L.R. and G.G.; investigation, M.M., S.A.R., L.R., D.P. and G.G.; resources, L.R., D.P. and G.G.; data curation, G.G. and M.M.; writing—original draft preparation, M.M., D.P. and G.G.; writing—review and editing, M.M., S.A.R., L.R., D.P. and G.G.; visualization, M.M.; supervision, D.P. and G.G.; project administration, M.M.; funding acquisition, M.M., L.R., D.P. and G.G. All authors have read and agreed to the published version of the manuscript.

Funding: This research was funded by Instituto Nacional de Tecnología Agropecuaria (INTA PEI#517), by Instituto Promoción Carne Vacuna Argentina (Convenio de Cooperación Técnica IPCVA-

INTA, 2017) and by Consejo Nacional de Investigaciones Científicas y Técnicas, Beca Doctoral Temas Estratégicos 2016–2021 (Res DN#5040).

Institutional Review Board Statement: Experimental procedures were approved by the Ethics Committee of the School of Veterinary Sciences of the University of Buenos Aires (Comité Institucional de Cuidado y Uso de Animales de Laboratorio, Facultad de Ciencias Veterinarias, protocol number 2016/42), Argentina. The assay was carried out as part of common zootechnical procedures and the animals did not suffer any intervention beyond those typical in such farming conditions. Muscle samples were collected for the analysis during the routine carcass cutting procedure at the meat processing plant.

Informed Consent Statement: Not applicable.

Data Availability Statement: The data presented in this study are available on request from corresponding author.

Acknowledgments: The authors express their gratitude to Karina Moreno, Marta Signorelli and Cecilia Barreto for their technical assistance. The authors are members of the Healthy Meat network funded by CYTED (119RT0568).

Conflicts of Interest: The authors declare no conflict of interest. The funders had no role in the design of the study; in the collection, analyses, or interpretation of data; in the writing of the manuscript, or in the decision to publish the results.

References

1. Araújo, W.A. Ethanol Industry: Surpassing Uncertainties and Looking Forward. In *Global Bioethanol*; Academic Press: Cambridge, MA, USA, 2016; pp. 1–33. ISBN 978-0-12-803141-4.
2. Renewable Fuels Association Annual Ethanol Production. Available online: https://ethanolrfa.org/markets-and-statistics/annual-ethanol-production (accessed on 3 January 2022).
3. Merayo, M.; Pighin, D.; Grigioni, G. Meat quality traits and feeding distillers grains to cattle: A review. *Anim. Prod. Sci.* **2020**, *60*, 1123–1135. [CrossRef]
4. Klopfenstein, T.J.; Erickson, G.E.; Bremer, V.R. Board-Invited Review: Use of distillers by-products in the beef cattle feeding industry. *J. Anim. Sci.* **2008**, *86*, 1223–1231. [CrossRef] [PubMed]
5. Faustman, C.; Sun, Q.; Mancini, R.; Suman, S.P. Myoglobin and lipid oxidation interactions: Mechanistic bases and control. *Meat Sci.* **2010**, *86*, 86–94. [CrossRef] [PubMed]
6. Domínguez, R.; Pateiro, M.; Gagaoua, M.; Barba, F.J.; Zhang, W.; Lorenzo, J.M. A comprehensive review on lipid oxidation in meat and meat products. *Antioxidants* **2019**, *8*, 429. [CrossRef] [PubMed]
7. Scollan, N.D.; Dannenberger, D.; Nuernberg, K.; Richardson, I.; MacKintosh, S.; Hocquette, J.-F.; Moloney, A.P. Enhancing the nutritional and health value of beef lipids and their relationship with meat quality. *Meat Sci.* **2014**, *97*, 384–394. [CrossRef]
8. Font-i-Furnols, M.; Guerrero, L. Consumer preference, behavior and perception about meat and meat products: An overview. *Meat Sci.* **2014**, *98*, 361–371. [CrossRef]
9. Faustman, C.; Suman, S.P. The Eating Quality of Meat: I-Color. In *Woodhead Publishing Series in Food Science, Technology and Nutrition, Lawrie ´s Meat Science*; Toldrá, F., Ed.; Woodhead Publishing: Sawston, UK, 2017; pp. 329–356. ISBN 9780081006979.
10. Testa, M.L.; Grigioni, G.; Panea, B.; Pavan, E. Color and Marbling as Predictors of Meat Quality Perception of Argentinian Consumers. *Foods* **2021**, *10*, 21. [CrossRef]
11. Mancini, R.A.; Hunt, M.C. Current research in meat color. *Meat Sci.* **2005**, *71*, 100–121. [CrossRef]
12. Holman, B.W.B.; Kerry, J.P.; Hopkins, D.L. Meat packaging solutions to current industry challenges: A review. *Meat Sci.* **2018**, *144*, 159–168. [CrossRef]
13. Descalzo, A.M.; Sancho, A.M. A review of natural antioxidants and their effects on oxidative status, odor and quality of fresh beef produced in Argentina. *Meat Sci.* **2008**, *79*, 423–436. [CrossRef]
14. Gobert, M.; Gruffat, D.; Habeanu, M.; Parafita, E.; Bauchart, D.; Durand, D. Plant extracts combined with vitamin E in PUFA-rich diets of cull cows protect processed beef against lipid oxidation. *Meat Sci.* **2010**, *85*, 676–683. [CrossRef]
15. Natalello, A.; Priolo, A.; Valenti, B.; Codini, M.; Mattioli, S.; Pauselli, M.; Puccio, M.; Lanza, M.; Stergiadis, S.; Luciano, G. Dietary pomegranate by-product improves oxidative stability of lamb meat. *Meat Sci.* **2020**, *162*, 108037. [CrossRef] [PubMed]
16. Mello, A.S.; Jenschke, B.E.; Senaratne, L.S.; Carr, T.P.; Erickson, G.E.; Calkins, C.R. Effects of feeding modified distillers grains plus solubles on marbling attributes, proximate composition, and fatty acid profile of beef. *J. Anim. Sci.* **2012**, *90*, 4634–4640. [CrossRef] [PubMed]
17. de Mello, A.S.; Jenschke, B.E.; Senaratne, L.S.; Carr, T.P.; Erickson, G.E.; Calkins, C.R. Effects of finishing diets containing wet distillers grains plus solubles on beef quality attributes and fatty acid profile. *Meat Sci.* **2018**, *136*, 16–22. [CrossRef] [PubMed]
18. Gill, R.K.; VanOverbeke, D.L.; Depenbusch, B.; Drouillard, J.S.; DiCostanzo, A. Impact of beef cattle diets containing corn or sorghum distillers grains on beef color, fatty acid profiles, and sensory attributes. *J. Anim. Sci.* **2008**, *86*, 923–935. [CrossRef]

19. Depenbusch, B.E.; Coleman, C.M.; Higgins, J.J.; Drouillard, J.S. Effects of increasing levels of dried corn distillers grains with solubles on growth performance, carcass characteristics, and meat quality of yearling heifers. *J. Anim. Sci.* **2009**, *87*, 2653–2663. [CrossRef]
20. Latimori, N.; Carduza, F.; Merayo, M.; Soteras, T.; Grigioni, G.; Garis, M. Efecto de la incorporación de burlanda de maíz en la dieta de bovinos para carne. In *Producción Bovinos para Carne (2013–2017) Programa Nacional de Producción Animal Alimentación de Bovinos Para Carne*; Pasinato, A., Grigioni, G., Alende, M., Eds.; INTA, Anguil: La Pampa, Argentina, 2018; ISBN 978-1-873671-00-9.
21. Merayo, M.; Pighin, D.; Latimori, N.; Kloster, A.; Cunzolo, S.; Grigioni, G. Utilización de subproductos de la industria de biocombustibles en la alimentación de bovinos para carne en argentina. In *Tecnologías Aplicadas en la Producción, Calidad y Competitividad de la Carne de Especies Pecuarias*; Dominguez Vara, I.A., Morales Almaraz, E., Ramirez Bribiesca, E., Eds.; Ediciones y Gráficos Eon: Toluca, Mexico, 2018; pp. 305–323. ISBN 9786078559671.
22. Buttriss, J.; Diplock, A. High-Performance Liquid Cromatography Methods for Vitamin E in Tissues. In *Methods in Enzymology*; Academic Press: Cambridge, MA, USA, 1984; Volume 105, pp. 131–138. ISBN 0-12-182005-X.
23. Descalzo, A.M.; Insani, E.M.; Biolatto, A.; Sancho, A.M.; García, P.T.; Pensel, N.A.; Josifovich, J.A. Influence of pasture or grain-based diets supplemented with vitamin E on antioxidant/oxidative balance of Argentine beef. *Meat Sci.* **2005**, *70*, 35–44. [CrossRef]
24. Ou, B.; Huang, D.; Hampsch-Woodill, M.; Flanagan, J.A.; Deemer, E.K. Analysis of Antioxidant Activities of Common Vegetables Employing Oxygen Radical Absorbance Capacity (ORAC) and Ferric Reducing Antioxidant Power (FRAP) Assays: A Comparative Study. *J. Agric. Food Chem.* **2002**, *50*, 3122–3128. [CrossRef]
25. Descalzo, A.M.; Rossetti, L.; Grigioni, G.; Irurueta, M.; Sancho, A.M.; Carrete, J.; Pensel, N.A. Antioxidant status and odour profile in fresh beef from pasture or grain-fed cattle. *Meat Sci.* **2007**, *75*, 299–307. [CrossRef]
26. Pensel, N.A. *Influence of Experimental Conditions on Porcine Muscle and Its Effect on Oxidation*; The Ohio State University: Columbus, OH, USA, 1990.
27. AMSA. *Meat Color Measurement Guidelines*; American Meat Science Association: Champaign, IL, USA, 2012.
28. Di Rienzo, J.; Casanoves, F.; Balzarini, M.; Gonzalez, L.; Tablada, M.; Robledo, C. InfoStat Versión 2020. Centro de Transferencia InfoStat, FCA, Universidad Nacional de Córdoba, Argentina, 2020. Available online: http://infostat.com.ar (accessed on 3 January 2022).
29. Li, M.; Tian, L.; Zhao, G.; Zhang, Q.; Gao, X.; Huang, X.; Sun, L. Formation of biogenic amines and growth of spoilage-related microorganisms in pork stored under different packaging conditions applying PCA. *Meat Sci.* **2014**, *96*, 843–848. [CrossRef]
30. Shin, E.-C.; Shurson, G.C.; Gallaher, D.D. Antioxidant capacity and phytochemical content of 16 sources of corn distillers dried grains with solubles (DDGS). *Anim. Nutr.* **2018**, *4*, 435–441. [CrossRef] [PubMed]
31. Loy, D.D.; Lundy, E.L. Nutritional Properties and Feeding Value of Corn and Its Coproducts. In *Corn*; Serna-Saldivar, S.O., Ed.; AACC International Press: Oxford, UK, 2019; pp. 633–659. ISBN 978-0-12-811971-6.
32. Koger, T.J.; Wulf, D.M.; Weaver, A.D.; Wright, C.L.; Tjardes, K.E.; Mateo, K.S.; Engle, T.E.; Maddock, R.J.; Smart, A.J. Influence of feeding various quantities of wet and dry distillers grains to finishing steers on carcass characteristics, meat quality, retail-case life of ground beef, and fatty acid profile of longissimus muscle. *J. Anim. Sci.* **2010**, *88*, 3399–3408. [CrossRef] [PubMed]
33. Blanco, M.; Ripoll, G.; Casasús, I.; Bertolín, J.R.; Joy, M. Carotenoids and tocopherol in plasma and subcutaneous fat colour to trace forage-feeding in growing steers. *Livest. Sci.* **2019**, *219*, 104–110. [CrossRef]
34. Pouzo, L.B.; Descalzo, A.M.; Zaritzky, N.E.; Rossetti, L.; Pavan, E. Antioxidant status, lipid and color stability of aged beef from grazing steers supplemented with corn grain and increasing levels of flaxseed. *Meat Sci.* **2016**, *111*, 1–8. [CrossRef]
35. Dunne, P.G.; Monahan, F.J.; O'Mara, F.P.; Moloney, A.P. Colour of bovine subcutaneous adipose tissue: A review of contributory factors, associations with carcass and meat quality and its potential utility in authentication of dietary history. *Meat Sci.* **2009**, *81*, 28–45. [CrossRef]
36. da Silva Hampel, V.; Poli, C.H.E.C.; Devincenzi, T.; Pötter, L. Feeding systems and tocopherol level in the diet and their effects on the quality of lamb meat: A meta-analysis. *Rev. Bras. Zootec.* **2019**, *48*, 48. [CrossRef]
37. Salami, S.A.; O'Grady, M.N.; Luciano, G.; Priolo, A.; McGee, M.; Moloney, A.P.; Kerry, J.P. Fatty acid composition, shelf-life and eating quality of beef from steers fed corn or wheat dried distillers' grains with solubles in a concentrate supplement to grass silage. *Meat Sci.* **2021**, *173*, 108381. [CrossRef]
38. Chao, M.D.; Domenech-Perez, K.I.; Senaratne-Lenagala, L.S.; Calkins, C.R. Feeding wet distillers grains plus solubles contributes to sarcoplasmic reticulum membrane instability. *Anim. Prod. Sci.* **2018**, *58*, 2215–2223. [CrossRef]
39. Saunders, C.; Wolf, F.; Sauber, T.; Owens, F. Method of Improving Animal Tissue Quality. US Patent No. 6,977,269 B1, 20 December 2005.
40. Merayo, M. Granos de Destilería en Alimentación Animal: Su Incidencia en Calidad de Carne Bovina, Universidad de Buenos Aires, 2020. Available online: http://hdl.handle.net/20.500.12123/10199 (accessed on 4 February 2022).
41. Chao, M.D.; Domenech-Pérez, K.I.; Voegele, H.R.; Kunze, E.K.; Calkins, C.R. Effects of dietary antioxidant supplementation of steers finished with 30% wet distillers grains plus solubles on fatty acid profiles and display life of strip loins. *Anim. Prod. Sci.* **2018**, *58*, 1949–1957. [CrossRef]
42. Tomasevic, I.; Tomovic, V.; Milovanovic, B.; Lorenzo, J.; Đorđević, V.; Karabasil, N.; Djekic, I. Comparison of a computer vision system vs. traditional colorimeter for color evaluation of meat products with various physical properties. *Meat Sci.* **2019**, *148*, 5–12. [CrossRef]

43. Prandl, O.; Fischer, A.; Schmidhofer, T.; Sinell, H. *Tecnología e Higiene de la Carne*; Ciencia y tecnología de la carne; Acribia: Zaragoza, Spain, 1994; ISBN 9788420007656.
44. Insani, E.M.; Eyherabide, A.; Grigioni, G.; Sancho, A.M.; Pensel, N.A.; Descalzo, A.M. Oxidative stability and its relationship with natural antioxidants during refrigerated retail display of beef produced in Argentina. *Meat Sci.* **2008**, *79*, 444–452. [CrossRef] [PubMed]
45. Salueña, B.H.; Gamasa, C.S.; Rubial, J.M.D.; Odriozola, C.A. CIELAB color paths during meat shelf life. *Meat Sci.* **2019**, *157*, 107889. [CrossRef] [PubMed]
46. Jose, C.; McGilchrist, P. Ageing as a method to increase bloom depth and improve retail colour in beef graded AUS-MEAT colour 4. *Meat Sci.* **2022**, *183*, 108665. [CrossRef] [PubMed]
47. McKenna, D.R.; Mies, P.D.; Baird, B.E.; Pfeiffer, K.D.; Ellebracht, J.W.; Savell, J.W. Biochemical and physical factors affecting discoloration characteristics of 19 bovine muscles. *Meat Sci.* **2005**, *70*, 665–682. [CrossRef]
48. Holman, B.W.B.; van de Ven, R.J.; Mao, Y.; Coombs, C.E.O.; Hopkins, D.L. Using instrumental (CIE and reflectance) measures to predict consumers' acceptance of beef colour. *Meat Sci.* **2017**, *127*, 57–62. [CrossRef]

MDPI

Article

Retail Packaging Affects Colour, Water Holding Capacity, Texture and Oxidation of Sheep Meat more than Breed and Finishing Feed

Minh Ha [1,*], Robyn Dorothy Warner [1], Caitlin King [1], Sida Wu [1] and Eric N. Ponnampalam [2]

1 School of Agriculture and Food, Faculty of Veterinary and Agricultural Sciences, The University of Melbourne, Parkville, VIC 3010, Australia; robyn.warner@unimelb.edu.au (R.D.W.); caitlink1@student.unimelb.edu.au (C.K.); sidaw1@student.unimelb.edu.au (S.W.)

2 Animal Production Sciences, Agriculture Victoria Research, Department of Jobs, Precincts and Regions, Bundoora, VIC 3083, Australia; Eric.Ponnampalam@agriculture.vic.gov.au

* Correspondence: minh.ha@unimelb.edu.au

Abstract: This study investigated the CIELab colour, water holding capacity, texture and oxidative stability of sheep meat from different breeds, finishing feeds, and retail packaging methods. Leg primal cuts from a subset of Composite wether lambs (n = 21) and Merino wether yearlings (n = 21) finished on a standard diet containing grain and cereal hay, a standard diet with camelina forage, or a standard diet with camelina meal, were used in this study. *Semimembranosus* and *Vastus lateralis* were packaged in vacuum skin packaging (VSP), or modified atmosphere packaging with 80% O_2 and 20% CO_2 (HioxMAP), or with 50% O_2, 30% N_2, and 20% CO_2 (TrigasMAP). Packaging had a greater effect ($p < 0.001$) on L*, a*, b*, hue, and chroma than the effects from breed and finishing feed. Purge loss was affected by packaging. Cooking loss was affected by breed for *Semimembranosus* and packaging for both muscle types. HioxMAP and TrigasMAP increased WBSF and Texture Profile Analysis hardness of the meat compared to VSP. Lipid oxidation, assessed by TBARS, were lower in camelina forage or camelina meal supplemented diets and TrigasMAP compared to standard diet and HioxMAP, respectively. Total carbonyl and free thiol content were lower in VSP. Thus, supplementing feed with camelina forage or meal and lowering oxygen content in retail packaging by TrigasMAP or VSP are recommended to ensure optimal sheep meat quality.

Keywords: Merino; composite; modified atmosphere packaging; trigas; camelina; lipid oxidation; colour stability; meat

Citation: Ha, M.; Warner, R.D.; King, C.; Wu, S.; Ponnampalam, E.N. Retail Packaging Affects Colour, Water Holding Capacity, Texture and Oxidation of Sheep Meat More than Breed and Finishing Feed. *Foods* **2022**, *11*, 144. https://doi.org/10.3390/foods11020144

Academic Editor: Rubén Domínguez

Received: 14 December 2021
Accepted: 27 December 2021
Published: 6 January 2022

Publisher's Note: MDPI stays neutral with regard to jurisdictional claims in published maps and institutional affiliations.

1. Introduction

In the sheep meat supply chain, quality traits such as colour, water holding capacity, texture, and oxidative stability are determined by various factors, including breed, finishing feed, and retail packaging method. High-oxygen modified atmosphere packaging (HioxMAP), using 70–80% O_2 and 20–30% CO_2, is a common meat retail packaging method due to its ability to maintain the "fresh" cherry red colour of meat. However, extensive research has shown an increased oxidation and lower eating quality of meat in HioxMAP compared with vacuum skin packaging (VSP) [1–3]. Trigas modified atmosphere packaging (TrigasMAP) is a more recently developed method in which oxygen is partially substituted with an inert gas, e.g., nitrogen, and has shown promising results in improving the eating quality and shelf life of meat [4].

The finishing feed of livestock is another factor affecting the quality of meat through altering the antioxidant activity in post-mortem muscles. The incorporation of antioxidants, such as vitamin E, or antioxidant-rich pasture crops were shown to result in reduced lipid oxidation and improved eating quality [5,6]. The use of oil crops and meals as animal feed supplements from the Brassica family, particularly camelina (Camelina sativa) has recently gained attention in improving animal productivity and carcass value [7]. Camelina seed is known to contain essential fatty acids, such as alpha-linolenic acid and different phenolic compounds such as flavonoids and proanthocyanidins, which provides an opportunity to improve the oxidative stability of meat [8].

Animal breed or the genetic background is also known to influence the sheep meat quality. For example, pure Merino sheep is believed to produce meat that is less tender or darker in colour compared with meat from crossbred sheep. The differences in texture and colour are associated with carcass fatness, muscle glycogen concentration, muscle iron concentration, and/or post-mortem chill effects [9,10].

Many studies have demonstrated the effect of the three supply chain factors of breed, feed and packaging individually. However, little is known about their interactive effect or the extent to which each factor affects meat quality. Thus, the aim of this study was to investigate the quality of sheep topsides (*Semimembranosus*) and knuckles (*Vastus lateralis*) from Merino yearlings and Composite wether lambs, finished on three different diets (standard diet; standard diet supplemented with camelina forage; or standard diet supplemented with camelina meal), and packed in three retail packaging methods (VSP, HioxMAP or TrigasMAP). The CIELab colour, water holding capacity, texture, lipid oxidation and protein oxidation were measured.

2. Materials and Methods

2.1. Animal Housing and Feeding

Feeding experiments were conducted at the Agriculture Victoria Research, Hamilton Centre, Hamilton, VIC, Australia. All animal procedures were conducted in accordance with the Australian code for the care and use of animals for scientific purposes (National Health and Medical Research Council 2013). Animal ethics approval was granted by the Department of Jobs, Precincts and Regions (DJPR) Agricultural Research and Extension Animal Ethics Committee (AEC Code No: 2016-17). Details of the experimental design, feeding of animals, and slaughter procedures were described previously [7]. In brief, a subset of maternal Composite wether lambs (n = 21) and Merino wether yearlings (n = 21) kept in different pens selected based on their final liveweights were used for this study. The primal cuts were from animals randomly allocated to three finishing feeds: a standard pelleted diet containing grain and cereal hay (SPD), a pelleted mixture diet containing 15% camelina forage hay (SCF), or a pelleted mixture diet containing 8% camelina meal (SCM). Diets were formulated using the ingredients available in the major sheep producing regions. The metabolizable energy (ME) and crude protein concentrations of the diets were managed to be 10–11 MJ ME/kg dry matter and 14–15% crude protein.

2.2. Slaughter Procedure and Collection of Sheep Primals

The animals were transported approximately 250 km using a semi-trailer to a commercial abattoir and slaughtered after 18 h in lairage. At 5 days post-mortem, legs from the left side of the animals were collected. Topsides and knuckles from the legs were boned from the legs. The muscles were vacuum-packed using a Multivac C450 (Sepp Haggenmüller GmbH & Co., Wolferschwenden, Germany) with Cryovac® vacuum pouches (PA/PE 70, Sealed Air, Fawkner, VIC, Australia) with an oxygen permeability less than 65 cc/m^2/24 h and water transmission less than 5 g/m^2/24 h. The vacuum-packed muscles were frozen at −20 °C for 6 weeks.

2.3. pH, Cutup, Packaging and Retail Display

Following thawing at 2 °C for 24 h, meat cutup and packaging were conducted at approximately 6 °C. Prior to cutup, the pH of the muscle was measured using a spear-head pH probe attached to a WP-80 pH-mV-temperature meter (TPS Pty Ltd., Brisbane, QLD, Australia). *Semimembranosus* and *Vastus lateralis* were extracted from the topsides and knuckles. The muscles were cut into three sections from the anterior and randomly allocated to VSP, HioxMAP, or TrigasMAP packaging treatments. All packaging was conducted using a Multivac T200 (Sepp Haggenmüller GmbH & Co., Wolferschwenden, Germany). Meat samples were placed on a cello soaker pad (130 mm × 90 mm; CBS, Carrum Downs, Australia) inside a black Cryovac® MAP packaging tray (TOD0901C 170 mm × 223 mm, Sealed Air, VIC, Australia). The trays were sealed with a Biaxially Oriented PolyAmide/Polyethylene/Ethylene vinyl alcohol-based film (OTR 15 cc/m^2/24 h). The gas composition in HioxMAP was 80% O_2 and 20% CO_2 while TrigasMAP had 50% O_2, 30% N_2, and 20% CO_2. Vacuum skin packaging was conducted using Cryovac® Darfresh® film (OTR 4 cc/m^2/24 h) and black Cryovac® trays (Sealed Air, Fawkner, VIC, Australia). Packaged samples were stored in a simulated retail display cabinet with LED lighting (~310 lux, Bromic Refrigeration, Ingleburn, NSW, Australia) at 4 °C for 10 days. The samples were rotated daily between shelves to minimise the effects of variations in illumination and temperature within the retail display cabinet on the samples.

2.4. Instrumental Colour Measurement

After 10 days in simulated retail display, CIE L* (lightness), a* (redness), and b* (yellowness) were measured on the meat surface using a Minolta chroma meter CR-300 (Minolta Co., Ltd., Osaka City, Japan), calibrated with a white plate (no. 20733120; Y = 84.9; x = 0.3171; y = 0.3240). The colour of vacuum skin-packed samples was measured after a 30 min blooming at 6 °C, whereas the colour of HioxMAP and TrigasMAP samples were measured immediately after the samples were removed from packaging. Duplicate colour measurements were taken on each sample. Hue angle (h*) and chroma (C*) were calculated using the following equations:

$$\text{Hue angle} = \arctan (b^*/a^*)$$

$$Chroma = \sqrt{(a^*)^2 + (b^*)^2}$$

2.5. Purge and Cooking Loss

Purge loss was expressed as the weight loss in packaging during retail display. Samples were weighed before packaging (initial weight) and after 10 days storage (final weight). Excess moisture was removed with paper towel before weighing. Purge loss was calculated using the following equation:

$$\text{Purge loss (\%)} = (\text{weight before pack} - \text{weight after pack})/(\text{weight before pack}) \times 100$$

Cooking loss was measured during the cooking procedure for Warner-Bratzler shear force and texture profile analysis. Each sample was weighed before cooking. After cooking, excess moisture on the meat surface was removed with paper towel before weighing. Cooking loss was calculated as:

$$\text{Cooking loss (\%)} = (\text{weight before cook} - \text{weight after cook})/(\text{weight before cook}) \times 100$$

2.6. Warner-Bratzler Shear Force

Warner-Bratzler shear force were measured according to Peng et al. [11]. Samples were individually placed in a plastic bag in temperature-equilibrated water baths (F38-ME, Julabo, Seelbach, Germany) set at 75 °C and cooked to internal temperature of 71 ± 0.5 °C. The internal temperature was monitored using T-type thermocouples inserted to the middle of meat samples and the thermocouples were connected to a Grant Squirrel Series 2020 datalogger (Grant Instruments Ltd., Cambridge, UK). After cooking, the samples were chilled in an ice water bath for 30 min and stored at 4 °C overnight. Six 4 cm long rectangular strips with 1 cm × 1 cm cross section area were obtained from each sample by cutting parallel to the muscle fibres. Each strip was sheared using a Lloyd Instruments LRX Materials Testing Machine (Lloyd Instruments Ltd., Hampshire, UK) equipped with a 5000 N load cell and a V-shape Warner-Bratzler shear force blade at an extension rate of

300 mm/min. The WBSF (N) of each sample was expressed as the average peak force of measurements from the six strips.

2.7. Texture Profile Analysis (TPA)

Texture profile analysis was measured using a double bite compression procedure outlined previously Peng et al. [11]. A piece of meat measuring 1 cm in thickness was obtained from each sample. The meat was compressed twice at the same position by a 6.3-mm diameter plunger which was driven 8 mm into the sample at a crosshead speed of 50 mm/min using Lloyd Instruments LRX Materials Testing Machine (Lloyd Instruments Ltd., Hampshire, UK) equipped with a 5000 N load cell. Hardness (N) (first bite compression), cohesiveness, and chewiness (N) were measured. TPA values for each sample were averaged from six measurements.

2.8. Lipid Oxidation

Lipid oxidation in meat was assessed by 2-thiobarbituric reactive substances (TBARS) procedure as reported by Sorensen and Jorgensen [12] with modifications. For each sample, duplicate (5 g) from each sample were finely chopped and homogenised in 12.5 mL of chilled (4 °C) trichloroacetic acid (TCA) solution (20% TCA (w/v) in 2 M phosphoric acid) at 12,000 rpm for 1.5 min using a Polytron PT 10–35 GT homogeniser (Thermo Fisher Scientific, VIC, Australia). The homogenate was then centrifuged at 1800× g using a Rotina 380R Hettich Centrifuge (LabGear, South Melbourne, VIC, Australia) for 10 min at 4 °C. The supernatant was filtered using Whatman filter paper no. 1. Equal volumes of the filtrate and 5 μM 2-thiobarbituric acid (TBA) were mixed and incubated in a test tube at 95 °C for 60 min. Following incubation, the tube was placed on ice for 15 min. Absorbance at 532 nm was measured for duplicate aliquots from each tube using a Multiskan spectrophotometer (Thermo Fisher Scientific, Scoresby, VIC, Australia). Malondialdehyde (MDA) was quantified against a standard calibration curve with 1,1,3,3-tetraethoxypropane (TEP). Results were expressed as mg MDA·kg^{-1} meat.

2.9. Total Carbonyl Content

Carbonyl content of the meat samples was determined according to Lund et al. [13] with modifications. Briefly, 1 g of meat samples were homogenised for 1 min at 15,000 rpm in 15 mL of homogenisation buffer (2.0 mM $Na_4P_2O_7$, 10 mM Tris-maleate, 2 mM EGTA, 100 mM KCl, pH 7.4) using a Polytron PT 10–35GT homogeniser (Thermo Fisher Scientific, VIC, Australia). Two equal aliquots (0.5 mL) from the homogenate were washed with a HCl:acetone (3:100 v/v) solution three times followed by washing with 10% (w/v) TCA twice. Out of the two identical samples, (i) 0.5 mL of DNPH dissolved in 2 M HCl was added to the first sample for carbonyl derivatisation and (ii) 0.5 mL of 2 M HCl was added to the other sample for protein concentration determination. Absorbance of the samples were measured at 280 nm to determine protein concentration against a standard curve with BSA (Sigma-Aldrich, Castle Hill, NSW, Australia); and, at 370 nm to determine total carbonyl content. Carbonyl concentration was determined by using the absorption coefficient at 370 nm for the hydrazones formed (22,000 M^{-1}·cm^{-1}) against the protein concentration and expressed as nmol·mg^{-1} protein.

2.10. Free Thiol Content

To determine the loss of thiol (sulfhydryl) groups, the 5,5′-Dithiobis (2-nitrobenzoic acid) (DTNB) method was used as described by Lund et al. [13]. Duplicates (2 g) of each sample were homogenised at 16,000 rpm in 40 mL of 5% (w/v) sodium dodecyl sulfate (SDS) in 0.1 M Tris buffer (pH 8) using a Polytron PT 10–35 GT homogeniser (Thermo Fisher Scientific, Scoresby, VIC, Australia). The homogenates were incubated at 95 °C for one hour in covered test tubes. The samples were cooled and centrifuges for 20 min at 1200× g using a Rotina 380R Hettich centrifuge (LabGear, VIC, Australia). The supernatants were filtered using Whatman filter paper no 1 and the protein concentration of filtrates was determined at 280 nm using a standard curve with bovine serum albumin (BSA) (Sigma-Aldrich, Castle Hill, NSW, Australia). The samples were diluted to a protein concentration of 1.5 mg.mL^{-1} using the SDS homogenisation buffer. The diluted samples were used to determine thiol group concentration by adding 2 mL of 0.1 M Tris buffer (pH 8) and 0.5 mL DTNB to 0.5 mL of sample. Samples were incubated for 30 min in the dark and absorbance at 412 nm was measured using a Multiskan spectrophotometer (Thermo Fisher Scientific, VIC, Australia). The concentration of thiol groups was analysed against a standard curve of L-cysteine prepared in 5% (*w*/*v*) SDS in 0.1 M Tris buffer (pH 8). Total thiol content was calculated and expressed as nmol·mg^{-1} protein.

2.11. Statistical Analysis

Data were analysed using restricted maximum likelihood (REML) with GenStat 16th Edition (VSN International, Hemel Hempstead, UK). For pH before packaging, breed (Composite and Merino), feed type (SPD, SCF and SCM), and muscles (*Semimembranosus* and *Vastus lateralis*) were fitted as fixed effects. Pen and carcass number (all nested within, i.e., pen/carcass number) were fitted as random effects. For all other quality traits, breed (Composite and Merino), feed type (SPD, SCF and SCM), packaging method (HioxMAP, TrigasMAP and VSP) were fitted as fixed effects. Pen and carcass number (all nested within) were fitted as random effects. Separate analyses were conducted for each muscle type (*Semimembranosus* and *Vastus lateralis*). $p < 0.05$ was used as the level for significant differences.

3. Results

3.1. pH and Colour

Figure 1 shows that the pH of meat prior to packaging significantly differed between the Composite and Merino sheep ($p = 0.004$) and between the *Semimembranosus* and *Vastus lateralis* muscles ($p < 0.001$). While the pH of *Vastus lateralis* was higher compared to *Semimembranosus* for both breeds, the difference was more substantial in meat from Merino compared to Composite sheep. There was no significant effect of feed ($p = 0.7$) on the pH.

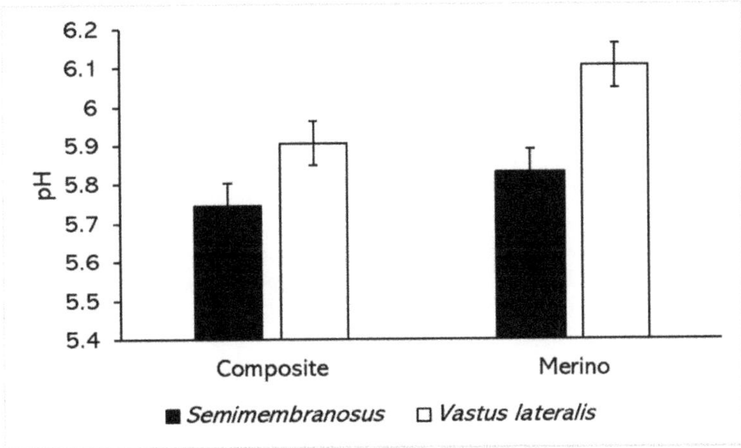

Figure 1. The pH of *Semimembranosus* and *Vastus lateralis* from Composite and Merino sheep before retail packaging. Values are predicted means ± standard error of differences (SED). p (breed × muscle) is 0.096.

Using the CIE L*, a*, and b*-values, the colour stability of lamb was evaluated (Table 1). Breed had differential effects on the lightness (L*) of *Semimembranosus* and *Vastus lateralis* muscles. Compared to Composite, Merino had a lower L* value for *Semimembranosus*, yet a higher L* value for *Vastus lateralis*. A significant effect of breed on a*, b*, and hue were also observed for *Vastus lateralis*, but not *Semimembranosus*. The finishing feed had no effect on any of the colour parameters. The packaging method had a greater influence on all colour parameters ($p < 0.001$ for all) compared to breed and feed effects. While there were small differences between HioxMAP and TrigasMAP, most colour parameters significantly differed between VSP and HioxMAP or between VSP and TrigasMAP for both muscles. Interestingly, when comparing VSP and HioxMAP, hue differed in *Semimembranosus*, but not *Vastus lateralis*. Together, these results show that the choice of packaging methods had a greater influence on the colour stability of sheep meat, compared to breed and feed, and the extent to which of HioxMAP negatively impacts meat colour was muscle dependent. A visual illustration of *Vastus lateralis* in the three packaging methods is presented in Figure 2.

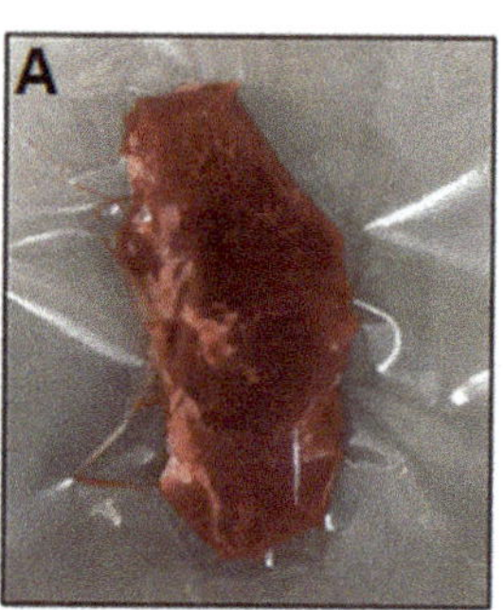

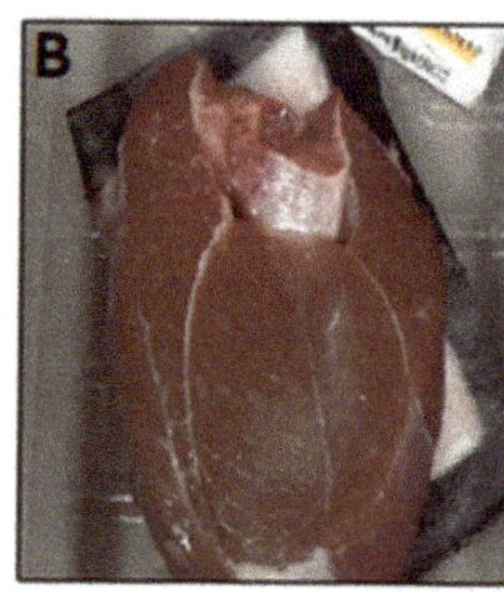

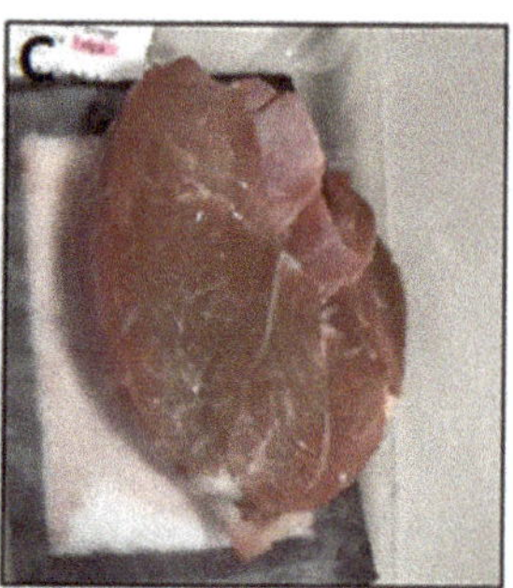

Figure 2. Visual comparison of *Vastus lateralis* from a Composite sheep finished on standard pelleted diet with grain and hay and packed in (**A**) vacuum skin packaging; (**B**) HioxMAP; or (**C**) TrigasMAP for 10 days at 4 °C.

3.2. Water Holding Capacity

Water holding capacity was measured as purge and cooking losses (Table 2). While the three supply chain factors (breed, feed, and packaging method) appear to influence purge loss to a similar extent, only packaging method had a significant effect ($p < 0.001$) on purge loss. It is worth noting that purge loss of *Semimembranosus* in TrigasMAP (5.7% ± 0.3 SED) was similar to VSP (5.7% ± 0.3 SED) and lower than HioxMAP (6.8% ± 0.3 SED). There was no difference in purge loss of *Vastus lateralis* in HioxMAP and TrigasMAP, indicating differences between the two muscles in their response to water holding capacity. A significant interaction between breed and packaging method was also observed for purge loss (Figure 3). While the purge loss did not differ across the three packaging methods for composite sheep *Semimembranosus* and *Vastus lateralis*, TrigasMAP reduced the purge loss in Merino *Semimembranosus* compared to HioxMAP (Figure 3A). This reduction in purge loss by TrigasMAP was not observed for *Vastus lateralis* (Figure 3B). Packaging method had a significant effect on purge loss of *Semimembranosus* and *Vastus lateralis* from Merino, but not those from Composite sheep.

Merino *Semimembranosus* had a lower cooking loss compared to the same muscle type from Composite sheep (Table 2). Finishing feed did not affect cooking loss for either *Semimembranosus* or *Vastus lateralis*. The difference in cooking loss between the three packaging methods were significant with the lowest cooking loss in TrigasMAP (30.9% ± 0.5 SED), followed by HioxMAP (32.2% ± 0.5 SED) and VSP (35.1% ± 0.5 SED). No significant interactions were observed for cooking loss.

Table 1. Effect of breed, feed and packaging method on the CIELab colour parameters of sheep *Semimembranosus* and *Vastus lateralis* after 10-day retail display.

Effect	Treatment	L* (Lightness)		a* (Redness)		b* (Yellowness)		h* (Hue Angle)		C* (Chroma)	
		Coeff	*p*-Value	Coeff	*p*-Value	Coeff	*p*-Value	Coeff	*p*-Value	Coeff	*p*-Value
	Semimembranosus Constant [1]	37.5 ± 1.37		20.41 ± 1.02		19.63 ± 0.98		43.67 ± 2.33		28.55 ± 1.16	
Breed	Merino	−2.4 ± 2.56	<0.001	1.87 ± 1.96	0.383	−3.6 ± 1.93	0.278	−8.35 ± 4.63	0.402	−1.14 ± 2.28	0.927
Feed	SCF [2]	2.7 ± 2.44	0.194	−1.89 ± 1.82	0.418	−1.70 ± 1.78	0.712	1.86 ± 4.27	0.671	−2.93 ± 2.12	0.29
	SCM [3]	−2.3 ± 2.44	0.194	0.46 ± 1.82	0.418	1.01 ± 1.78	0.712	2.14 ± 4.27	0.671	0.68 ± 2.12	0.29
Packaging	HioxMAP [4]	9.91 ± 1.57	<0.001	−11.71 ± 1.48	<0.001	−6.46 ± 1.24	<0.001	13.3 ± 3.24	<0.001	−12.64 ± 1.63	<0.001
	TrigasMAP [5]	9.72 ± 1.57	<0.001	−11.39 ± 1.48	<0.001	−7.66 ± 1.24	<0.001	9.6 ± 3.24	<0.001	−13.44 ± 1.63	<0.001
	Vastus lateralis Constant [6]	38.99 ± 1.52		19.33 ± 1.05		18.37 ± 0.84		42.96 ± 2.5		26.74 ± 1.07	
Breed	Merino	0.62 ± 2.48	<0.001	−2.94 ± 1.74	0.005	−5.18 ± 1.37	0.043	−2.75 ± 4.11	0.003	−5.33 ± 1.75	0.08
Feed	SCF [2]	1.38 ± 2.49	0.63	−0.89 ± 1.72	0.104	−0.22 ± 1.39	0.174	1.52 ± 4.05	0.452	−0.78 ± 1.76	0.053
	SCM [3]	1.58 ± 2.49	0.63	−0.35 ± 1.72	0.104	2.07 ± 1.39	0.174	4.19 ± 4.05	0.452	1.21 ± 1.76	0.053
Packaging	HioxMAP [4]	9.74 ± 1.48	<0.001	−6.88 ± 0.98	<0.001	−5.99 ± 0.90	<0.001	2.9 ± 2.14	<0.001	−9.1 ± 1.09	<0.001
	TrigasMAP [5]	10.45 ± 1.48	<0.001	−9.25 ± 0.98	<0.001	−6.59 ± 0.90	<0.001	6.72 ± 2.14	<0.001	−11.19 ± 1.09	<0.001

Coefficients ± standard error of differences (Coeff ± SED) and level of significance (*p*-values) are presented. [1] For *Semimembranosus* from a Composite lamb, finished on a standard pelleted diet containing grain and cereal hay, and retail displayed in vacuum skin packaging for 10 days. [2] SCF = standard pelleted diet containing 15% camelina forage hay. [3] SCM = standard pelleted diet containing 8% camelina meal (SCM). [4] HioxMAP = high-oxygen modified atmosphere packaging with 80% O_2 and 20% CO_2; [5] TrigasMAP = trigas modified atmosphere packaging with 50% O_2, 30% N_2 and 20% CO_2; [6] For *Vastus lateralis* from a Composite sheep, fed with standard pelleted diet containing grain and cereal hay, and packaged in vacuum skin packaging.

Table 2. Effect of breed, feed and packaging method on water holding capacity and texture measurements of sheep *Semimembranosus* (topside) and *Vastus lateralis* (knuckle).

Effect	Treatment	Purge Loss (%)		Cooking Loss (%)		WBSF (N)		Hardness (N)		Cohesiveness		Chewiness (N)	
		Coeff	p-Value	Coeff	p-Value	Coeff	p-Value	Coeff	p-Value	Coeff	p-Value	Coeff	p-Value
	Semimembranosus Constant [1]	6.3 ± 0.5		35.3 ± 0.9		25.3 ± 2.7		33.6 ± 1.5		0.16 ± 0.01		1.78 ± 0.32	
Breed	Merino	−0.6 ± 1	0.506	−2.3 ± 1.7	0.038	−0.4 ± 4.9	0.703	2.8 ± 4.3	0.001	−0.01 ± 0.03	<0.001	0.16 ± 0.8	<0.001
Feed	SCF [2]	−0.3 ± 0.8	0.984	0.5 ± 1.3	0.566	3.2 ± 4.1	0.18	1.7 ± 2.3	0.952	−0.004 ± 0.02	0.334	0.21 ± 0.49	0.508
	SCM [3]	0.5 ± 0.8	0.984	0.9 ± 1.3	0.566	4.6 ± 4.1	0.18	0.8 ± 2.3	0.952	0.02 ± 0.02	0.334	0.37 ± 0.49	0.508
Packaging	HioxMAP [4]	0.4 ± 0.6	0.014	−3 ± 1.1	<0.001	3.8 ± 3.1	0.001	3.0 ± 3.3	0.07	0.05 ± 0.02	<0.001	1.14 ± 0.63	<0.001
	TrigasMAP [5]	−0.5 ± 0.6	0.014	−3.3 ± 1.1	<0.001	5.1 ± 3.1	0.001	6.6 ± 3.3	0.07	0.03 ± 0.02	<0.001	1.07 ± 0.63	<0.001
	Vastus lateralis Constant [6]	6.0 ± 0.5		35.3 ± 0.9		20.9 ± 1.2		32.9 ± 2		0.18 ± 0.02		2.04 ± 0.56	
Breed	Merino	−0.8 ± 1	0.101	−1.4 ± 1.6	0.348	0.3 ± 2.5	0.688	−4.8 ± 4.5	0.756	−0.01 ± 0.05	0.005	−0.08 ± 1.22	0.01
Feed	SCF [2]	0.6 ± 0.8	0.936	1.3 ± 1.3	0.742	2.9 ± 1.9	0.117	−0.9 ± 2.9	0.034	−0.02 ± 0.03	0.926	−0.31 ± 0.85	0.551
	SCM [3]	0.2 ± 0.8	0.936	1.1 ± 1.3	0.742	1 ± 1.9	0.117	−1.2 ± 2.9	0.034	−0.02 ± 0.03	0.926	−0.23 ± 0.85	0.551
Packaging	HioxMAP [4]	−0.1 ± 0.7	0.031	−2.7 ± 0.9	<0.001	2.1 ± 1.7	0.763	0.3 ± 3.2	0.695	0.05 ± 0.03	<0.001	1.36 ± 0.82	<0.001
	TrigasMAP [5]	0.3 ± 0.7	0.031	−4.9 ± 0.9	<0.001	1.4 ± 1.7	0.763	0.7 ± 3.2	0.695	0.04 ± 0.03	<0.001	1.11 ± 0.82	<0.001

Coefficients ± standard error of differences (Coeff ± SED) and level of significance (*p*-values) are presented. [1] For *Semimembranosus* from a Composite lamb, finished on a standard pelleted diet containing grain and cereal hay, and retail displayed in vacuum skin packaging for 10 days. [2] SCF = standard pelleted diet containing 15% camelina forage hay. [3] SCM = standard pelleted diet containing 8% camelina meal (SCM). [4] HioxMAP = high-oxygen modified atmosphere packaging with 80% O_2 and 20% CO_2; [5] TrigasMAP = trigas modified atmosphere packaging with 50% O_2, 30% N_2 and 20% CO_2; [6] For *Vastus lateralis* from a Composite sheep, fed with standard pelleted diet containing grain and cereal hay, and packaged in vacuum skin packaging.

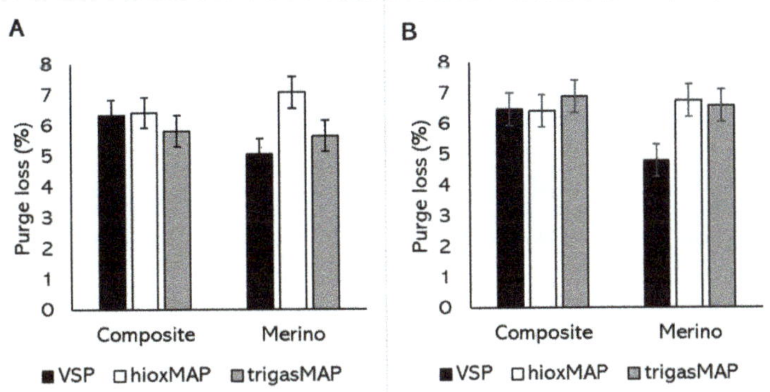

Figure 3. Purge loss of (**A**) *Semimembranosus* and (**B**) *Vastus lateralis* from two sheep breeds (Composite or Merino) in three retail packaging methods (VSP = vacuum skin packaging; HioxMAP = high-oxygen modified atmosphere packaging with 80% O_2 and 20% CO_2; or TrigasMAP = trigas modified atmosphere packaging with 50% O_2, 30% N_2 and 20% CO_2). Values are predicted means ± standard error of differences (SED). p (breed × packaging method) values are 0.014 for *Semimembranosus* (**A**) and 0.015 for *Vastus lateralis* (**B**).

3.3. WBSF and Texture Profile Analysis

Breed or finishing feed had no effect on WBSF for either of the two muscles (Table 2). Differences in WBSF between the three packaging methods were only found for *Semimembranosus*, which were tougher in HioxMAP and TrigasMAP, compared to VSP. No significant interactions were observed for WBSF in either muscle types.

The effect of the three supply chain factors on Texture Profile Analysis hardness differed between the two muscles (Table 2). Within the *Semimembranosus* samples, hardness was affected by breed only, with *Semimembranosus* from Merino having a higher hardness value compared to *Semimembranosus* from Composite sheep. The hardness of *Vastus lateralis* was only affected by finishing feed, with SCM having a lower hardness value than that of SPD and SCF. Cohesiveness and chewiness were affected by breed and packaging method in both muscle types. Cohesiveness and chewiness were lower in VSP compared to HioxMAP and TrigasMAP for both muscle type, suggesting a softer texture in a low oxygen packaging environment. No significant interactions were observed for hardness, cohesiveness, and chewiness in either muscle types.

3.4. Lipid Oxidation

Lipid oxidation in meat was assessed using TBARS assay and the levels were expressed as mg MDA/kg of meat. Breed did not affect lipid oxidation in either of the two muscle types (Table 3). However, supplementation of feed with either camelina forage or camelina meal led to a reduction in TBARS values compared to the standard pelleted diet containing cereal hay and grains. Packaging type not only had a significant effect but also to a greater extent (substantially higher coefficients) than feed on TBARS values. TrigasMAP was able to reduce lipid oxidation compared to HioxMAP for *Semimembranosus*, but not *Vastus lateralis*. There was also a significant interaction between finishing feed and packaging method for the *Semimembranosus* samples. Figure 4 shows that the TBARS value of meat in HioxMAP were substantially greater in the control diet (SPD) compared to the two camelina supplemented diets (SCF and SCM), especially for *Semimembranosus*. These results further emphasise the need for sheep meat to be packaged in a lower oxygen environment when sheep feed is not supplemented with antioxidants.

Table 3. Effect of breed, feed and packaging method on lipid and protein oxidation measurements of sheep *Semimembranosus* (topside) and *Vastus lateralis* (knuckle).

Effect	Treatment	TBARS (mg MDA·kg^{-1} Meat)		Total Carbonyl (nmol·mg^{-1} Protein)		Free Thiol Content (nmol·mg^{-1} Protein)	
		Coeff	*p*-Value	Coeff	*p*-Value	Coeff	*p*-Value
	Semimembranosus Constant [1]	0.51 ± 0.33		1.52 ± 0.58		53.48 ± 3.14	
Breed	Merino	−0.13 ± 0.46	0.323	−0.11 ± 0.77	0.781	−14.87 ± 6.85	0.661
Feed	SCF [2]	−0.4 ± 0.47	0.034	−0.66 ± 0.81	0.395	0.85 ± 5.02	0.37
	SCM [3]	−0.28 ± 0.47	0.034	−0.50 ± 0.81	0.395	0.02 ± 5.02	0.37
Packaging	HioxMAP [4]	4.05 ± 0.45	<0.001	1.75 ± 0.58	<0.001	−10.62 ± 2.52	<0.001
	TrigasMAP [5]	1.80 ± 0.45	<0.001	1.74 ± 0.58	<0.001	−8.92 ± 2.52	<0.001
	Vastus lateralis Constant [6]	0.29 ± 0.36		1.2 ± 0.29		54.62 ± 3.15	
Breed	Merino	0.14 ± 0.50	0.716	0.28 ± 0.41	0.404	−13.21 ± 6.51	0.424
Feed	SCF [2]	−0.23 ± 0.54	0.06	−0.36 ± 0.42	0.162	0.36 ± 5	0.382
	SCM [3]	−0.11 ± 0.54	0.06	−0.23 ± 0.41	0.162	−1.62 ± 5	0.382
Packaging	HioxMAP [4]	2.59 ± 0.41	<0.001	2.06 ± 0.39	<0.001	−10.29 ± 2.43	<0.001
	TrigasMAP [5]	2.67 ± 0.41	<0.001	1.96 ± 0.39	<0.001	−8.22 ± 2.43	<0.001

Coefficients ± standard error of differences (Coeff ± SED) and level of significance (*p*-values) are presented. [1] For *Semimembranosus* from a Composite lamb, finished on a standard pelleted diet containing grain and cereal hay, and retail displayed in vacuum skin packaging for 10 days. [2] SCF = standard pelleted diet containing 15% camelina forage hay. [3] SCM = standard pelleted diet containing 8% camelina meal (SCM). [4] HioxMAP = high-oxygen modified atmosphere packaging with 80% O_2 and 20% CO_2; [5] TrigasMAP = trigas modified atmosphere packaging with 50% O_2, 30% N_2 and 20% CO_2; [6] For *Vastus lateralis* from a Composite sheep, fed with standard pelleted diet containing grain and cereal hay, and packaged in vacuum skin packaging.

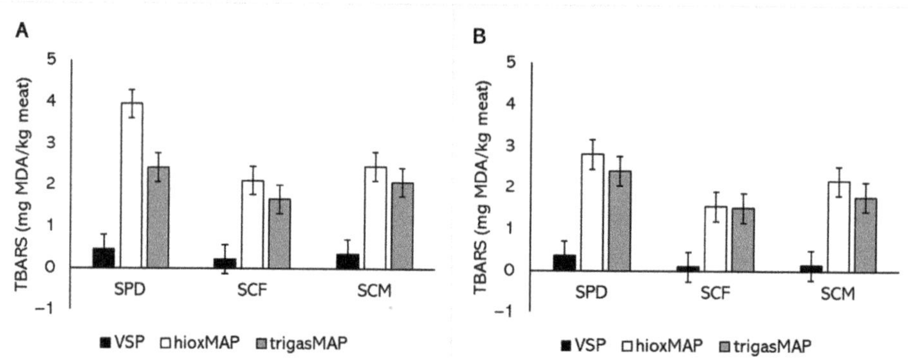

Figure 4. Thiobarbituric acid reactive substances (TBARS) values of (**A**) *Semimembranosus* and (**B**) *Vastus lateralis* from sheep finished on three diets (SPD = standard pelleted diet containing grain and cereal hay; SCF = pelleted mixture diet containing 15% camelina forage hay; or SCM = pelleted mixture diet containing 8% camelina meal) and retail displayed in three packaging methods (VSP = vacuum skin packaging; HioxMAP = high-oxygen modified atmosphere packaging with 80% O_2 and 20% CO_2; or TrigasMAP = trigas modified atmosphere packaging with 50% O_2, 30% N_2 and 20% CO_2). Values are predicted means ± standard error of differences (SED). p (feed × packaging method) values are 0.011 for *Semimembranosus* (**A**) and 0.243 for *Vastus lateralis* (**B**).

3.5. Protein Oxidation

There were no differences between breed and finishing feed treatments on total carbonyl and free thiol content in either *Semimembranosus* or *Vastus lateralis* (Table 3). On the other hand, protein oxidation significantly differed between the three packaging methods in both muscle types. Total carbonyl was significantly lower in VSP compared to HioxMAP and TrigasMAP. There was a small but significant difference between total carbonyl of HioxMAP and TrigasMAP with TrigasMAP inducing a lower carbonyl generation. The free thiol content values were significantly lower in VSP compared to either HioxMAP or TrigasMAP for both muscle types, suggesting that minimising protein oxidation in sheep meat can be achieved by the use of oxygen at a level below 50%. Within SPD treatment, Merino *Semimembranosus* or *Vastus lateralis* had lower free thiol contents compared to equivalent muscles from Composite sheep, indicating the importance of cameline in sheep finishing diets for Merino sheep to reduce protein oxidation. This difference was not observed for total carbonyl.

4. Discussion

4.1. Colour and pH

Differences in Instrumental CIELab parameters due to breed was more apparent in *Vastus lateralis* compared to *Semimembranosus*. Merino *Vastus lateralis* had higher L*, and lower a*, b*, hue, and chroma than the same muscle from Composite sheep. These results suggest that the meat from Merino sheep was less colour stable, which coincided with a substantially higher pH compared to meat from Composite sheep. This agrees with a previous study which found the fastest drop in oxy-/met-myoglobin ratio in meat from Merino, compared to other crossbreeds [14]. The lower colour stability in Merino sheep meat was also reported in other studies [9,10]. Furthermore, significant differences in fatty acid composition and vitamin E concentration were found in the *Longissimus* of Merino and crossbred sheep [7,10]. These differences are likely to result in variation in the oxidation of myoglobin, thus affecting the colour stability of meat.

Breed and pH have been shown to be among the most important predictors in sheep meat colour stability. Meat from Merino often has a different ultimate pH to meat from crossbreed sheep [10,14]. A study on lamb from the Australian Cooperative Research Centre for Sheep Industry Innovation showed that Merino *Longissimus* with a higher pH had the least colour stability in overwrap [14]. The link between breed, ultimate pH, and colour stability is complex. Meat pH has been linked to myoglobin autooxidation, changes in enzymatic activities, iron molecule oxidation, and light scattering, all of which affect the appearance of the meat [15]. It is worth noting that most studies on sheep meat have focused mainly on the *Longissimus*, which is known to differ from *Semimembranosus* and *Vastus lateralis* in muscle fibre type, contributing to differences in colour [15].

Myoglobin oxidation and oxygenation status is affected by the level of oxygen during retail packaging. This study shows that the storage of lamb under high (80%) and moderate (50%) oxygen environments for 10 days significantly reduces the colour stability of both *Semimembranosus* and *Vastus lateralis*, when compared to lamb stored in VSP. The significant decrease in L*, a*, b*, hue and chroma in HioxMAP, compared to VSP, are consistent with the results of previous studies [3,16,17]. Lower chroma and higher hue values are undesirable in red meat as it represents paler and duller meat [18]. The colour results from TrigasMAP in the present study suggest that after 10 days of storage, TrigasMAP does not offer enhanced colour stability, similar to results of Resconi et al. [16] in which beef were displayed in different O_2 levels for up to 8 days. However, it is possible that meat in TrigasMAP with a shorter retail display time may have better colour than in HioxMAP. Zakrys et al. [19] suggested that 50% O_2; 30% N_2; 20% CO_2 may provide opportunity for improved shelf life by enhancing the a* value of beef, compared to HioxMAP after a 3-day storage. Meat surface colour has been shown to deteriorate after three days of storage in HioxMAP [20]. This was attributed to the reduction of metmyoglobin reducing activity during prolonged storage, thereby favouring the oxidative process of oxymyoglobin to metmyoglobin. The study of Khliji et al. [21] indicated that consumers discriminate against red meat with a* values below 14.5. The a* values for both muscles in HioxMAP and TrigasMAP were well below this threshold for both muscles in this study. Thus, retail displaying of sheep meat in HioxMAP or TrigasMAP for 10 days is not recommended for colour enhancement.

4.2. Water Holding Capacity and Texture

Contradictory results have been reported for the purge loss of meat in different packaging treatments, while others reported an increase in the purge of meat under vacuum [22]. Taylor et al. [17] showed that the weight loss of vacuum-packed beef and pork was less than MAP (75%O_2/25%CO_2)-packed samples after storage. Similar results were reported in other studies [23,24]. However, it was suggested in a review by McMillin [25] that the purge loss of meat displayed under vacuum packaging is higher than MAP, partly attributed to the negative pressure. In the current study, the two muscles responded differently to the effect of packaging. HioxMAP led to a higher purge loss for *Semimembranosus* but lower purge loss for *Vastus lateralis* compared to VSP, emphasising that future purge loss investigations should consider muscle differences.

Unlike purge loss, similar cooking losses were observed for both muscle types with a higher cooking loss found for both HioxMAP and TrigasMAP compared to VSP, in agreement with previous studies [26]. On the other hand, cooking loss due to breed differences appear to be muscle-specific, with *Semimembranosus* from Merino having a lower cooking loss compared to the same muscle from Composite. While the underlying mechanisms of water holding capacity remains an ongoing research area, previous studies showed that variations in muscle fibre type and connective tissue composition play a role in cooking loss differences [27–29].

Packaging appeared to affect the two muscles differently. WBSF of *Semimembranosus* in HioxMAP and TrigasMAP were higher than VSP. Similar results have been reported for beef topside and beef round, where the beef topside is a more likely response to ageing than beef round muscle after storage [30]. Numerous studies have shown the negative effect of HioxMAP on sheep meat eating quality. Frank et al. [1] showed a significantly lower sensory tenderness of lamb in HioxMAP compared to VSP. Similar results on various texture measurements were also found for meat from other species [11,13,31]. Previous studies found while the WBSF of beef *Longissimus* did not differ between oxygen levels from 40–80%, sensory panellists preferred beef in lower O_2 environments 40–50% [19,32,33]. Furthermore, various studies have established that the toughening of meat in HioxMAP is caused by increased protein oxidation resulting in more disulfide bond formation between actomyosin complexes, less degradation of structural proteins, e.g., desmin and troponin T, and deactivation of calpain [13,32,34–36]. It is worth noting that the exact mechanisms appeared to be muscle- and species-specific [2,3,34]. Our results on texture are consistent with the protein oxidation results, which showed that significant differences were only found when VSP was compared to HioxMAP and TrigasMAP. Together, these results indicate VSP is the preferred packaging method for lamb regardless of breed and finishing feed treatments.

4.3. Lipid Oxidation

Lipid oxidation is a key quality determinant in meat, as it causes the development of off-flavours and rancidity in meat. Free radical formation from lipid oxidation has also been linked to increased myoglobin oxidation and thus discolouration [37]. Feeding strategy of livestock can play a significant role in manipulating lipid oxidation of meat. The present study found a reduction of lipid oxidation of *Semimembranosus* and *Vastus lateralis* from sheep finished on diets supplemented with camelina forge or camelina meal. These findings compliment previous studies which found significant decreases in TBARS for forage fed animals when compared to grain-fed animals [5,38,39]. Lamb muscles finished on diets supplemented with camelina cake has been shown to have a different fatty acid composition compared to those on the standard pelleted diet without camelina supplementation [40]. Furthermore, we have reported in a separate study that both camelina hay- and camelia meal-supplemented diets reduced ($p < 0.001$) arachidonic acid concentration of *Longissimus* from these animals compared with the SPD diet [7]. In addition, the SCM diet significantly increased alpha linolenic acid (ALA) concentration of the *Longissimus* compared to SPD and SCF, resulting in an increase in total omega-3 concentration and the decrease in the ratio of $n-6/n-3$ in meat [7].

The packaging results in this study show that TrigasMAP is an effective method to reduce lipid oxidation in packaging, regardless of breed and finishing feed treatments, consistent with previous studies on the effect of varying oxygen content on lipid oxidation [19,34,41]. Reducing oxygen content in retail packaging is even more important when sheep is not finished on supplemented diet.

Consumers discriminate against the off-flavour of beef when TBARS reaches the 2.28 mg MDA/kg meat threshold [42]. While similar investigations are needed for sheep meat, the present findings suggest that retail display of sheep meat in HioxMAP for 10 days leads to unacceptable flavour regardless of breed, feed or muscles. Supplementation of finishing feeds with camelina forage and TrigasMAP offers the potential to reduce TBARS values to below this threshold, thus reducing the economic loss for the industry. It should be noted that VSP provided consistently minimal lipid oxidation regardless of breed, feed or muscle treatments.

4.4. Protein Oxidation

Protein oxidation during retail display has been shown to lead to changes in protein aggregation and degradation, with implication for meat tenderisation. Carbonyl content substantially increased after 10 days of retail display in both TrigasMAP and HioxMAP (Table 3). Interestingly, TrigasMAP reduced the extent of formation of carbonyl groups compared to HioxMAP. This is similar to that observed in lipid oxidation, and agrees with previous studies [19] which reported increases in carbonyl content with increases in oxygen concentration. This would suggest that reducing the oxygen concentration in the packaging system to 50% reduced the extent of post-mortem oxidative processes.

Morzel et al. [43], using an ·OH radical generating system from pig *Longissimus*, showed oxidation induced formation of disulfide bridge and protein polymerisation led to a reduction in proteolysis susceptibility of myofibril proteins. Free thiol groups (sulfhydryl) are susceptible to oxidation; therefore, the quantification is a useful measure to determine the extent of protein oxidation in muscle foods. The present study showed the free thiol content of both *Semimembranosus* and *Vastus lateralis* did not differ between breed and finishing feed treatments. However, significant differences were observed between VSP and HioxMAP and TrigasMAP treatments. Bao and Ertbjerg [34] reported no difference in free thiol content between 80% O_2 and 60% O_2 in HioxMAP packaged beef. The underlying mechanisms behind differences in free thiol content between Composite and Merino on SPD is not understood. However, differences in muscle fibre type, lipid content and composition, and antioxidant capacities between breeds are likely to be involved [44].

5. Conclusions

By examining the colour, water holding capacity, texture, and oxidative stability of sheep meat from different breeds, finishing feed, and retail packaging methods, this study demonstrated the complexity in how different sheep breeds and muscles respond to variations in finishing feeds and packaging methods. Packaging of sheep meat in low, moderate, or high oxygen environments affected the colour to a greater extent than breed and finishing feeds. However, supplementation of the finishing feed with either camelina forage or camelina meal significantly reduced the lipid oxidation of sheep meat. Understanding how and to which extent supply chain factors affect the quality of sheep meat enables sheep producers and processors to prioritise intervention strategies to ensure optimal quality.

Author Contributions: Conceptualization, E.N.P. and M.H.; methodology, E.N.P. and M.H.; validation, M.H., S.W. and C.K.; formal analysis, M.H.; investigation, S.W. and C.K.; resources, E.N.P. and R.D.W.; data curation, M.H.; writing—original draft preparation, M.H., C.K. and S.W.; writing—

review and editing, M.H., R.D.W. and E.N.P.; supervision, M.H. and R.D.W.; project administration, E.N.P. and M.H.; funding acquisition, E.N.P. All authors have read and agreed to the published version of the manuscript.

Funding: The funding for the animal feeding experiment was provided by the Victorian Department of Jobs, Precincts and Regions (Agriculture Victoria Research), grant number CMI 105393.

Institutional Review Board Statement: The study was conducted according to the Australian Code of Practice for the Care and Use of Animals for Scientific Purposes, and approved by the Agricultural Research and Extension Animal Ethics Committee of the Department of Jobs, Precincts and Regions (AEC Approval No: 2016-17).

Informed Consent Statement: Not applicable.

Data Availability Statement: The data presented in this study are available on reasonable request.

Acknowledgments: The authors gratefully acknowledge the technical contributions of staff from Hamilton and Bundoora during animal feeding, slaughter, and primal cuts collection. The camelina forage and camelina meal were purchased from Yarrock Pty Ltd., Kaniva, VIC 3419, Australia. Technical and operational support from the Faculty of Veterinary and Agricultural Sciences, The University of Melbourne is also acknowledged for the collection of muscle cuts, biochemical analyses, and meat quality measurements.

Conflicts of Interest: The authors declare no conflict of interest. The funder had no role in the design of the study; in the collection, analyses, or interpretation of data; in the writing of the manuscript, or in the decision to publish the results.

References

1. Frank, D.C.; Geesink, G.; Alyarenga, T.I.R.C.; Polkinghorne, R.; Stark, J.; Lee, M.; Warner, R. Impact of high oxygen and vacuum retail ready packaging formats on lamb loin and topside eating quality. *Meat Sci.* **2017**, *123*, 126–133. [CrossRef] [PubMed]
2. Geesink, G.; Robertson, J.; Ball, A. The effect of retail packaging method on objective and consumer assessment of beef quality traits. *Meat Sci.* **2015**, *104*, 85–89. [CrossRef]
3. Kim, Y.H.; Huff-Lonergan, E.; Sebranek, J.G.; Lonergan, S.M. High-oxygen modified atmosphere packaging system induces lipid and myoglobin oxidation and protein polymerization. *Meat Sci.* **2010**, *85*, 759–767. [CrossRef] [PubMed]
4. Torngren, M.A.; Darre, M.; Gunvig, A.; Bardenshtein, A. Case studies of packaging and processing solutions to improve meat quality and safety. *Meat Sci.* **2018**, *144*, 149–158. [CrossRef] [PubMed]
5. Luciano, G.; Monahan, F.J.; Vasta, V.; Pennisi, P.; Bella, M.; Priolo, A. Lipid and colour stability of meat from lambs fed fresh herbage or concentrate. *Meat Sci.* **2009**, *82*, 193–199. [CrossRef] [PubMed]
6. Ponnampalam, E.N.; Dunshea, F.R.; Warner, R.D. Use of lucerne hay in ruminant feeds to improve animal productivity, meat nutritional value and meat preservation under a more variable climate. *Meat Sci.* **2020**, *170*, 108235. [CrossRef]
7. Ponnampalam, E.N.; Butler, K.L.; Muir, S.K.; Plozza, T.E.; Kerr, M.G.; Brown, W.G.; Jacobs, J.L.; Knight, M.I. Lipid oxidation and colour stability of lamb and yearling meat (muscle longissimus lumborum) from sheep supplemented with camelina-based diets after short-, medium-, and long-term storage. *Antioxidants* **2021**, *10*, 166. [CrossRef]
8. Quezada, N.; Cherian, G. Lipid characterization and antioxidant status of the seeds and meals of Camelina sativa and flax. *Eur. J. Lipid. Sci. Technol.* **2012**, *114*, 974–982. [CrossRef]
9. Cloete, J.; Hoffman, L.; Cloete, S. A comparison between slaughter traits and meat quality of various sheep breeds: Wool, dual-purpose and mutton. *Meat Sci.* **2012**, *91*, 318–324. [CrossRef] [PubMed]
10. Mortimer, S.I.; van der Werf, J.H.; Jacob, R.H.; Hopkins, D.L.; Pannier, L.; Pearce, K.L.; Gardner, G.E.; Warner, R.D.; Geesink, G.H.; Edwards, J.E.; et al. Genetic parameters for meat quality traits of Australian lamb meat. *Meat Sci.* **2014**, *96*, 1016–1024. [CrossRef]
11. Peng, Y.L.; Adhiputra, K.; Padayachee, A.; Channon, H.; Ha, M.; Warner, R.D. High oxygen modified atmosphere packaging negatively influences consumer acceptability traits of pork. *Foods* **2019**, *8*, 567. [CrossRef] [PubMed]
12. Sorensen, G.; Jorgensen, S.S. A critical examination of some experimental variables in the 2-thiobarbituric acid (TBA) test for lipid oxidation in meat products. *Z. Lebensm. Unters. Forsch.* **1996**, *202*, 205–210. [CrossRef]
13. Lund, M.N.; Lametsch, R.; Hviid, M.S.; Jensen, O.N.; Skibsted, L.H. High-oxygen packaging atmosphere influences protein oxidation and tenderness of porcine longissimus dorsi during chill storage. *Meat Sci.* **2007**, *77*, 295–303. [CrossRef]
14. Jacob, R.H.; D'Antuono, M.F.; Gilmour, A.R.; Warner, R.D. Phenotypic characterisation of colour stability of lamb meat. *Meat Sci.* **2014**, *96*, 1040–1048. [CrossRef] [PubMed]
15. Bekhit, A.; Morton, J.D.; Bhat, Z.F.; Kong, L. Meat color: Factors affecting color stability. In *Encyclopedia of Food Chemistry*; Varelis, P., Melton, L., Shahidi, F., Eds.; Elsevier Inc.: Amsterdam, The Netherlands, 2019; Volume 2, pp. 202–210.
16. Resconi, V.C.; Escudero, A.; Beltran, J.A.; Olleta, J.L.; Sanudo, C.; Campo, M.D. Color, lipid oxidation, sensory quality, and aroma compounds of beef steaks displayed under different levels of oxygen in a modified atmosphere package. *J. Food Sci.* **2012**, *77*, S10–S18. [CrossRef]

17. Taylor, A.; Down, N.; Shaw, B. A comparison of modified atmosphere and vacuum skin packing for the storage of red meats. *Int. J. Food Sci. Technol.* **1990**, *25*, 98–109. [CrossRef]
18. Renerre, M. Oxidative processes and myoglobin. In *Antioxidants in Muscle Foods*; Decker, E., Faustman, C., Lopez-Bote, C.J., Eds.; John Wiley & Sons, Inc.: Toronto, ON, Canada, 2000; Volume 2000, pp. 113–133.
19. Zakrys, P.I.; Hogan, S.A.; O'Sullivan, M.G.; Allen, P.; Kerry, J.P. Effects of oxygen concentration on the sensory evaluation and quality indicators of beef muscle packed under modified atmosphere. *Meat Sci.* **2008**, *79*, 648–655. [CrossRef]
20. Sorheim, O.; Nissen, H.; Nesbakken, T. The storage life of beef and pork packaged in an atmosphere with low carbon monoxide and high carbon dioxide. *Meat Sci.* **1999**, *52*, 157–164. [CrossRef]
21. Khliji, S.; van de Ven, R.; Lamb, T.A.; Lanza, M.; Hopkins, D.L. Relationship between consumer ranking of lamb colour and objective measures of colour. *Meat Sci.* **2010**, *85*, 224–229. [CrossRef]
22. Cayuela, J.M.; Gil, M.D.; Banon, S.; Garrido, M.D. Effect of vacuum and modified atmosphere packaging on the quality of pork loin. *Eur. Food Res. Technol.* **2004**, *219*, 316–320. [CrossRef]
23. Bağdatli, A.; Kayaardi, S. Influence of storage period and packaging methods on quality attributes of fresh beef steaks. *CyTA-J. Food* **2015**, *13*, 124–133. [CrossRef]
24. Lagerstedt, A.; Lundstrom, K.; Lindahl, G. Influence of vacuum or high-oxygen modified atmosphere packaging on quality of beef M. longissimus dorsi steaks after different ageing times. *Meat Sci.* **2011**, *87*, 101–106. [CrossRef]
25. McMillin, K.W. Where is MAP Going? A review and future potential of modified atmosphere packaging for meat. *Meat Sci.* **2008**, *80*, 43–65. [CrossRef] [PubMed]
26. Clausen, I.; Jakobsen, M.; Ertbjerg, P.; Madsen, N.T. Modified atmosphere packaging affects lipid oxidation, myofibrillar fragmentation index and eating quality of beef. *Packag. Technol. Sci.* **2009**, *22*, 85–96. [CrossRef]
27. Vaskoska, R.; Ha, M.; Naqvi, Z.B.; White, J.D.; Warner, R.D. Muscle, ageing and temperature influence the changes in texture, cooking loss and shrinkage of cooked beef. *Foods* **2020**, *9*, 1289. [CrossRef] [PubMed]
28. Vaskoska, R.; Ha, M.; Ong, L.; Chen, G.; White, J.; Gras, S.; Warner, R. Myosin sensitivity to thermal denaturation explains differences in water loss and shrinkage during cooking in muscles of distinct fibre types. *Meat Sci.* **2021**, *179*, 108521. [CrossRef]
29. Vaskoska, R.; Venien, A.; Ha, M.; White, J.D.; Unnithan, R.R.; Astruc, T.; Warner, R.D. Thermal denaturation of proteins in the muscle fibre and connective tissue from bovine muscles composed of type I (masseter) or type II (cutaneous trunci) fibres: DSC and FTIR microspectroscopy study. *Food Chem.* **2021**, *343*, 128544. [CrossRef] [PubMed]
30. Stolowski, G.D.; Baird, B.E.; Miller, R.K.; Savell, J.W.; Sams, A.R.; Taylor, J.F.; Sanders, J.O.; Smith, S.B. Factors influencing the variation in tenderness of seven major beef muscles from three Angus and Brahman breed crosses. *Meat Sci.* **2006**, *73*, 475–483. [CrossRef] [PubMed]
31. Fu, Q.Q.; Liu, R.; Zhang, W.G.; Li, Y.P.; Wang, J.; Zhou, G.H. Effects of different packaging systems on beef tenderness through protein modifications. *Food Bioprocess Technol.* **2015**, *8*, 580–588. [CrossRef]
32. Zakrys-Waliwander, P.I.; O'Sullivan, M.G.; O'Neill, E.E.; Kerry, J.P. The effects of high oxygen modified atmosphere packaging on protein oxidation of bovine M. longissimus dorsi muscle during chilled storage. *Food Chem.* **2012**, *131*, 527–532. [CrossRef]
33. Zakrys-Waliwander, P.I.; O'Sullivan, M.G.; Walsh, H.; Allen, P.; Kerry, J.P. Sensory comparison of commercial low and high oxygen modified atmosphere packed sirloin beef steaks. *Meat Sci.* **2011**, *88*, 198–202. [CrossRef] [PubMed]
34. Bao, Y.L.; Ertbjerg, P. Relationship between oxygen concentration, shear force and protein oxidation in modified atmosphere packaged pork. *Meat Sci.* **2015**, *110*, 174–179. [CrossRef] [PubMed]
35. Chen, L.; Zhou, G.H.; Zhang, W.G. Effects of high oxygen packaging on tenderness and water holding capacity of pork through protein oxidation. *Food Bioprocess Technol.* **2015**, *8*, 2287–2297. [CrossRef]
36. Jongberg, S.; Wen, J.; Tørngren, M.A.; Lund, M.N. Effect of high-oxygen atmosphere packaging on oxidative stability and sensory quality of two chicken muscles during chill storage. *Food Packag. Shelf Life* **2014**, *1*, 38–48. [CrossRef]
37. Faustman, C.; Sun, Q.; Mancini, R.; Suman, S.P. Myoglobin and lipid oxidation interactions: Mechanistic bases and control. *Meat Sci.* **2010**, *86*, 86–94. [CrossRef] [PubMed]
38. O'Sullivan, A.; Galvin, K.; Moloney, A.; Troy, D.; O'Sullivan, K.; Kerry, J. Effect of pre-slaughter rations of forage and/or concentrates on the composition and quality of retail packaged beef. *Meat Sci.* **2003**, *63*, 279–286. [CrossRef]
39. Descalzo, A.M.; Rossetti, L.; Grigioni, G.; Irurueta, M.; Sancho, A.M.; Carrete, J.; Pensel, N.A. Antioxidant status and odour profile in fresh beef from pasture or grain-fed cattle. *Meat Sci.* **2007**, *75*, 299–307. [CrossRef]
40. Cieslak, A.; Stanisz, M.; Wojtowski, J.; Pers-Kamczyc, E.; Szczechowiak, J.; El-Sherbiny, M.; Szumacher-Strabel, M. Camelina sativa affects the fatty acid contents in M-longissimus muscle of lambs. *Eur. J. Lipid Sci. Technol.* **2013**, *115*, 1258–1265. [CrossRef]
41. Spanos, D.; Torngren, M.A.; Christensen, M.; Baron, C.P. Effect of oxygen level on the oxidative stability of two different retail pork products stored using modified atmosphere packaging (MAP). *Meat Sci.* **2016**, *113*, 162–169. [CrossRef]
42. Campo, M.M.; Nute, G.R.; Hughes, S.I.; Enser, M.; Wood, J.D.; Richardson, R.I. Flavour perception of oxidation in beef. *Meat Sci.* **2006**, *72*, 303–311. [CrossRef]
43. Morzel, M.; Gatellier, P.; Sayd, T.; Renerre, M.; Laville, E. Chemical oxidation decreases proteolytic susceptibility of skeletal muscle myofibrillar proteins. *Meat Sci.* **2006**, *73*, 536–543. [CrossRef] [PubMed]
44. Prache, S.; Schreurs, N.; Guillier, L. Review: Factors affecting sheep carcass and meat quality attributes. *Animal* **2021**, 100330. [CrossRef] [PubMed]

MDPI
St. Alban-Anlage 66
4052 Basel
Switzerland
Tel. +41 61 683 77 34
Fax +41 61 302 89 18
www.mdpi.com

Foods Editorial Office
E-mail: foods@mdpi.com
www.mdpi.com/journal/foods

www.ingramcontent.com/pod-product-compliance
Lightning Source LLC
LaVergne TN
LVHW070850170726
843515LV00005B/610
* 9 7 8 3 0 3 6 5 3 8 7 9 2 *